Languages of Science in the Eighteenth Century

Languages of Science in the Eighteenth Century

Edited by
Britt-Louise Gunnarsson

De Gruyter Mouton

ISBN 978-3-11-025505-8
e-ISBN 978-3-11-025506-5

Library of Congress Cataloging-in-Publication Data

Languages of science in the eighteenth century / edited by Britt-Louise Gunnarsson.
p. cm.
Includes bibliographical references and index.
ISBN 978-3-11-025505-8 (alk. paper)
1. Science – Europe – History – 18th century. 2. Technical writing – Europe – History – 18th century. 3. Scientists – Europe – Intellectual life – 18th century. I. Gunnarsson, Britt-Louise.
Q127.E8L36 2011
509.4'09033–dc23

2011017257

Bibliographic information published by the Deutsche Nationalbibliothek

The Deutsche Nationalbibliothek lists this publication in the Deutsche Nationalbibliografie; detailed bibliographic data are available in the Internet at http://dnb.d-nb.de.

Typesetting: PTP-Berlin Protago-TEX-Production GmbH, Berlin
Printing: Hubert & Co. GmbH & Co. KG, Göttingen
∞ Printed on acid-free paper

Printed in Germany

www.degruyter.com

Contents

Introduction

Section 1. The forming of scientific communities

Section 2. The emergence of new languages of science

Section 3. The spread of scientific ideas

Section 4. The development of scientific writing

Contributors

Charles Bazerman
University of California, Santa Barbara
Department of Education
3208 Education Building
Santa Barbara
CA 93106
USA
bazerman@education.ucsb.edu

Han-liang Chang
Fudan University
Department of Chinese
220 Handan Road
Shanghai 200433, China
National Taiwan University
1 Roosevelt Road, Section 4
Taipei, Taiwan 10674
changhl@ntu.edu.tw

Palmira Fontes da Costa
Universidade Nova de Lisboa
Unit of History and Philosophy of Science and Technology
Faculty of Sciences and Technology
2829-516 Campus da Caparica
Portugal
pfc@fct.unl.pt

Rosemarie Gläser
University of Leipzig
Faculty of Philology
Institute of British Studies
Private address:
Am Staffelstein 6
01328 Dresden
Germany

Gunilla Gren-Eklund
Uppsala University
Department of Linguistics and Philology
Private address:
Salagatan 12B
753 30 Uppsala
Sweden
gunilla.gren-eklund@lingfil.uu.se

Britt-Louise Gunnarsson
Uppsala University
Department of Scandinavian Languages
Box 527
751 20 Uppsala
Sweden
britt-louise.gunnarsson@nordiska.uu.se

Anna Helga Hannesdóttir
University of Gothenburg
Department of Swedish
Box 200
405 30 Göteborg
Sweden
anna.hannesdottir@svenska.gu.se

Ann-Mari Jönsson
Uppsala University
Department of Linguistics and Philology
Box 635
751 26 UPPSALA
Sweden
ann-mari.jonsson@lingfil.uu.se

Kenneth J. Knoespel
Georgia Institute of Technology
School of Literature, Communication and Culture
221 Bobby Dodd Way
Atlanta
GA 30332-0165,
USA
kenneth.knoespel@iac.gatech.edu

Andreas Nord
University of Gothenburg
Department of Swedish
Box 200
405 30 Göteborg
Sweden
andreas.nord@svenska.gu.se

Päivi Pahta
University of Tampere
School of Language, Translation and Literary Studies
33014 University of Tampere
Finland
paivi.pahta@uta.fi

Bo Ralph
University of Gothenburg
Department of Swedish
Box 200
405 30 Göteborg
Sweden
bo.ralph@svenska.gu.se

Renata Schellenberg
Mount Allison University
Modern Languages and Literature
Crabtree Building
49A York Street
Sackville
N.B. E4L 1C7
Canada
rschelle@mta.ca

Philippe Selosse
Université Lumière Lyon 2
Faculté des Lettres, Sciences du Langage et Arts
18 Quai Claude-Bernard
69365 Lyon cedex 07
France
selosse.philippe@wanadoo.fr

Richard Sörman
Uppsala University
Department of Modern Languages
Box 636
751 26 UPPSALA
Sweden
richard.sorman@moderna.uu.se

Ulf Teleman
Lund University
Centre for Languages and Literature
Box 201
221 00 Lund
Sweden
ulf.teleman@nordlund.lu.se

Lars Wollin
Åbo Akademi University
Department of Swedish
Finland
Private address:
Noreens väg 21
752 63 Uppsala
Sweden
lars.wollin@abo.fi

Editor's acknowledgements

Most of the chapters in this volume were originally presented as plenary lectures or section papers at the Symposium on Languages of Science in the Time of Linnaeus, held in Uppsala in June 2007. This international symposium, co-organized by Hans Helander and myself, was arranged on the initiative of the Faculty of Languages at Uppsala University as part of the commemoration of the tercentenary of the birth of Carl Linnaeus.

As editor of this volume, I wish to thank Urban Örneholm for technical assistance at an initial stage in the editing process, and Martin Naylor for his careful scrutiny of the English used in chapters written by non-native speakers of that language. Finally, I gratefully acknowledge the financial support provided for the editing and publication of this book by the Faculty of Languages at Uppsala University.

Uppsala, August 2011 *Britt-Louise Gunnarsson*

Introduction

Introduction: Languages of science in the eighteenth century

Britt-Louise Gunnarsson

The eighteenth century is an important period in both the history of science and the history of languages. Interest in science, and especially in the useful sciences, exploded and a new, modern approach to scientific discovery and the accumulation of knowledge emerged. It was during this century, too, that ideas on language and language practice began to change more widely, including in northern Europe. There, Latin had been more or less the only written language used for scientific purposes, but gradually the vernaculars became established as fully acceptable alternatives for scientific writing.

The period is of interest, moreover, from a genre-historical point of view. Encyclopedias, dictionaries and also correspondence played a key role in the spread of scientific ideas. Scientific nomenclatures were established for the key areas of the time, and concepts and names were discussed in letters, journal articles and encyclopedias. The textual embedding of this discussion, however, showed considerable variation. Writing on scientific matters was not as distinct from fiction, poetry or religious texts as it is today, a fact which also gave a creative liberty to individual writers. It also meant that a few important scientists came to play a role in the development both of their areas of expertise and its nomenclature and of scientific language and prose.

In this volume, seventeen authors explore, from a variety of angles, the construction of a scientific language and discourse.[1] The chapters are thematically organized into four sections, each contributing to our understanding of this dynamic period in the history of science: their themes are the forming of scientific communities, the emergence of new languages of science, the spread of scientific ideas, and the development of scientific writing. There is a particular focus in this book on the Swedish botanist Carl Linnaeus (1707–1778). The writing

1. The theoretical framework that can be said to unite the chapters is that of social constructivism. This perspective has been associated with both the sociology of knowledge (e.g. Berger and Luckmann 1967) and the sociology of science (Latour and Woolgar 1986, Bazerman 1988).

and correspondence of this internationally recognized scientist are highlighted and taken as an example of how the scientists of the eighteenth century played an active part in the construction of both a scientific nomenclature and scientific prose. In this introductory chapter, I will begin with a general background to these four thematic sections, before presenting the authors and contents of the individual chapters.

1. The forming of scientific communities

Careful and detailed observation of natural phenomena was at the heart of the sciences during this period. Man and nature were to be studied empirically. Scientists gathered specimens of plants, insects, animals and minerals, but also of machines, fabrics and household implements. They established collections of these organisms and objects, made meticulous drawings of them, and set about naming them and grouping them into systems and families. Scientific discovery and classification also had a practical purpose. It was hoped that new useful plants could be introduced and hardy species for medicinal use cultivated. Botanical gardens were created and glasshouses built to learn how best to improve and multiply the fruits of nature.

The eighteenth-century view of science and progress was in fact imbued with utilitarian thinking.[2] In many European countries, politicians were inspired by the doctrine of mercantilism, which prescribed that as much as possible should be exported and as little as possible imported. Foreign trade was seen as the primary source of a country's wealth. The state was to be actively involved in the economy, subsidizing industry and promoting both commerce and agriculture. But a nation's resources also included its population: the more people there were, the better it was for the country. One of the useful sciences was medicine, and a growing number of doctors observed the progress of diseases in individual patients. Different cures and treatments were also tried. By meticulous observation and description of individual cases, this branch of science, too, sought to combine empirical precision with benefits to humanity.

Great store was set by both the economic and the natural sciences during this period. To bring together practitioners of different useful sciences, academies and societies inspired by classical models were established in the second half of

2. Among literature with a focus on the climate of science and thought in the eighteenth century, I would mention Heckscher (1953), Lindroth (1978) and Johannisson (1988). A detailed account of the early history of Uppsala University can be found in Annerstedt (1912).

the seventeenth and the first half of the eighteenth century, many of them with a king or emperor as their patron. In England the *Royal Society of London* was founded in 1660, in France the *Académie des sciences* in 1666, in Schweinfurt, Germany, the *Akademie der Naturforscher* in 1652, in Russia the *Saint Petersburg Academy of Sciences* in 1724, in Sweden the *Royal Swedish Academy of Sciences* in 1739, and in Denmark the *Royal Danish Academy of Sciences and Letters* in 1742.

The economic welfare of the country was one of the guiding stars of these institutions, and to achieve the greatest possible benefits they were anxious to disseminate useful scientific findings. As part of their activities, therefore, periodicals were launched: the Royal Society of London published its *Philosophical Transactions* and the Royal Swedish Academy of Sciences its *Handlingar* [Transactions].

Despite the creation of these scientific societies, though, it has to be said that the scientists of the eighteenth century worked in relative isolation.[3] Universities were concerned with teaching, and scientific experiment and observation were not among their undertakings. Professors' salaries were modest, moreover, so even if a scientist was attached to a university he still had to rely on his ability to secure his own funding. If he was not born rich and had been unable to marry rich, he was forced to seek good relations with individuals in positions of wealth and power. For many scientists, relations with royal or imperial families were important, as heads of state were likely to support useful experiments and expeditions.

With only a small number of scientists in each country, scientific communities were not at all specialized. An individual scientist would have few fellow countrymen working in the same subject area as himself. University professors had to cover broad fields in their teaching. When Carl Linnaeus, for example, assumed the chair of medicine at Uppsala University, his teaching duties encompassed not only dietetics and *materia medica*, but natural history as well (Broberg 2007:23). Similarly, the academies and societies consisted of individuals with different scholarly backgrounds. Scientific discoveries and findings were thus discussed among groups of learned men who, though interested, were not specialists in the same area.

Thus, in the eighteenth century, scientific communities were not divided into specialities in the way we are now used to. Nor was there any clear dividing line between research and science on the one hand and family life on the other. Carl Linnaeus's students, for example, were often guests at his dinner table.

3. For a more in-depth discussion of the sociohistorical construction of scientific discourse, see Gunnarsson (1997 and 2005).

By modern standards, then, the scientific communities of the period were small and non-specialized. The scientists of the time were men of society, taking an interest in everything that could be useful to the country in which they lived. As scientific communities, eighteenth-century groupings of scientists thus represent what can be termed a pre-establishment stage of science.

2. The emergence of new languages of science

As several studies have shown, a rich tradition of vernacular scientific writing in southern and central European countries can be found as early as the Middle Ages. This tradition gained considerable strength in the sixteenth and seventeenth centuries, culminating in the first scientific journals in England and France in the mid-1660s, which mostly appeared in English and French (Gotti 1996; Crossgrove 1996; Taavitsainen and Pahta 2004). In northern Europe, however, Latin remained the language of the learned throughout the seventeenth century. The eighteenth century can thus be said to be a period of language development and language change more generally. The total dominance of Latin as the language of scholarship gradually ceded ground to a more varied language practice. The Reformation, with its Bible translations, had of course paved the way for the use of the vernacular in northern Europe back in the sixteenth century, but in science Latin retained its hold into the eighteenth.[4] If science was to have the desired practical benefits, however, scientists now had to write about their discoveries in the vernacular. The choice of the latter would prove to be a conscious step in the direction of language planning. When a chair in economic sciences was established at Uppsala University in 1741, it was stipulated that the subject was to be taught in Swedish. And it was with practical benefits in mind that Swedish rather than Latin was chosen as the language of the Transactions of the newly founded Swedish Academy of Sciences.

Of course, the shift from classical languages to the vernacular was not universal. To reach an international audience, many scientists still preferred to write their major works in Latin. Nor was it absolute, as Latin and Greek were often used for names and terms. Not uncommonly, though, we find terminology in other languages as well.

Naming and terminology were an important aspect of science at this time. Scientists were discovering new species and describing new relationships, for

4. Cf. Wendt's (2005: 1353–1354) discussion of writing for scientific purposes in Denmark and Sweden in the sixteenth, seventeenth and eighteenth centuries.

which it was necessary to devise names that fitted into a system. Naming, in other words, proceeded in parallel with discovery and classification. Encyclopedias and dictionaries served to standardize these processes, which of course also attracted debate and criticism.

The eighteenth century is an interesting period from the standpoint of language history, partly because of the gradual transition from Latin to the vernacular, and partly in view of the considerable effort devoted to creating nomenclature and terminology. The study of languages at the universities, however, offered no real basis for that endeavour. There, textual interpretation remained the primary concern, i.e. students were taught to read theological and classical works rather than to compose texts of their own.

3. The spread of scientific ideas

In parallel with the concern to reach an interested public, there was of course also a desire that the new discoveries should have an impact internationally. And scientific ideas and findings did indeed spread, across Europe and to other continents as well.

Scientific travel in the form of expeditions was part of the exploration of the world – and, for that matter, of nature. It was in the eighteenth century that Captain James Cook made his round-the-world voyages. And it was during the same century that the botanist Carl Linnaeus undertook his expeditions to different provinces of Sweden, before sending his "apostles" to explore plant life around the globe. The travelogues and journals that resulted were printed and read.

These expeditions also gave rise to international contacts. Among those who accompanied Cook were, on his first voyage, the naturalist Daniel Solander and, on his second, Anders Sparrman, both of them disciples of Linnaeus. It was also common at this time for young men to travel abroad to engage in academic studies. The universities they chose varied over time, partly depending on the professors teaching there. Educational travel of this kind was of course important in disseminating scientific ideas. It provided scholars with an international network which they could later maintain by means of correspondence. Letter writing played a crucial role in the spread of ideas, but also in sustaining networks among scientists.

The eighteenth century, moreover, was a time when encyclopedias and dictionaries were written and printed. In France, Diderot and d'Alembert edited the *Encyclopédie ou Dictionnaire raisonné des sciences, des arts et des métiers*, and in Britain the *Encyclopaedia Britannica* appeared. The printer's art had conquered the world, and it was no longer difficult to produce and distribute books. The *Philosophical Transactions* of the Royal Society in London were also read by an international audience (Atkinson 1999: 27).

4. The development of scientific writing

To the modern-day reader, eighteenth-century texts show little sign of standardization and uniformity. Of course, the strict control of genre conformity nowadays undertaken by editors and established in style sheets and instructions to authors did not exist at that time. The construction of scientificality therefore took place in a freer arena of different models and genres. Nor was the degree of specialization among scientists comparable to what we find today. The academic field of medicine covered botany, zoology, pharmacology, treatment of diseases, health issues and sexology, as is shown for instance by the various topics in the works of Carl Linnaeus. His writing also illustrates how the modern boundary between science and popular science was less clear in those days; scientists knew how to write both for the public and for learned colleagues. The dividing line between science and literature was also less clear; scientists could convey their scholarly findings in a poem (Haskell 2007) or surround them with literary associations. Nor could a clear line be drawn between scientific and religious writing: many scientists, among them Linnaeus, placed the exploration of nature in a religious context, and religious text models also made themselves felt in scientific writing.

In genre terms too, then, the eighteenth century represents a pre-establishment stage. Careful and detailed observation was a common ideal, but as far as textual form was concerned, wide variation was permitted. The latter could of course also be the subject of debate.

5. Europe in the eighteenth century

Many historically oriented studies of texts have focused on scientific writing in English. Here, mention may be made for example of Bazerman (1988), who examined how scientists reported their experiments in the *Philosophical Transactions* of the Royal Society between 1665 and 1800; Atkinson (1999), who analysed the development of scientific writing in the same publication from 1675 to 1975; and Valle (1999), who studied the scientific discourse in the life sciences within the Royal Society from 1665 to 1965. I would also like to mention Grund (2009) on alchemical texts, Gotti (1996) on Robert Boyle's writing, and Taavitsainen and Pahta (2004) on medical and scientific writing in late medieval English. The present volume, however, takes a broader European approach to the emergence of a scientific discourse. Several chapters explore scientific language and texts in eighteenth-century Sweden. Others focus on Germany, Russia, France, Portugal or Britain.[5]

The political geography of the eighteenth century was different from that of today. At the dawn of the century, Sweden was a European power, with dominions in Poland, the Baltic states, Finland and part of Norway, but the ensuing decades saw major changes taking place. Sweden lost its territories in Poland and the Baltic states to Russia. With the economy in ruins after a long period of war, economic growth became the main political aim, and the mercantilist doctrine was declared the great saviour of the country. In the academic sphere, the period saw a flourishing of all the useful sciences, and scientists like Carl Linnaeus, Nils Rosén von Rosenstein and Anders Celsius brought fame to Uppsala University.

6. Carl Linnaeus – a scientist of his time

In this volume, a particular focus is placed on the Swedish botanist Carl Linnaeus (1707–1778). From the point of view of the natural sciences, Linnaeus is renowned for his principles for defining genera and species of organisms and his creation of a uniform system for naming them. From the standpoint of this volume, however, he is also of interest as an example of a European scientist of the eighteenth century. In many ways, the story of his life can be seen as an illustration of what it meant to be a successful scientist at that time. Below, I will begin with a brief biography of Linnaeus.[6]

> Carl Linnaeus was born in 1707 in a province of southern Sweden. His father was a Lutheran pastor and also an avid gardener. In 1727 Linnaeus began to study medicine. After a short period at Lund University, he transferred to Uppsala. At the time, training in botany was part of the medical curriculum, and most of Linnaeus's time was spent collecting and studying plants. Despite being in hard financial straits, in 1731 Linnaeus mounted a botanical and ethnographical

5. Analyses of German, French and English texts from the eighteenth century can also be found in Gross, Harmon and Reidy (2002).
6. This account is an abridged version of the "Biography of Linnaeus" section of the article "Carl Linnaeus (1707–1778)" found at www.ucmp.berkeley.edu/history/linnaeus.html. For more detailed biographies, see Frängsmyr (2004) and Broberg (2007).

expedition to Lapland. This expedition to the north of the country aroused great interest in learned circles in Sweden, and in 1734 he was able to finance another expedition, to the central Swedish province of Västergötland.

As a typical scholar of his day, Linnaeus wished to spend some time abroad. He fell in love with and became engaged to a young girl from a wealthy family. Partly with financial support from his future father-in-law, he managed to spend some important years in the Netherlands. In 1735 he completed his medical degree at the University of Harderwijk, and then enrolled at the University of Leiden. That same year, he published the first edition of his classification of living things, the *Systema Naturae*. During these years abroad, he met or corresponded with Europe's great botanists and continued to develop his classification scheme, which was published for example in *Genera Plantarum* (1737). Returning to Sweden in 1738, he practised medicine (specializing in the treatment of syphilis) and lectured in Stockholm, before being awarded a professorship at Uppsala in 1741. At Uppsala, he restored the university's botanical garden (arranging the plants according to his system of classification), made three more expeditions to various parts of Sweden, and inspired a generation of students. He was also one of the founders of the Royal Swedish Academy of Sciences.

Linnaeus was instrumental in having his students sent out on voyages of trade and exploration to all parts of the world. Meanwhile, he himself continued to revise his *Systema Naturae*, which grew from a slim pamphlet into a multi-volume work. He first published his sexual system in the *Systema Naturae* in 1735, later applying it to every known species in the *Species Plantarum* (first edition 1753).[7] Linnaeus was also deeply involved with ways to make the Swedish economy more self-sufficient and less dependent on foreign trade, either by acclimatizing valuable plants to grow in Sweden or by finding native substitutes. He also found time to practise medicine, eventually becoming personal physician to the Swedish royal family. In 1758 he bought the estate of Hammarby, outside Uppsala, where he built a small museum for his extensive personal collections. In 1761 he was raised to the nobility, becoming Carl von Linné. His later years, however, were marked by increasing depression and pessimism. He probably suffered from a series of small strokes. In 1778 he died at the age of 71. His son, also named Carl, took over the professorial chair in botany, although he was not noteworthy as a botanist. When Carl the Younger died five years later with no heirs, his mother and sisters sold the elder Linnaeus's library, manuscripts and natural history collection to an English natural historian Sir James Edward Smith, who founded the Linnean Society of London to take care of them.

To sum up, Linnaeus was in many ways a scientist of his time. He played an important role in the development of a Swedish scientific community, and unites in an interesting way the connections with university, royal family and state

7. See Broberg (2007: 43).

that were so crucial in his day. Linnaeus held a chair at the prestigious University of Uppsala and was one of the founders of the Royal Swedish Academy of Sciences. He received funding for his expeditions from the state (i.e. from Parliament). He was also personal physician to the royal family. In addition, as the son of a clergyman, he had his roots in the Protestant church. He himself managed to create a synthesis between religion and the natural sciences. Broberg (2007: 33) writes: "The Linnaean project was a combination of themes religious and secular. It was man's duty to wonder at Creation in all its diversity and in doing so to give thanks to the Creator for His generosity. Linnaeus never tires of praising the deity, but as a Creator, not as a Saviour."

As a scientist, Linnaeus devoted himself to collecting plants, minerals and animal species. He made drawings of them, but above all he classified them and ordered them into systems. Naming was of course an important aspect of collecting, and he gave plants names in both Latin and Swedish. He was also typical in his practical, utilitarian thinking. He collected plants not just for scientific purposes, but also as basis for cultivating hardy species that could be used for food and medicine. Linnaeus restored the university botanical garden and had glasshouses built there. He also saw to it that new, exotic plants were brought back to Sweden, all for the purpose of enhancing the country's wealth. In true eighteenth-century spirit, he was interested in horticulture and plant breeding.

Linnaeus's pioneering work became widely known internationally even in his own lifetime. Through an extensive correspondence, he maintained the contacts he had established during his few years in Holland. As his fame grew, his network of international connections became ever larger and his works were cited, translated and debated the world over. Knowledge of his system was also disseminated by his "apostles". To explore nature and acquire new plants, he sent his best students off on expeditions around the world.

Linnaeus wrote his major scientific works in Latin. For his travel writings and letters, however, he used Swedish, often interspersed with Latin words and phrases. He was no purist as far as language was concerned, but he was a remarkably precise observer. His creative use of the Swedish language was not only something new to his contemporaries, but would also prove significant in the subsequent development of Swedish non-literary prose. This is how Broberg (2007: 30) describes his travelogues: "We find ourselves travelling in the company of an all-seeing eye, a horseman continually dismounting to scrutinise the flowers at the road side, making notes and gathering material."

7. The contents of the volume

The seventeen chapters of this volume are organized into four sections. The first section includes three studies which examine, from different vantage points, the forming of scientific communities in the eighteenth century. **Charles Bazerman** explores the conditions for early scientific publication in Europe from a broad, socio-constructivist perspective. He describes how competing political, religious and economic players created conditions that fostered the freedom and growth of empirical science and the emergence of autonomy in scientific writing. Throughout the eighteenth century, the European universities remained under church control, and domains of study followed the traditional, church-regulated faculties of liberal arts, theology, medicine and law. Scientific research and publication thus came to develop outside the university, frequently also beyond the reach of church or state control. As Bazerman writes, this independence from large institutions paved the way for heterodox publication and forced scientists to seek patronage and support from multiple sponsors. Societies of learned people were formed, often under royal patronage. In England, the Royal Society of London was founded in 1660, and in Sweden the Royal Society of Sciences at Uppsala in 1710 and the Royal Swedish Academy of Sciences in 1739. To underscore the importance of this multiple sponsorship, Bazerman compares the European situation with that in the centralized Chinese state four centuries earlier. In his concluding discussion, he also establishes a link to the modern research university.

Gunilla Gren-Eklund provides a general survey of philology in eighteenth-century Europe and Sweden. By focusing on the history of linguistics, and some important intellectuals of the Enlightenment, she introduces a new perspective on scientific discourse and writing. From a discussion of the material background in the history of scholarship, she summarizes what the learned at the time knew and thought about the function, nature and genesis of languages, and what methods they used to study them. She then turns to how academic philology was pursued at Nordic universities in the eighteenth century and gives an inspired presentation of Johan Ihre, professor of Oriental languages at Uppsala University. Ihre, who was contemporary with Carl Linnaeus, enjoyed an international reputation for his comparisons of languages. Among Swedish philologists at the time, however, he was fairly unique in his focus on the theoretical side of scholarship. Gren-Eklund also reflects on practical knowledge of languages at the time, i.e. what languages were taught and which ones scholars knew. She claims that it is clear that the professors at Uppsala University had a mastery of European languages. It is also clear that this was a result not of teaching at the university, but of their international contacts and early studies abroad. German and Dutch universities were of importance for Swedish professors, and many had also visited the universities of England and France, and even of Italy and Spain.

Ulf Teleman, in his chapter, considers the language policy and language practices of the Royal Swedish Academy of Sciences. He focuses on the role of this Academy, founded in 1739, in the development of scientific writing in Swedish. Although no linguist was a member of the Academy, the cultivation of the Swedish language was one of its primary aims. The Academy supported essential language projects, and its members debated various language-related issues. More important for the development of a Swedish for scientific purposes, however, was the Academy's decision to use Swedish in its Transactions. In the second part of his chapter, Teleman turns to an analysis of these publications. He discusses the results of a comparison of the style and language of transactions from a few volumes from the early years with those of transactions published around 1775. With the emphasis on the transition between Latin and Swedish, or between traditional academic writing and speech and writing outside the learned world, he compares textual organization, objectivity, rhetorical devices, lexical apparatus, syntactic machinery, techniques of presentation and standardization in the texts. In his conclusion, Teleman writes that, according to the principle of utility, the new style chosen for the Academy's Transactions was based on traditional Latin and non-academic Swedish prose. He claims that this entailed an early move towards standardization of the language.

The second section comprises five chapters exploring the emergence of new discourses of science in the eighteenth century. **Renata Schellenberg** examines the evolution of scientific literacy in eighteenth-century Germany. As she claims, German intellectual culture flourished during this century, establishing a national presence and a strong tradition of academic thought. According to Schellenberg, the most significant effect of the German Enlightenment was on language itself, as writers of the time increasingly endorsed German as a language of learned discourse. As there was no consensus regarding the status of German as a scientific language among the new German academies, successful efforts to standardize German came instead from individual scientists and from private organizations and gatherings. Schellenberg also stresses the importance of periodicals, such as the well-known *Berlinische Monatsschrift*. As a popular and new genre, the periodical created a new dynamic network of communication between readers and writers. The many scientific disputes were also important for the spread of ideas and, in addition, for the development of a precise articulation of ideas. As an example of a debate which contributed to scientific literacy in Germany, she mentions the famous one on preformation vs. epigenesis, which involved a number of famous scholars: Albrecht von Haller and

Caspar Friedrich Wolff, as well as Johann Friedrich Blumenbach and Immanuel Kant.

Anna Helga Hannesdóttir suggests a fresh approach to the problem of the transition from vernacular Swedish to a fully-developed medium of scientific discourse. She argues that the sociolinguist Einar Haugen's notion of "language planning" can also serve as a model to analyse and explain language change from a sociolinguistic point of view. In her chapter, she applies the four phases of this LP model, i.e. selection, codification, implementation and elaboration, to describe the progression which Swedish underwent, from a poorly codified vernacular to a developed, standardized language. With reference to these four phases, she elaborates on the role played by the Swedish botanist Carl Linnaeus. According to Hannesdóttir, Linnaeus supported the selection and implementation phases by promoting Swedish rather than Latin for the lectures and transactions of the Royal Swedish Academy of Sciences. Linnaeus also contributed directly to the elaboration and codification phases by creating a Swedish vocabulary appropriate to his scientific findings.

Lars Wollin sheds light on the decline of Latin and the emergence of Swedish in scientific writing, a process that was promoted by the Swedish Academy and the Swedish dictionaries which it commissioned. From a discussion of the overall proportions of Swedish and Latin in early book printing in Sweden, he turns to an analysis of the frequency of loanwords with eight particular Latin suffixes and prefixes. By means of this analysis, he is able to compare the relative distribution of Latin words in common and professional language (*fackspråk*). Wollin finds it useful to summarize the relationships observed in chronologies specific to each type of language. His conclusions are further presented as a hypothesis concerning the character of the gradual integration of loan words into a receiving language.

The following two chapters have Carl Linnaeus and his writing as a starting point. **Richard Sörman's** contribution offers a picture of the scientific controversies and debates of the eighteenth century. The two influential scientists Georges Louis Leclerc de Buffon and Carl Linnaeus never corresponded with each other, and Linnaeus is not mentioned in the *Encyclopédie française*. Buffon published his main work *Histoire naturelle* from 1749 onwards, and his chief objections to Linnaeus were the alleged arbitrariness of the system which he imposed on nature (i.e. the deductive principle) and the creation of artificial terms. As the analysis presented in this chapter reveals, Buffon's opinions on writing largely relate to the aesthetics of French seventeenth-century classicism. In the midst of the French Enlightenment, he expresses an anti-modern view of scientific research. Sörman's conclusion is that Buffon's critique of modern scientists' use of language is far from out of date. Instead it raises the general and ahistorical question of the effective value of abstract language as a tool for understanding and describing reality.

In the next chapter, **Philippe Selosse** broadens the perspective on Linnaeus's botanical nomenclature to include a philosophical level. Selosse claims that Linnaeus's nomenclature can be seen as a point of view on the world, and relates his ideas to philosophers such as Gottfried Wilhelm Leibniz and Francis Bacon. According to Selosse, Linnaeus's work constitutes a genuine synthesis of earlier theories and those that followed, at the same time as his nomenclature is "self-conscious" in a new way. The Linnaean nomenclature is "conscious" of being a system as such, a perspective that was a novelty at the time. In order to explore Linneaus's theory of nomenclature, Selosse chooses to focus on his use of the Latin word "succedaneum". This word (almost a synonym for "substitute") was common in pharmacopoeias in the eighteenth century, where it applied to a drug substituted for another because they shared some common properties. In Linnaeus's aphorisms, however, "succedaneum" frequently occurs with conceptual rather than medicinal applications. In his chapter, Selosse finds Linneaus's use of "succedaneum" relevant to an illustration of how his various taxonomic and linguistic concepts can be conceived in a homogeneous epistemic frame.

Turning next to the third section of the volume, we find four studies dealing with the spread of scientific ideas in the eighteenth century. **Ann-Mari Jönsson** presents a study of Linnaeus's correspondence with leading botanists in Holland, England, Germany, Switzerland and France. Jönsson begins her chapter with an overview of this international correspondence. Linnaeus had some 600 correspondents, and the total correspondence is estimated to have consisted of around ten thousand letters. Jönsson's main claim is that the Linnaean correspondence is not only an important source for an understanding of his scientific work, but also reveals how he disseminated his new ideas and handled criticism, that is, how he spread his "scientific revolution". Jönsson discerns three stages in this revolution. The first is seen in 1735, when Linnaeus published his *Systema naturae*. As the correspondence from this period shows, his ideas were well received in a personal circle of Dutch botanists, but met with compact resistance from botanists in the German-speaking countries, England, France and Italy. The second stage can be dated to the 1740s. During this period many of his correspondents made it clear to him that he was now in the middle of an open war. The revolution began to spread to wider circles, including botanists in New York and Zürich. The men of the second stage regarded themselves as reformers and missionaries. The third stage comes in the 1750s, when his correspondents wrote that he had finally and decisively won his war. Linnaeus had laid the theoretical ground for his revolution in *Philosophia botanica* in 1751, and it can be seen as culminating in the publication of his *Species plantarum* in 1753. The men of the third stage finally confirmed Linnaeus's revolution.

In the following chapter, **Rosemarie Gläser** deals with Linnaeus's influence on the *Encyclopaedia Britannica* of 1771. After a brief comment on the Swedish botanist's connections with English and Scottish scholars, and a general presentation of the first *Encyclopaedia Britannica*, Gläser focuses her discussion on some prominent articles in the fields of botany, zoology and medicine which reveal the influence of Linnaeus's work. The breakthrough of his system of classification and nomenclature for plants and animals, she writes, came with the decision of the editors and authors of the *Encyclopaedia Britannica* of 1771 to prefer Linnaeus's system to similar attempts by competing contemporaries. The repercussions of the Swede's work as a biologist and physician are exemplified by instructive passages drawn from articles and treatises included in this national work of reference. As Gläser's analysis shows, Linnaeus's ideas were well received in the British Isles.

Kenneth J. Knoespel then shifts our focus to Russia. He discusses the commitment of Linnaeus to the Swedish expeditions to Siberia that began in 1724. Linnaeus's multifaceted response to information from these expeditions allows us to follow his interaction with the newly founded Imperial Academy of Sciences at St Petersburg (1725) through his botanical research at the University of Uppsala. The works of Linnaeus's disciples inspired him to create a plot of Siberian plants in his garden in Uppsala. Moreover, he was interested in how plants were named in different languages – Swedish, Russian, German and Latin – and how they could be transferred from one environment to another. Using extracts from the "Russian dissertations", which in line with the academic tradition of the time were written by the professor, i.e. Linnaeus himself, Knoespel elaborates on ways in which his work led to the stabilization of codes and the development of strategies that could be used to share information across nations. According to Knoespel, Linnaeus's interaction with Russian correspondents – in Latin and German as well as through drawings, diagrams and physical specimens – showed a growing awareness of how a universally shared language of natural history could replace a political empire with an "empire of knowledge".

Linnaeus's work also had a major impact on terminology and nomenclature in countries in southern Europe. In her chapter, **Palmira Fontes da Costa** describes how the Linnaean language of nature was influential not only in Portugal, but also in the Portuguese Empire. A particular emphasis is placed on Domingos Vandelli's dictionary of technical terms of natural history (published in Portuguese in 1788), which were in fact extracted from the works of Linnaeus. This dictionary propagated the Linnaean system of classification and its nomenclature. As da Costa writes, the publication of a Portuguese flora using the Linnaean system was associated with important national aims, and the Linnaean method of classification contributed to a reshaping of botanical education in Portugal. To demonstrate the variety of works that contributed to the diffusion of Linnaean ideas of classification in Portugal, da Costa writes about the Marquesa de Alorna's poem *Botanical Recreations.* This work is mentioned as an example of the influence of presentations of the Linnaean classification in texts which crossed the boundaries between science and literature.

The fourth section of the volume, finally, contains five chapters on the development of writing in the eighteenth century. Here, texts on botany and medicine are studied from different angles and using different methodologies. **Bo Ralph**, in his chapter, emphasizes the role of Carl Linnaeus in Swedish language history. The modern reader of Linnaeus's travel writings might be struck by the numerous passages in Latin that are interspersed throughout the Swedish texts, but certainly also by his keen-sighted observations formulated using concrete and illustrative Swedish words and expressions. As Ralph notes, Linnaeus's particular style of prose reveals both his acquaintance with the rhetorical tradition of Latin and his unbiased creativity. Neither Linnaeus himself nor his students took an active part in the language debate as such. Linnaeus's importance lies, rather, in his establishment of a "model" for scientific writing in Swedish. According to Ralph, the particular kind of prose Linnaeus used in the diaries from his journeys to different parts of Sweden in the 1730s and 1740s was taken as a model by his "apostles". A homogeneous type of factual prose for scholarly purposes was thus developed, characterized by Linnaeus's stylistic ideals and his manner of detailed reporting. Ralph argues that, by using Swedish for scientific purposes, the Linnaeans came to contribute to the standardization process. He further claims that Linnaeus's influence can also be extended to Swedish literature in general and to the modern breakthrough of impressionistic Swedish prose. Thus, the opening passage of the novel *Röda rummet* (1879) by the Swedish author August Strindberg echoes Linnaeus's rapid and expressive use of Swedish in his diaries.

Han-Liang Chang's chapter also deals with the writing of Carl Linnaeus, in this case with a focus on two Latin texts: *Philosophia Botanica* (1751) and *Fundamenta Botanica* (1736). Chang elaborates on the generic conventions of "calendar" and "aphorism", thus tracing the generic sources of Linnaeus's style. In order to picture the two faces of Linnaeus, he compares the strict classification found in *Systema Naturae*, which was encoded in scientific Latin, with the narrative – and sometimes even poetic – style of Linnaeus's travelogues, which were written in Swedish. From a stylistic point of view, Chang claims that *Philosophia Botanica,* which was developed from *Fundamenta Botanica*, is unique, as Linnaeus makes ample use in these calendars of the discursive device of the aphorism. As Chang writes, the calendars exhibit stylistic similarities with his travelogues, in that they are "random and sketchy, discursive but inconclusive". In this respect, they follow the essayistic and aphoristic tradition of Renaissance authors, and of Francis Bacon in particular.

Botany is also the topic of the texts analysed by **Andreas Nord**. Using a textlinguistic approach, Nord examines style and textual patterns in Swedish gardening literature from the eighteenth century. The material analysed comprises excerpts from eight popular handbooks on gardening aimed at laymen (private garden owners) from different socio-economic classes or at practising professionals (gardeners or garden apprentices). In his analysis, Nord uses the notions of social semiotics, appraisal theory, and the concept of the "model reader". Among other things, his study revolves around whether the texts are oriented towards "action" or "knowledge", and goes on to reveal the occurrence of two different patterns, or "model readers": one action-oriented and one more critical or reflective. The gardening texts examined orient towards different ideal readerships and different views of knowledge. It may be assumed that both types of reader position were necessary in raising the level of knowledge about practical cultivation and gardening techniques in eighteenth-century Sweden.

The last two chapters of the book deal with Swedish and English texts on medicine. **Britt-Louise Gunnarsson** explores how eighteenth-century medical professionals textually created scientificality in their writing in Swedish. Her corpus comprises twelve articles on smallpox and cataract, published in the Transactions of the Royal Swedish Academy of Sciences between 1750 and 1769. Applying her socio-constructivist framework, which relates language and textual patterns to three different levels of texts – cognitive, social and societal – she explores how the authors of the time constructed a trustworthy description of their findings. The detailed analysis of the texts includes a categorization of their content in terms of five "cognitive worlds", a description of how the content is structured, and the way medical cases are introduced in the texts, as well as a description of the use of references, names of colleagues, personal pronouns, terms and figures. At a cognitive level, the proportion of different "worlds" in the texts and the structuring of the content suggest a pre-establishment stage of medical science. At a social level, the authors' use of references, colleagues' names and personal pronouns reveals a small scientific community with a manifest connection between one individual and another. At a societal level, finally, the use of terms and figures, as well as the way the medical cases are introduced, suggests a close relationship between the author-scientist-doctor and his patients, and also between scientists-experts and other groups in society.

Päivi Pahta, for her part, deals with medical writing in English. She examines eighteenth-century English medical texts and discourses on reproduction in their socio-historical contexts. The texts discussed are part of the *Corpus of Early English Medical Writing*, which includes texts from 1375 to 1800. This study takes in theoretical sources on embryology, as well as practically oriented texts on human reproduction, obstetrics and gynaecology. In her analysis, which is mainly qualitative and contextualized, Pahta focuses on the cognitive layer of discourse, the concepts writers use to construct knowledge, and also how they indicate the sources of knowledge and the certainty of their claims. This study confirms a clear trend, from authority-dependent and low-modality modes of discourse to a more complex and layered view in seventeenth- and eighteenth-century texts. The old discourse patterns continue to dominate in texts aimed at general audiences and female professionals, while new patterns emerge in the upper strata of the discipline. In the latter, Pahta concludes, professionals are no longer just transmitters of knowledge constructed by earlier generations, but are themselves actively engaged in constructing new knowledge by their experiments and observations and drawing on their own cognitive processes.

To conclude, this volume is unique both in its broad linguistic approach – including studies on textlinguistics, stylistics, sociolinguistics, lexicon and nomenclature – and in its combination of language studies, philosophy of language, history and sociology of science. The book covers writing in different European languages: Swedish, German, French, English, Latin and Portuguese. With its focus on the history of scientific language and discourse during a dynamic period in Europe, the book promises to contribute to new insights both for readers interested in language history and for those with an interest in the history of ideas and thought.

References

Annerstedt, Claes
1912 *Upsala universitets historia. Bihang III. Handlingar 1695–1749* [The history of Uppsala University. Appendix III. Transactions 1695–1749]. (Uppsala universitets årsskrift 1912.) Uppsala: Universitetet.

Atkinson, Dwight
1999 *Scientific Discourse in Sociohistorical Context. The Philosophical Transactions of the Royal Society of London, 1675–1975*. Mahwah, New Jersey & London: Lawrence Erlbaum.

Bazerman, Charles
1988 *Shaping Written Knowledge. The Genre and Activity of the Experimental Article in Science*. Madison, Wisconsin: The University of Wisconsin Press.

Berger, Peter and Thomas Luckmann
1967 *The Social Construction of Reality*. Harmondsworth: Penguin.

Broberg, Gunnar
2007 *Carl Linnaeus.* Stockholm: Swedish Institute.

Crossgrove, William
1998 Introduction. *Early Science and Medicine: A Journal for the Study of Science,Technology and Medicine in the Pre-modern Period* 3/2, Special issue: *The Vernacularization of Science, Medicine, and Technology in Late Medieval Europe: 81–87.*

Frängsmyr, Tore (ed.)
2004 *Linnaeus. The Man and His Work* (Second printing). (Uppsala Studies in History of Science, Volume 18.) Sagamore Beach, MA: Science History Publications.

Gotti, Maurizio
1996 *Robert Boyle and the Language of Science.* Milano: Guerini.

Gross, Alan G., Joseph E. Harmon and Michael Reidy
2002 *Communicating Science. The Scientific Article from the 17th Century to the Present.* Oxford and New York: Oxford University Press.

Grund, Peter
2009 Textual alchemy: The transformation of pseudo-Albertus Magnus's *Semita Recta* into the *Mirror of Lights*. *Ambix: The Journal for the Society for the History of Alchemy and Chemistry* 56(3): 202–225.

Gunnarsson, Britt-Louise
1997 On the sociohistorical construction of scientific discourse. In: Britt-Louise Gunnarsson, Per Linell and Bengt Nordberg (eds.), *The Construction of Professional Discourse*, 99–126. London and New York: Longman.

Gunnarsson, Britt-Louise
2005 Medical Discourse: Sociohistorical Construction. In: Keith Brown (ed.), *Encyclopedia of Language and Linguistics* (2nd Ed.), Vol. 7, Article 2360, 709–716. Oxford: Elsevier.

Haskell, Yasmin
2007 Advancement of Science or Self-Promotion? A Dutch Doctor's 'Grand Tour' in Latin Verse. Paper presented at the Symposium on the Languages of Science in the Time of Linnaeus. Uppsala University, Sweden, June 7–9, 2007.

Heckscher, Eli F.
1953 *Merkantilismen* [Mercantilism] (Andra upplagan). Stockholm: Norstedt.

Johannisson, Karin
1988 *Det mätbara samhället: Statistik och samhällsdröm i 1700-talets Sverige* [A measurable society: Statistics and the societal dream in eighteenth-century Sweden]. Stockholm: Norstedt.

Latour, Bruno and Steve Woolgar
1986 *Laboratory Life. The Construction of Scientific Facts*. Princeton, NJ: Princeton University Press.

Lindroth, Sten
1978 *Svensk lärdomshistoria. Frihetstiden* [The history of science in Sweden. The Age of Liberty]. Stockholm: Norstedt & Söner.

Porter, Roy (ed.)
2003 *The Cambridge History of Science. Volume 4. Eighteenth-century Science.* Cambridge, UK: Cambridge University Press.

Roger, Jacques
1963, 1997 *The Life Sciences in Eighteenth-Century French Thought.* Stanford, California: Stanford University Press.

Taavitsainen, Irma and Päivi Pahta
2004 *Medical and Scientific Writing in Late Medieval English.* Cambridge, UK: Cambridge University Press.

Valle, Ellen
1999 *A Collective Intelligence. The Life Sciences in the Royal Society as a Scientific Discourse Community 1665–1965.* Anglicana Turkensia No. 17. Department of English, University of Turku, Finland.

Wendt, Bo A.
2005 The development of the types of text in the Nordic languages from the 16th to the end of the 18th century. In: Oskar Bandle et al. (eds.): *The Nordic Languages. An International Handbook of the History of the North Germanic Languages.* Volume 2, 1346–1356. Berlin, New York: Walter de Gruyter.

Section 1.
The forming of scientific communities

Church, state, university, and the printing press: Conditions for the emergence and maintenance of autonomy of scientific publication in Europe[1]

Charles Bazerman

In early modern Europe the multiplicity of competitive political, religious, and economic players created conditions of both support and freedom that seeded the free flow of knowledge, the flourishing of competing knowledge claims, and the growth of science. Yet, surprisingly from a modern perspective, the university was not a central part of this story. When the new state-sponsored research university emerged in the nineteenth century it maintained elements of autonomy for both scholars and scholarly publishers that fostered scientific freedom. It is not clear, however, how the contemporary reconfigurations of what has been called the triple helix of state, industry and science may restrict the university, i.e. science and scientific publication, diminishing its autonomy to support the free growth of knowledge.

In Europe, universities, from their medieval invention through the eighteenth century, had remained largely under church control. Domains of study followed the church regulated traditional faculties of liberal arts, theology, medicine and law. Empirical science was little pursued or studied within university walls. During the seventeenth and eighteenth centuries, scientific research and publication developed largely outside of the university, frequently outside the reach of church or state control. The complex fracturation of power in Europe from the sixteenth through the eighteenth centuries, meant that scientific inquiry and publication could escape the domination of a unified political or religious authority.

As this period developed there was some state patronage for individuals and state charter for scientific societies, but this reflected more the desire of the state to enlist the emerging value and prestige of science rather than to exert authority over it. Rather it was individuals acting as entrepreneurs, privately

1. This essay draws heavily on chapters nine and ten of the Handbook of Research on Writing. I co-wrote these chapters with Paul Rogers, whom I thank for all his assistance and collaboration.

organized societies, and especially printers and publishers who were responsible for the production, communication, discussion and development of sciences prior to the nineteenth century. This independence from large institutions, and especially state or church control allowed the license for exploration and heterodox publication. At the same time this independence allowed scientists and their compatriot publishers to seek patronage and support from multiple sponsors, each of whom saw different values and opportunities in the new forms of knowledge.

Eighteenth century Scotland, Germany, and Sweden, however, foreshadowed a change in university role in the production and dissemination of knowledge, bringing the university into the center of new alliances with the state and publishers. These new alliances, reflecting the value of knowledge to the state rather than the state's fear of the uncontrolled proliferation of knowledge, were accompanied by ideologies and arrangements that fostered academic independence as well as practicality – ideas that would develop in the secular research university of the nineteenth century. These new arrangements changed the character and conditions of scientific publication, as well as the sponsorship of scientists and science.

1. The printing press and changing networks of knowledge in Europe

In the early European Middle Ages classical knowledge was limited to a few Latin texts and compendia derived from them. The modern university was born out of a curiosity about texts arriving during the 11th and 12th centuries in Europe through contact with Islamic scholarship held in the libraries of the Umayyad courts of Spain. Scholars in the monasteries and larger cities of Europe began translating and studying such texts as Ptolemy's synthesis of the work of Greek astronomers, known through its Arabic title *al-Majisti* or *Almagest* (Ridder-Symoens 1991). As available texts increased, students and scholars gathering in greater numbers organized themselves in guilds to form the bases of universities. Monastic and commercial copyists were of course important in providing texts for the libraries that were at the heart of these universities. By the end of the twelfth century, universities existed at Salerno, Bologna and Reggio, and soon others emerged at Vicenza, Palencia, Paris, Oxford, Montpelier, Arrezo, Salamanca, Padua and Naples. By 1500 over sixty universities were active throughout Europe: from Uppsala in the north to Catania in Sicily in the south; from Lisbon in the west to Cracow in the east (Verger 1991). From the middle of the fourteenth until the start of the sixteenth century, approximately three quarters of a million students matriculated throughout Europe (Schwinges 1991). Two forces served to organize and support this movement, shaping its destiny until the reforms of the nineteenth century. The contemporary economic system of guilds provided the internal organization, defining structures of faculty governance, student rights, and protection of the interests of guild members. The church provided sponsorship and curricular regulation – around the four faculties of Liberal Arts, Theology, Law and Medicine.

During the Middle Ages, the close nexus of the universities, the church, scriptoria, and education for church careers kept universities at the center of the knowledge maintenance, dissemination, and production. The Protestant Reformation and the accompanying religious struggles of the sixteenth and seventeenth centuries, moreover, did not fundamentally change the church-bound character of the universities, although changing some allegiances and disrupting the Vatican's universal curricular authority.

In the fifteenth century, however, knowledge moved out into the world. The moveable type printing press, along with related inventions and social arrangements made books available in increasing quantity (Eisenstein 1979), accelerating and transforming a process that had already begun in the scriptoria (McKetterick 2003). Increasingly, scholars were freed from the university or monastery library and from church supervision. Even more, the printing houses proliferating across Europe no longer came under a single religious jurisdiction and therefore could not be uniformly censored or controlled, nor did they serve a single international organization. Separate states had neither wealth nor jurisdictional reach to keep the production of texts subservient to their needs. To underscore the importance of this multiple sponsorship for creating substantial autonomy for printers and scholars, it is useful to compare the European situation to those in the centralized Chinese state four centuries earlier when printing was first invented, but with very different consequences as it became an instrument of state power.

2. The centralization of power, knowledge, and printing in China

In China, long before the introduction of printing, the national order was administered by an elite trained in classical learning and its ideals. Knowledge and its production was regulated through a system of rewards and controlled dissemination among government officials. Valued learning was institutionally regulated by the imperial civil service examinations which lasted over two millennia, until the final collapse of Imperial power in the early twentieth century. The Han dynasty (206 BCE–220 CE), to repair the destruction of books by the preceding Qin dynasty (221–207 BCE), fostered bureaucratic expertise in philology and lexicography through instituting scholarly examinations. For the next two millennia those exams defined the aims of schooling, what texts were valued, and the literate development of every individual seeking power and place. Further, the need for objectivity of evaluation lead to a narrowing of the canon of texts concerned, a formalization of the questions and a ritual patterning of expected answers in the notorious eight-legged essay based on eight matched pairs of opposing concepts. As the most valuable knowledge was that which would provide advantage on the examinations, much scholarly production was summary, commentary and interpretation of the classic texts. Some of these commentaries in turn became part of the examined canon (Lee 2000).

Throughout a two thousand year period, there was great consistency in the ethical, philological, literary and aristocratic knowledge valued in the civil service, the exams and the schools that prepared candidates. Learning in the law, medicine, astronomy, mathematics and military arts was also supported in preparation for appropriate civil service roles for the control and maintenance of the economy and national welfare (Ronan and Needham 1981; Needham and Lu 1970a, 1970b).

However, while these knowledge domains, useful to the state, had some coherent development and expanding literature, other areas of knowledge were sporadic with little organized distribution of texts or institutional support. The many technological advances made in agriculture, textile manufactures, mining, fishing, construction, weaponry, explosives, mechanical and civil engineering, ship-building and other arts and crafts were developed largely by artisans, workers, crafts people or people in the lowest rungs of the state bureaucracy. Higher level administrators trained in the classics had at most a supervisory role in the development of these practical arts. Thus the makers of practical knowledge neither were educated and highly literate nor had they access to the means of publication and text distribution. Practical work tended to be atheoretic and did not depend on the dominant educated thought systems of Confucianism, Taoism and Buddhism. Sometimes inventions and discoveries remained local and sporadic because of the lack of textual transmission. When this practical knowledge did spread, it was through objects and practices. It was thus in these concrete forms that much of this knowledge was diffused to India, the Islamic world and Europe (Needham 1970).

While in Europe the invention of the printing press was to foster novel texts, new communities of knowledge seekers and producers, and new disciplines of learning, in China the much earlier invention of printing – block printing by the eighth century CE and movable type circa 1041–1048 (Carter 1955) – led to much less diversity. The control of the press remained largely in the hands of the state and monasteries (Luo 1998). As a result most mass-produced and widely circulated print documents reflected bureaucratic functions of the state, the literary classics and commentaries associated with examination, religious scriptures and government issued paper money. Sometimes leisured elites used the government press for publication of special interest limited-editions of their poetry and avocations, reflecting their educated tastes steeped in the classics. When private printing flourished (often based in private academies), it too was dominated largely by the culture of the classically-based examination system. Only during the Ming (1368–1644) and Qing (1644–1911) dynasties did private printing of vernacular texts (such as popular novels and tales, books on crafts and technology, and gazetteers) appear on a large scale. However, most private printing remained devoted to such ritual artifacts as New Year pictures and funerary money. Thus the printing press largely supported and participated in the same world of knowledge fostered by the government civil service and examinations.

3. Learning as a competitive force in Europe

In Renaissance and early modern Europe, however, learning became a competitive force that could enhance the status and power of monarchs, starting with the great merchant princes of Italy who patronized such scholars as da Vinci and Galileo (Biagioli 1993). Monarchs throughout Europe patronized scholars and brought them to court to bring grandeur and luster, if not the vision of a new world, as in the court of Rudolph of Austria (Evans 1973). In the free city of Magdeburg, Otto von Guericke rose to power in part on his demonstrations of learning, which he then turned to the benefit of the state (Bazerman 1993). Printing houses saw themselves as beyond the force of any state and began to fashion themselves as a Republic of Letters, spreading cosmopolitan thoughts and ideals (Eisenstein 1979). Gaining knowledge of each other through books, scholars across Europe engaged in lively correspondence networks.

Science, previously called natural philosophy, has been closely associated with consequences of the printing press, i.e. with easier access to classic texts, with wide and rapid dissemination of new data, observations and theories, with the reproduction of exact descriptions, tables, illustrations and maps that allowed the comparison and aggregation of astronomic, geographic, botanic, zoological and anatomic data, with the impetus to criticism, commentary, taxonomy and theory based on the access to multiple sources which then could be compared to new results, and with the impetus for improved maps, illustrations, tables and taxonomies to meet the book-buying market (Eisenstein 1979). Publishers were instrumental in creating cultures of trust that allowed readers to rely on the authority of editions untainted by piracy and other forms of immorality and amorality (Johns 1998). While universities, scriptoria and monasteries formed communities of trust within which books could be selected, shared-interpreted and evaluated, the proliferation of copies of printed books seemed to set them free of social context, which needed to be re-created around the networks of publishers, authors, collectors and sponsors. These new communities of knowledge, communicating across national and religious boundaries, challenged the authority and legitimacy of at least one state, England, in the seventeenth century (Jacob 1976; Shapin and Schaffer 1985). The restored monarchy in England needed to position itself warily with respect to natural philosophic inquiry, which it sequestered apart from public discourses of faith and royal legitimacy. In the eighteenth century new philosophy, knowledge and rationalism formed the ideology of American and French revolutions, the Napoleonic empire, and the consequent nineteenth century remaking of the European political/administrative landscape.

In urban areas where new learning thrived outside the walls of universities or government, societies of learned people formed to share their readings, thoughts, and discoveries, as well as to support and criticize their new claims to knowledge. These societies, often enjoying patronage of rich families or royalty, became the centers of learning. The Scholarly Societies Project (www.scholarly-societies.org) has identified thirty such societies prior to 1600. The earliest that specifically turned its attention to natural philosophy appears to be the Accademia dei Segreti founded by Giambattista della Porta in 1560 in Naples and lasting twenty years until shut down by ecclesiastical opposition. Among the other early scientific societies were the Accademia dei Lincei (1603–1630 in Rome), Accademia degli Investiganti (circa 1650–1670 in Naples), and the Accademia del Cimento (1657–1667 in Florence). In 1660 the Royal Society of London, the oldest scientific society in continuous existence, was organized from a series of informal meetings. As the first in Scandinavia, the Royal Society of Sciences at Uppsala (Kungl. Vetenskapssocieteten i Uppsala) was founded in 1710 and the Royal Swedish Academy of Sciences (Kungliga Svenska Vetenskapsakademien) in Stockholm in 1739. At first communication among scientists across Europe was facilitated by active letter writing with some individuals becoming the centers of correspondence, such as Marin Mersenne (whose correspondents were to form the basis of the Académie Royale des Sciences) and Henry Oldenburg (who was secretary of the Royal Society of London). Out of these two networks were to form in 1665 the first scientific journals *Le Journal des Sçavans* and the *Philosophical Transactions of the Royal Society*. While the earliest journal issues carried the trappings of letter correspondence, this was to rapidly evolve into distinctive authored articles.

By 1790 over 1000 scientific journals had appeared, at least briefly, of which three quarters presented original contributions and/or were society proceedings (Kronick 1976). Currently the Scholarly Societies Project indexes over 4000 societies.

The interest in nature was coupled with a desire for language appropriate for communicating about nature. The wide availability of detailed descriptions and illustrations of botanic species, for example, vexed prior taxonomy, as principles were needed to aggregate and organize these many species in collections. Bacon (1603), in *The Advancement of Learning,* argued that we often mistake words for things and lose sight of the things themselves: words come to us filled with unconsidered and unsubstantiated associations, and words sometimes name things that do not exist or that are ill-defined. Bacon expressed a desire for a method of notation that would not be deluded by what he called the Idol of the Marketplace. His critique inspired projects for universal languages which could be used to record and organize all knowledge in its true form – the best known of which is Bishop Wilkins *Essay towards a Real Character and a Philosophic Language*. Bacon's (1620) description of Solomon's house in the *Novum Organum* set out a communal project for the gathering, inscription and interpreting of knowledge of nature that inspired the Royal Society. Thomas Sprat's (1667) hyperbolic description of *The History of the Royal Society* sees language purification at the heart of the society's project. Despite hopes for a language that transcended rhetoric, scientific writing was always to remain persuasive and argumentative, but the grounds of the argument were to shift to accounts of empirical experience. A plainer style, less reliant on ornaments, was to influence pages of the new scientific journals. Nonetheless, figures of speech and thought (such as antithesis, series and repetition) were to remain an essential part of scientific writing (Fahnestock 1999).

Journal publication and society meetings created new forums for scientific arguments that had previously been published in books that were only publicly contestable years later in new books (Bazerman 1988). Further books contained such a myriad of details and claims that it would be difficult to focus a specific disagreement across books. At Royal Society meetings, however, the heart of the argument was a physical demonstration of an empirical reality (Dear 1985; Shapin and Schaffer 1985). Issues of detail could be directly debated. Further, the rapid response available in journals allowed for controversies to be argued with many rounds of responses. But as journals could contain only accounts of demonstrations, to be read by distant audiences, the credibility of the witnesses and the impressiveness of the described apparatus carried persuasive value. At first, credibility drew on earlier social resources for gentlemanly credibility, but, over time, scientific expertise became the source of credibility (Shapin 1994).

Credibility also came to be enhanced by the scientific credibility of the editor of the journal and the persons who were to assist in the evaluation, criticism and selection of the articles in what emerged as a system of referees by the middle of the eighteenth centuries. These social changes were accompanied by transformation of a more gentlemanly style for a more overtly contestative and professional one (Atkinson 1999; Gross et al. 2002), expressing evaluations through facts, use of the literature and irony rather than overt first-person judgments (Gunnarsson 2001; Myers 1989, 1990b). This professional discourse had unique features that set it apart from languages in other social domains and made it increasing difficult for non-specialist and amateur reading (Halliday and Martin 1993; Battalio 1998). Differing historical, social, cultural and economic circumstances in different countries lead to distinct kinds of journals and forms of articles (Gunnarsson 1997; Gross et al. 2002).

Controversy was to erupt on the pages of the journals as natural philosophers questioned each other's results. More detailed accounts of the conditions and actions that led to the results soon followed, as did quantification and precision in reporting the results. More extensive reasoning connecting theory and research design and results led to theoretical claims being supported through experimental and other methodologically focused empirical evidence (Bazerman 1988). Changing ideological beliefs about the value of collective experiences along with the mounting accumulation of empirical results led to the development of modern practices of citation and reviews of literature in the latter part of the eighteenth century (Bazerman 1991). Many of the rewards and values associated with participation in science developed in conjunction with journal publication and served to reinforce participation within the journal system (Merton 1973; Bazerman 1988). Recurrent violation of these values in terms of misrepresentation of parts of the experiments and results, plagiarism, lack of supervision, collusion, or self-delusion serves to illustrate how strongly rewards are tied to values. The periodic scandals and calls for self-policing indicate how much hangs on the reliability of the system threatened by such acts (Broad and Wade 1982; LaFollette 1992).

The systems of publication and authorship grew hand in hand with the formation of modern science. The work of scientists to contribute to knowledge was directed and focused for publication in the emerging journals. Scientists adopted roles of editors, critical readers, and referees as they became engaged in journal production. Communal values of criticism, shared production of knowledge, and objectivity became formulated around the conflicts of the publication process. And the published literature came to stand for the accumulated accomplishment of the sciences. Within that simultaneously cooperative and agonistic social system, the concept of the individual scientific authorship and credit for discoveries arose as a reward along with accountability and responsibility for claims (Merton 1973), although authorship has in recent decades been transformed through the emergence of large collaborative science (Biagioli and Galison 2003). Further, within the social organization of reviewing, criticism, publication, and uptake, even the singly authored article is a social accomplishment (Myers 1990a).

4. Systems of worldly knowledge

Although the emergence of modern science is seen as paradigmatic of the growth of knowledge, many other systems of knowledge were also developing in the renaissance and early modern Europe, including commercial, journalistic, technical, colonial governmental and military. Each of these developed somewhat separately from the other. Each had their own documentary systems, different uses for print media, and restrictions on the free flow of information. Eventually, however, they were all to find common interest in the modern research university, ultimately putting pressures on what the university should be producing and how its knowledge should be circulated or restricted in access.

Commercial information was and remains in large part proprietary financial information maintained through the Renaissance technologies of accounting (Littleton 1933). As commercial enterprises grew and became geographically dispersed, particularly in the last two centuries, new technologies from typewriter and filing cabinets to electronic storage and computing were invented to produce and keep track of the growing information needed to manage (Yates 1989, 2005). Expanding commerce also required information about foreign markets and trade – giving rise to newspapers, market reporting, financial and industrial journalism, and other databases that are part of business decision making (Raymond 1996; Andrews 1968; Bourne 1887; Sommerville 1996). Financially valuable market and commercial information particularly motivated information technologies, whether nineteenth century telegraphy or current internet.

Knowledge of the specific arts upon which commerce was based also became of great value. The origins of technical writing have been traced to the printed books of instruction in practical arts such as silkworm production, beekeeping, and cooking that appeared in the Renaissance (Tebeaux 1997; Brockmann 1998). Some of the arts were so complex as to require extensive documents closely held among the adept, such as apothecaries and herbalists, lens makers, and alchemists. Today's technological enterprises are even more deeply tied to the production and use of new knowledge. Patents and their publication

(Federico 1929; Bugbee 1967) were until recently a knowledge system carried out almost entirely separate from the university. With the industrial revolution and the formation of large corporations technological and industrial development became closely intertwined. (Currently about 85 % of patents are granted to corporations.) Recent partnerships between universities and corporations, particularly in the biotech industry, however, have raised questions about restrictions of scientific publication, hampering scientific advance, and sheltering embargoed work from peer criticism and evaluation of the work (Lievrouw 2004; Etzkowitz, Webster, and Healey, 1998).

Today information and knowledge themselves are commercially valuable commodities. The economic value of texts was established by the extension of patent monopoly to copyright in the eighteenth century. As the length of the copyright monopoly has been extended, more extended ownership of the knowledge instantiated in texts has been made possible, and ownership has aggregated in publishing houses. As modern society has become more dependent on knowledge, the economic value of many sorts of information, and the texts that bear them, has increased, particularly with the advent of electronic communication and the internet. This means that the purchaser may only gain transient use of the purchased knowledge product, while the permanent and authoritative copy still resides solely in the possession of the owner. The consequences of these arrangements have tempted a few corporations to try to gain ownership of large segments of the knowledge our society depends on, knowledge now largely produced by the university.

Another related driver of knowledge production that is now influencing the future of the university has been national interest. At first national interest was expressed through exploration and colonialism, then through nationalism and national identity, and in the twentieth century through military technology and national security concerns (Ruegg 1996). During the period of exploration and colonialism, knowledge of the resources and economies of foreign holdings and the internal wealth of the home nations became matters for internal circulation within governments and more broadly within society as entrepreneurship and citizen patriotism became part of the enterprises (Eisenstein, 1979). By the eighteenth century knowledge of standardized national languages and then a century later knowledge of national literatures became means and markers of participation in the enterprises of the nation (Anderson 1983; Helgerson 1992; McArthur 1986). Texts of political and social philosophy became widely circulated controversial documents, as societies sought for the grounds of order outside church doctrine or monarchical authority. Hobbes, Locke, Hume, Montaigne and Rousseau, among others, pervaded a new public sphere which sought explicit rational justifications and designs for their constitutions, most notably during the American and French Revolutions. Each of these new political formations created institutions for the advance of knowledge, as well as the collection and distribution of texts (Fliegelman 1993; Warner 1990). Though this age of political thoughts was fostered in an international climate of freedom and exchange, this movement towards cosmopolitan democratic rationalism was to become fractured by national identities and national languages. Consequently, distinctive national traditions, affecting what scholars were likely to read, developed in philosophy, humanities, and social thought – and even to some degrees in the natural sciences (see, for example, Guerlac 1981). Further, insofar as scholarship remained international, national languages competed to be the dominant in each area of study, with French and German each having domains of dominance until the general dominance of English from the middle of the twentieth century on. This language situation, in turn, led to an expectation that any person of learning (even in areas of little language contact, as in the U.S.) needed familiarity with several European languages. Gradually in the nineteenth century, the universities began to accommodate their curricula to include more instruction in contemporary foreign languages, in the local vernacular and in local history and culture, particularly as the Napoleonic and Humboldtian reforms reorganized universities and new subjects and disciplines.

The military has long seen knowledge as providing strategic advantage, but only in the middle of the twentieth century has the university been seen as a provider of that knowledge. Treatises on military knowledge were produced in ancient China, India and Rome. At the time when printing emerged in Europe, however, the political conditions were particularly unstable with nations in frequent conflict on economic, national, and religious grounds. These conditions created a rich market for technical military books on fortifications, shipbuilding, gunnery and ballistics. As science demonstrated its military potential, governments began to enlist it to produce new weapons. Over the ensuing centuries, advances in cartography, communication and transportation (such as telegraphy and rail), propulsion (steam and internal combustion), armaments (such as the machine gun) and shipbuilding (ironclads and steampower) were of military interest. Chemistry, physics and information technologies were central to the efforts of both sides in the two world wars of the twentieth centuries. Aeronautical and aerospace engineering along with bio- and nano-technologies were added to the mix in the latter part of the century. The knowledge produced in developing each of these military technologies was a complex of secret, bureaucratic, field operational and open scientific knowledge, with increasing involvement of the university as the century progressed. Currently most academic research in the United States is funded by the federal government. (On average, 60 % of it is defense related.) Much of those funds are administered by the Department of Defense, which has developed an elaborate congressionally-regulated system for developing projects, calling for and receiving proposals, and forming contracts with academic and industrial vendors. This system forms tight communicative relations among universities, corporations, and the military (Van Nostrand 1997), and exerts a strong though quiet influence on the growth and operations of universities.

5. The modern research university

While some creators of knowledge in the sixteenth through eighteenth centuries were university trained and held university posts, the main advances occurred outside universities and were largely disseminated outside university networks. Galileo is a case in point; although he studied medicine at the University of Pisa, he left without a degree to study mathematics under a military engineer. He then taught mathematics, astronomy, mechanics and fortification in the cities of Siena, Pisa and Padua, but only in part at universities. He left universities entirely when he gained the patronage of the Medicis.

Gradually some universities made some curricular adjustments and hosted chairs in new specialisms (such as the Lucasian Chair in Mathematics that Newton occupied at Cambridge). Yet the university curriculum generally remained conservative, aimed at the moral formation and intellectual discipline of leadership classes, principally clergy, lawyers and physicians. The Reformation did not bring secularization, autonomy, or research to the university, but only changed the religious auspices, to which national sponsorship was sometime added. Sweden only in part followed this model. At Uppsala the pro-chancellor was regularly the archbishop of Sweden, and at other Swedish universities at Abo (Turku) and Dorpat (Tartu) clergy also were pro-chancellors (Ridder-Symoens 1991). Nonetheless, the Swedish royalty also seemed to understand the relationship between free knowledge and the prosperity and power of the state. King Gustav II Adolph appointed his personal advisor Johan Skytte, chancellor of Uppsala in 1622, only two years after a generous Royal gift put the University on solid financial grounds (Ridder Symoens 1991). Skytte was also to be appointed Chancellor at Abo and Dorpat. He apparently took an active role in the administration of these universities. Gustav Adolf took a strong interest in all of the Swedish universities, including the philosophic curricula, and he provided a large donation to expand the library. Linnaeus, (according to Rausing 2003) saw his botanic project as part of a Christian economy and stewardship of nature, placing his work at the intersection of church and state – making it a candidate for sponsorship within the Swedish university. Sweden also forged a novel arrangement among crown, university, and the printer Lars Salvius. Swedish academic interest in the intersection of scientific and practical knowledge distinguished it from the more clerical concerns of most European Universities. This practical interest made the 18th century Swedish universities in some part similar to the Scottish universities that with secular charters were the most overt exceptions to academic traditionalism during that period.

Of course in the next century major university reforms in France – in the wake of the Enlightenment, the Revolution and the Napoleonic reorganization – abolished the colleges of the *ancien régime*, and forming new secular professional schools. Research was, nonetheless, supported in non-university institutes and centers, such as the botanic and zoological gardens. This model of reform held some influence over mid-nineteenth century universities elsewhere in Europe. Prussia, following the ideas of Kant, Fichte, Schliermacher and Humboldt, developed another model of university reform at Göttingen, Halle, and Berlin, based on scholarly research professorships and advanced research seminars and degrees. While the professorships initially were in philosophy and theology, these soon became differentiated into philology, history, economics, and the sciences. This model spread to the rest of Germany, particularly after its unification in the nineteenth century, as well as to Austria, Russia and the United States. By the turn of the twentieth century the German model influenced the more traditional systems of England and southern Europe, as well as the French bureaucratic system

Even though universities had become the primary center of scientific research by the end of the nineteenth century, scientific publishing had remained largely in the hands of the independent printers and publishers of books and journals. As societies formed and published journals, they also worked with commercial publishers and printers. The few existing university publishers such as Cambridge and Oxford were devoted to history, the humanities and theology. With the rise of the research university, by the turn of the twentieth century, university presses became more common, often with a special responsibility for the work of their faculty.

The changing nature of the university also affected the role and collections of the university libraries. The early medieval university was mainly devoted to the study of the classic canon, and the purpose of the university library was to make canonical texts available to faculty and students. But libraries changed and took on a new importance as science developed into a highly intertextual, cooperative system in the late eighteenth and nineteenth centuries, moving hand in hand with the development of modern citation practices. Libraries needed to collect the most up-to-date material and not just be a storehouse of canonical texts. Thus today, a researcher cannot publish in science without positioning his or her work against a rapidly evolving literature, knowledge of which requires an up-to-date library, or now its virtual extension. Again in Sweden there seems to have been a productive relationship among crown, scientists, and university and national libraries.

Sweden's and Scotland's unusual eighteenth century arrangements of worldly and knowledge institutions were in some ways precursors to the general university reforms and scientific reorganizations of the nineteenth century. They might now be worth looking at afresh to find clues as to how to maintain a degree of independence of knowledge in the face of forces reconfiguring the power, interests, and alliances surrounding scientific and university institutions.

The success of the university in producing and disseminating knowledge has increased the value of that knowledge as well as the interest of its sponsors in wanting more direct benefit. Higher degrees of political and economic organization have exposed the university to increasing external influence even as its success has increased its size and resources. The clients of university-produced knowledge are willing to pay very high prices for knowledge valuable for commerce, medicine, military, law and other practical uses. These clients do not necessarily share the view of knowledge as a public good to be shared internationally; for various competitive reasons they are just as happy to keep knowledge flow restricted. These same clients have entered into tighter relations with research universities so as to foster the kinds of knowledge they perceive as most useful and to gain proprietary edges against competitors, again with pressures for the restriction of knowledge to the higher paying patrons.

The publishers at the same time have been taken up in larger corporate conglomerates and have replaced the traditional ideology of the publishing industry which was culturally aligned with the university and other knowledge institutions with the monopolistic financial logics of mega-corporations. They are aware that information and knowledge are valuable commodities with customers willing to pay the cost.

The value of information has also fostered new technologies for its dissemination that are disrupting traditional routes of distribution, access and storage. These new technologies change the function of the university libraries, which from the beginning had been the central collecting place and local distribution point for the knowledge essential to the university – whether in the traditional church or the modern research version. The core collections now can reside in servers owned by the external providers rather than in the physical copies held in the library. The corporate publishers are taking the opportunity of this moment to restructure the market in their favor basing their pricing on their highest paying clients, and putting the squeeze on the vulnerability of the university library in transition. Insofar as they succeed, they limit the flow of knowledge (maintaining its high price) and hamper the cooperative enterprise of knowledge production and restrict the wide distribution of knowledge for the public good, including those forms of knowledge aiming at social reform that are not of interest to the higher paying clients.

These same technologies that are providing this commercial monopolistic opportunity, however, provide low cost opportunities to bypass the marketplace logic of the most well-heeled clients and the conglomerate sellers. That is the story of the growing open access movement.

All these forces are creating tensions and destabilizations within the current models of university-based scientific knowledge production and distribution. All threaten the independence of knowledge production carved out in a complex landscape of fractionated power over the last six centuries. While the more stable and unified world that has been emerging in the last half-century is a great blessing, the forces of centralization through alliances of capital and government threaten to harness knowledge production and dissemination more directly to the needs of state, national security, and economy, as was the case in the stable, hierarchical Chinese empire whose knowledge evolved only slowly for two thousand years.

The conditions for autonomy of scientific publication in early modern Europe have clearly changed but they have left a remarkable legacy of secular inquiry and open distribution of knowledge. The wonder of the modern research university is transient and fragile – a conjunction of historical forces. The independence and dynamic growth of knowledge depends on creating new arrangements that allow and even encourage scholars and students to pursue new truths that do not seem to have immediate pay-offs for the state, military, and industry who pay the bills – truths that may even seem heterodox to the sponsoring powers. Looking back to the earlier configurations and the forces that led to our current arrangements is more than a matter of historical celebration, it is a matter of understanding what our world has been constructed of so we can continue in its constant reconstruction.

References

Anderson, Benedict
1983 *Imagined Communities.* London: Verso.
Andrews, Alexander
1968 *The History of British Journalism* (Vols. 1–2). London: Haskell House.

Atkinson, Dwight
1999 *Scientific Discourse in Sociohistorical Context: The Philosophical Transactions of the Royal Society of London, 1675–1975*. Mahwah, NJ: L. Erlbaum Associates.

Bacon, Francis
1603 *The Advancement of Learning*. London.

Bacon, Francis
1620 *Novum Organum.* London.

Battalio, John T.
1998 *The Rhetoric of Science in the Evolution of American Ornithological Discourse*. Stanford: Ablex.

Bazerman, Charles
1988 *Shaping Written Knowledge: The Genre and Activity of the Experimental Article in Science*. Madison: University of Wisconsin Press.

Bazerman, Charles
1991 How natural philosophers can cooperate. In: Charles Bazerman and James Paradis (eds.), *Textual Dynamics of the Professions*, 13–44. Madison, Wisconsin: University of Wisconsin Press.

Bazerman, Charles
1993 Forums of validation and forms of knowledge: The magical rhetoric of Otto von Guericke's sulfur globe. *Configurations*, 1: 201–228.

Biagioli, Mario
1993 *Galileo Courtier: The Practice of Science in the Culture of Absolutism.* University of Chicago Press.

Biagioli, Mario and Peter Galison (eds.)
2003 *Scientific Authorship: Credit and Intellectual Property in Science*. New York: Routledge.

Bourne, Henry Richard Fox
1887 *English Newspapers: Chapters in the History of Journalism* (vols. 2). London: Chatto and Windus.

Broad, William and Nicholas Wade
1982 *Betrayers of the Truth.* New York: Simon and Schuster.

Brockmann, R. John
1998 *From Millwrights to Shipwrights to the Twenty-first Century.* Cresskill NJ: Hampton.

Bugbee, Bruce W.
1967 *Genesis of American Patent and Copyright Law.* Washington, D.C.: Public Affairs Press.

Carter, Thomas Francis
1955 *The Invention of Printing in China and its Spread Westward*. NY: Ronald Press.

Dear, Peter
1985 Totius in verba: Rhetoric and authority in the early royal society. *Isis 76*: 145–161.
Eisenstein, Elizabeth L.
1979 *The Printing Press as an Agent of Change.* Cambridge: Cambridge University Press.
Etzkowitz, Henry, Andrew Webster and Peter Healey (eds.)
1998 *Capitalizing Knowledge: New Intersections of Industry and Academia*. Albany, NY: State University of New York Press.
Evans, Robert J. W.
1973 *Rudolf II and his World*. Oxford: Oxford University Press.
Fahnestock, Jeanne
1999 *Rhetorical Figures in Science*. New York: Oxford University Press.
Federico, Pasquale J.
1929 Origin and early history of patents. *Journal of the Patent Office Society, 11:* 292–305.
Fliegelman, Jay
1993 *Declaring Independence: Jefferson, Natural Language, and the Culture of Performance.* Palo Alto: Stanford University Press.
Gross, Alan G., Joseph Harmon and Michael Reidy
2002 *Communicating Science: The Scientific Article from the Seventeenth Century to the Present*. New York: Oxford University Press.
Guerlac, Henry
1981 *Newton on the Continent.* Ithaca, New York: Cornell University Press.
Gunnarsson, Britt-Louise
1997 On the sociohistorical construction of scientific discourse. In: Britt-Louise Gunnarsson, Per Linell and Bengt Nordberg (eds.), *The Construction of Professional Discourse,* 99–126. London: Longman.
Gunnarsson, Britt-Louise
2001 Expressing criticism and evaluation during three centuries. *Journal of Historical Pragmatics, 2*, 115–139.
Halliday, Michael A. K. and James Martin
1993 *Writing Science: Literacy and Discursive power*. London: Falmer Press.
Helgerson, Richard
1992 *Forms of Nationhood: The Elizabethan Writing of England*. Chicago: University of Chicago Press.
Jacob, Margaret C.
1976 *The Newtonians and the English Revolution, 1689–1720*. Ithaca, New York: Cornell University Press.
Johns, Adrian
1998 *The Nature of the Book: Print and Knowledge in the Making*. Chicago: University of Chicago Press.

Kronick, David A.
1976 *A History of Scientific and Technical Periodicals: The Origins and Development of the Scientific and Technical Press, 1665–1790*. Metuchen, N.J.: Scarecrow Press.

LaFollette, Marcel
1992 *Stealing into Print: Fraud, Plagiarism and Misconduct in Scientific Publishing.* Berkeley: University of California Press.

Lee, Thomas H. C.
2000 *Education in Traditional China, A History*. (Handbook of Oriental Studies vol 13.) Leiden: Brill.

Lievrouw, Leah A.
2004 Biotechnology, intellectual property, and the prospects for scientific communication. In: Sandra Braman (ed.), *Biotechnology and Communication: The Meta-Technologies of Communication, 145–172*. Mahwah NJ: Erlbaum.

Littleton, Ananias Charles
1933 *Accounting Evolution to 1900.* New York: American Institute Publishing.

Luo, Shubao
1998 *An Illustrated History of Printing in Ancient China*. Hong Kong: City University Press.

McArthur, Tom
1986 *Worlds of Reference: Lexicography, Learning, and Language from the Clay tablet to the Computer.* Cambridge: Cambridge University Press.

McKitterick, David
2003 *Print, Manuscript and the Search for Order, 1450–1830*. Cambridge: Cambridge University Press.

Merton, Robert K.
1973 *The Sociology of Science: Theoretical and Empirical Investigations.* Chicago: University of Chicago Press.

Myers, Greg
1989 The pragmatics of politeness in scientific articles. *Applied Linguistics, 10:* 1–35.

Myers, Greg
1990a *Writing Biology.* Madison: University of Wisconsin Press.

Myers, Greg
1990b The rhetoric of irony in academic writing. *Written Communication, 7*: 419–455.

Needham, Joseph
1970 The Unity of Science; Asia's Indispensible Contribution. In: Joseph Needham et al. (eds.), *Clerks and Craftsmen in China and the West,* 14–29. Cambridge: Cambridge UP.

Needham, Joseph and Gwei-Djen Lu
1970a Medicine and Chinese culture, In: Joseph Needham et al. (eds.), *Clerks and Craftsmen in China and the West*, 263–293. Cambridge: Cambridge University Press.

Needham, Joseph and Gwei-Djen Lu
1970b China and the origin of qualifying examinations in Medicine In: Joseph Needham et al. (eds.), *Clerks and Craftsmen in China and the West*, 379–395. Cambridge: Cambridge University Press.

Rausing, Lisbet
2003 Underwriting the oeconomy: Linnaeus on nature and mind. *History of Political Economy Annual Supplement* 35: 173–203.

Raymond, Joad
1996 *The Invention of the Newspaper: English Newsbooks, 1641–1649.* Oxford: Clarendon Press.

Ridder-Symoens, Hilde de
1991 Mobility. In Hilde de Ridder-Symoens (ed.) *A History of the University*, I, 280–304. Cambridge: Cambridge University Press.

Ronan, Colin and Joseph Needham
1981 *The Shorter Science and Civilization in China*. II. Cambridge: Cambridge UP.

Ruegg, Walter
1996 Themes. In Hilde de Ridder-Symoens (ed.) *A History of the University, II*, 3–42. Cambridge: Cambridge University Press.

Scholarly Societies Project
2006 www.scholarly-societies.org. Waterloo, Canada: University of Waterloo.

Schwinges, Ranier Christoph
1991 Admission. In Hilde de Ridder-Symoens (ed.) *A History of the University* 1, 171–194. Cambridge: Cambridge University Press.

Shapin, Steven
1994 *A Social History of Truth: Civility and Science in Seventeenth-century England.* Chicago: University of Chicago Press.

Shapin, Steven and Simon Schaffer
1985 *Leviathan and the Air-pump: Hobbes, Boyle, and the Experimental Life.* Princeton, N.J.: Princeton University Press.

Sommerville, C. John
1996 *The News Revolution in England.* New York: Oxford University Press.

Sprat, Thomas
1667 *History of the Royal Society*. London.

Tebeaux, Elizabeth
1997 *The Emergence of a Tradition: Technical Writing in the English Renaissance, 1475–1640.* Amityville NY: Baywood

Verger, Jacques
1991 Patterns. In Hilde de Ridder-Symoens (ed.) *A History of the University* 1, 35–74. Cambridge: Cambridge University Press.

Warner, Michael
1990 *The Letters of the Republic: Publication and the Public Sphere in Eighteenth-Century America*. Cambridge MA: Harvard University Press.

Wilkins, John
1668 *Essay towards a Real Character and a Philosophic Language*. London.

Yates, JoAnne
1989 *Control through Communication*. Baltimore MD: Johns Hopkins University Press.

Yates, JoAnne
2005 *Structuring the Information Age: Life Insurance and Information Technology in the 20th Century*. Baltimore MD: Johns Hopkins University Press.

Philology in the eighteenth century: Europe and Sweden

Gunilla Gren-Eklund

1. Introduction

There are certain difficulties in choosing a word relevant to the activity of "studies of language" at the academies in the centuries preceding the nineteenth, i.e. before such terms as "linguistics" and "Sprachwissenschaft" appeared and assumed their modern meaning. The word "philology" might best suggest the main idea of language studies in earlier times. Philology, in the sense in which it is still used in German and Swedish, seems to be quite an apposite term to describe the undertakings of professors of languages at the universities for many centuries from the Middle Ages on, since the emphasis then was exclusively on the interpretation of texts, Biblical and classical.

Though little represented among those professionally engaged in language studies, certain theoretical ideas about language were explored from ancient times. But language as an object of reflection in itself was not an independent discipline of research, and such was the situation until the nineteenth century, when studies of language came to be separated from the discipline of philosophy, both as subject of academia and as far as ideas were concerned.

At the outset, it may be said of the first half of the eighteenth century in this field that, unlike the seventeenth century to some extent, it was by no means an innovative or epoch-making time as far as ideas about language are concerned – the renewal was still to come. Regarding the situation at Swedish academies, there were very few philologists deserving mention as contributors to scholarship on language.

In order to rightly represent the topic of language studies in the eighteenth century, when scholarship in general developed a growing interest in the natural sciences, it is necessary to give a historical perspective on views and studies of language during earlier centuries and, to a certain extent, what came later might also be considered.

An interest in language has always followed humankind, and certain issues have been discussed at all times. Two main topics with a theoretical bearing are constantly discernible when man expresses his thoughts on language. They are traceable even in the most ancient documentation available, the texts of the Greeks, but also in the Indian and Arabic cultures. These two concerns are the *function* and the *nature* of language.

First, there has been an incessant interest in language as function, i.e. as an expression of thoughts, a question clearly put by Aristotle and extensively discussed during the Middle Ages and, in the vein of Aristotle, even into the eighteenth century. Such a question would be inevitable in the discipline of logic – was natural language of use to logic?

The second theme, the nature of language, might be appreciated as a question of ontology, which was also in focus during the post-medieval centuries. The question, expressed, answered and solved in different ways, more precisely regarded the primary genesis of language and languages.

These two aspects will be treated below in the sections on theories of language and on the genesis of languages, respectively. But it is necessary to start with an inventory of the general understanding of language and the material available for language studies during our period of interest.

2. The material background in the history of scholarship

Philology and ideas of language did not change very much when the eighteenth century set in – the lively debates of the seventeenth century in this field seem to a certain extent to have frozen in a rather fixed position.

To begin with, the system of learning inherited from the Middle Ages, based on the *trivium* – grammar, dialectics and rhetoric – was still dominant within the basic education of the century. The academic institutions of Europe, and also very much so the rather conservative academia of the Nordic countries, were still primarily theological. The chairs that were concerned with language were found not only in faculties of philosophy, to which the chairs in Greek and Oriental languages as well as rhetoric (to wit, Latin) were affiliated, but also to a considerable extent in faculties of theology. In the latter part of the century, many new chairs were established in new and quite different fields, when the natural sciences were promoted and the educational ambitions of the universities were widened to other professions than the clerical. But professorships in other languages than the Oriental and classical ones were not established until another century later.

On the other hand, since as early as the sixteenth century there had been projects of learning, not necessarily within the universities, that were initiated in order to gain a knowledge of other languages and to try to explore other kinds of grammars besides that of Latin. Grammars of such European languages as French, Italian, Spanish, Polish and Church Slavonic saw the light of day already during the fifteenth and sixteenth centuries (Arens 1969: 62). Even such a special language as Basque was treated as early as 1562 in a book by the Spaniard Franciscus Sanctius.[1]

A natural consequence of certain recording enterprises was that languages could be put side by side and compared. The first author with such an ambition was Postellus,[2] who in 1538 edited a work presenting 12 different languages with their genuine alphabets (Postellus 1538).

The work of Postellus was one of the earliest manifestations of an interest in collecting material from languages which, as time went on, led to a gradual broadening of the documentation available. A special genre was developed, the so-called polyglottic collection, which came to be an important material for language studies. *Mithridates* by Conrad Gesner[3] (Gesner 1555) was an early work in the genre. The title came to apply to all works of this kind, emanating from King Mithridates of Pontus in the first century BC, who was known to have conquered 22 nations and learnt the languages of all of them. Gesner's work was published in 1555 and the first edition includes the Lord's Prayer in exactly 22 languages, besides other material from 130 languages. In course of time, more and more languages were added and the genre developed into real dictionaries; around 1800 such a collection could comprise up to 500 languages or even more; the final standard work was *Mithridates* 1806–1817. Such collections came to be a rich material for use also in the more modern comparative linguistics emerging in the nineteenth century. A scholar from Uppsala in the first half of the eighteenth century, Philip Johan Strahlenberg (1676–1747), a geographer and cartographer with an interest in ethnography, made a rather special contribution to this genre. He mainly touched upon "Tatarian" languages and peoples in his work (Strahlenberg 1730), in which he also collected 32 languages in a *Tabula polyglotta*, recording in particular the numerals.

In other words, the horizon had widened already in the sixteenth century, when colonization and also the Christian mission outside Europe had begun.

1. Sanctius' Spanish name was Sanchez and his work was entitled *Minerva seu de Latinae linguae causis et elegantia.* on Latin syntax (Arens 1969: 1965). For the core languages at that time, the "classical" ones, i.e. Greek, Hebrew and Arabic, new grammars and dictionaries were written during the sixteenth century.
2. Guillaume Postellus Barentonius (1510–1581), professor Linguarum Orientalium in Paris.
3. Conrad Gesner, polyhistor (1516–1565).

New political and economic contacts gave rise to knowledge about other cultures. Some of the people involved in such projects – not least the Jesuit missionaries – had a special interest in languages, which led to the publication even in this early century of grammars of what could be termed "exotic" languages, to the extent that even American Indian languages were recorded (Arens 1969: 64). As far as concerns Asia, grammars and glossaries were written for Japanese, Vietnamese, certain Indian languages and also for Persian in the seventeenth century, and in addition this was a time which had a special significance for the discussion of the genesis of languages. Such works are to be found in the collections of the library of Uppsala University, notably a Malayan dictionary printed in 1631.

Availability of texts was not a problem, but it was almost only the Biblical and the classical ones that were studied and published. There was little interest in publishing texts from other languages. After the sixteenth century, however, ecclesiastical texts also came to include those translated into the vernacular languages. Such a practical interpretative enterprise could of course also raise certain theoretical questions. Reflections at least about what form or expression was right or wrong in the particular language concerned could be expressed, but they were not connected to any theories about language.

The material available after the Middle Ages could be used for both the basic questions about language mentioned, that of function and that of genesis.

3. Theories of language

The possibility of a common structure was the subject of discussion, with the most advanced one occurring in the field of logic and philosophy, and elaborated in particular in the so-called *grammaire raisonné*, the "rational grammar", from the French school of Port-Royal. The *Grammaire générale et raisonné*, published in 1660, was in fact based on the logic of the Middle Ages and there was also a link to certain ideas of Renée Descartes (1596–1650). Port-Royal was a scholarly, Jansenist and anti-Jesuitical school which produced works on dogmatics, logic and also grammatical theory over a number of decades in the middle of the seventeenth century (1637–1661). The severe Christian morals of the school and its Jansenist message, however, brought it into a conflict with the *sedes apostolica*, and it was closed down after some decades. The impact of the *grammaire générale* was not immediately obvious, but the idea of discussing such questions of grammar and philosophy was revived during the Enlightenment, and it was taken up a century later in the great Encyclopedia (Encyclopédie 1751–1780).

Whether there was any interest in general at the Nordic academies in discussing theoretical issues concerning language, and also other scholarly matters, little evidence of it is found. In academic philological works in Uppsala, mostly dissertations, and mostly written by the professors – there were few other, independent scholarly publications by the professors – the commonest reference to any scholar concerned with languages was for centuries, and still in the eighteenth century, to the philologist Joseph Scaliger[4] and his views on language. Already in the sixteenth century he had broken with long (from the thirteenth century) dominant ideas emanating from the intricate, Latin-based ideas of grammar of the *modistae* (Arens 1969: 66). Another person not without influence on the intellectual life of Sweden was the European J. Amos Comenius (1592–1670), who spent some years in Sweden in the 1640s. A distinguishing mark of that scholar was his influential thoughts on the pedagogy of language.

Although those students who later became professors in Uppsala, especially in the seventeenth century, made their peregrinations and stayed at various European universities, it is well known that it would be a long time before such general ideas as those of Descartes became a subject of discussion, and then also of conflict, at Uppsala University; as late as 1689, Cartesian anti-biblical interpretations were forbidden there, nearly 40 years after the death of Descartes (Lindroth 1976: 75). There seems to have been little familiarity in Sweden with a European intellectual tradition which to a certain extent also discussed questions of language and which was upheld by the English empiricism of Francis Bacon (1561–1626) (Brekle 1975: 281–289) and, in the latter part of the seventeenth century, was represented by the philosopher John Locke (1632–1704) (Brekle 1975: 296–300).

This limited interest in theoretical matters should also be seen in the light of the fact that the universities in Sweden were, from the beginning and also during these centuries, mainly looked upon as educational institutes, providing professional skills rather than serving as arenas for theoretical discussions and research institutions.

During the seventeenth century academia was held in check by the church. This state of affairs changed significantly in the eighteenth century, when the academies, at least in Sweden, were to be defined, rather, as governed by the state and developing into political instruments. There was a new predominance of education for civil professions over those of the church; but this neither promoted theoretical discussions of scholarly matters, nor changed the general status of the universities as mainly educational establishments.

4. Joseph Justus Scaliger (1540–1609) was a scholar at the academy in Leiden.

4. Genesis of languages

The second enduring question about languages concerned their genesis. More specifically, for centuries it was discussed which language was the original one and thus the language of greatest authority. As long as the ideas of scholarship were dominated by theology and the message of the Bible, a monogenesis not only of the whole world as such, but of all its manifestations, including language, was assumed. Hebrew was maintained as being the primeval language even after the Middle Ages, by Postellus[5] and by many scholars during the centuries that followed.

Of course such a discussion, which was ontologically relevant, was also dependent on certain basic facts of language, though it utilized the material in a special way. It must, however, be stressed that what was in focus was not the history of language or languages, neither was the internal development of languages reflected upon. Nor was there a widespread concern to group languages together, especially not according to a genealogy. Historical linguistics in our sense was hardly possible before the nineteenth century, without the impetus of that time from Darwinian ideas of evolution; it also required a Ranke to be able to judge changes in language as historical processes, as well as a Saussure to look at language as structure. Furthermore, a new view of phonetic facts was necessary to finding tenable sound laws. All of this did not happen until the nineteenth century, when systemic grouping of languages also gave rise to such terms as "Indo-European" and an earlier term, "Semitic", came into use as its complement.

During the centuries after the Middle Ages the dominant answer about which was the original language was given on the basis of the Christian conception of the Bible. According to biblical interpreters, languages arose from Hebrew in their great number when the sons of Noah spread all over the world.[6] This was in fact commonly taken for granted in dissertations on language, in Uppsala as in Europe generally, but without any depth of ontological support. It was held, rather, as an unreflected and ahistorical truth, based on an idea of the divine and the myths of creation.

In certain settings and in the works of individual scholars, however, the idea of Hebrew as the original language was challenged by certain other theories (Droixhe et al. 2000). More secular ideas inherited from classical antiquity

5. Cf. footnote 2, above.

6. Genesis 11: 1 concerning the one and only language on earth, which according to Genesis 11: 7 had to be confused as a punishment, resulting in a mutual lack of understanding.

about the genesis of languages were floating around in Europe and they were quite influential on Swedish academic minds. Many scholars, especially those who worked on the glossaries and to a certain extent also on the morphology of different languages, held that "Scythian" was the original language. Helpful evidence for Scythian was to be found in comparisons between European languages and Persian. This rather flexible idea was not least of interest to Nordic scholars, who could connect their languages more easily to Scythian than to Hebrew. The notion of Scythian, inherited from classical/Greek antiquity, was at the time geographically identified as comprising two parts, the European, north of the Danube, and the Asiatic, all of Asia north of India from Persia to China, then Cathay (Zedler 1732–1754, s.v. Scythien.) The idea was also flexible in the sense that, even within the thesis, it was permissible to refer to Hebrew as a still more genuine language, or sometimes even to integrate it as a relative. Another reference sometimes made to an ancestral language was to "Egyptian", a term likewise inherited from the texts of classical antiquity.

One of the effects of this – or perhaps, rather, a reason for the impact of the idea – was that when a more secular kind of genesis was assumed, there was ample scope for various manifestations of "Goropianism", in Swedish known as "Rudbeckianism". In Sweden there was, already at the beginning of the seventeenth century, an interest in the possibility of another original language than Hebrew. Such an idea started with Johannes Bureus (1568–1652), who by his recording of the runes initiated an interest in the Nordic linguistic heritage. To this was added an interest in the Gothic language, which emerged in the seventeenth century through the edition of the *Codex Argenteus*, prepared in 1665 by Franciscus Junius (the Younger, 1589–1677).[7]

Junius positioned the Gothic language within a group of Germanic languages, but he did not assume it to be the original language. Georg Stiernhielm (1598–1672) was a Swedish cultural personality well known for his vast education, acquired through journeys abroad and studies of languages. In 1671 he published his edition of the *Codex Argenteus*[8] in four languages, and in the introduction he mentioned the Japhetic language group, derived from one of the

7. The Codex Argenteus was discovered in 1563 in Germany, but its existence was forgotten until the manuscript found its way to Sweden through Queen Christina in 1648. It acquired international fame through the edition of Franciscus Junius, whose Gothic glossary was included in the edition. The glossary was also printed separately in Amsterdam in 1684.
8. In Stiernhielm's edition, the Gothic text was transcribed with Roman script, together with the texts of the Vulgate and of the Icelandic and Swedish Bibles.

three sons of Noah, Japheth, a group of languages that he expressly equated to the language of the Scythians:

> Scythen sind die Einwohner beyder Scythien, nehmlich des Europäischen und Asiatischen Scythiens welche Nachkommen des Magogs, eines Sohns des Japhets, gewesen seyn. (Zedler 1732–1754, s.v. Scythen).

He thereby also established a vital position for the Swedes and the Swedish language in the hierarchy of languages. The same connections were later also acknowledged by the two Olof Rudbecks, father and son[9], although both still reverently identified Hebrew as the most important language. Erik Benzelius the Younger (1675–1743) (Agrell 1954: 134–143), who had been a pupil of Leibniz, took the same line.

5. A comparative method in nuce

When it comes to the method used to compare languages in the 17th century, in the form that was still dominant in the first half of the eighteenth century, it had its basis in an idea about "permutations" of parts of words. The method implied a rather primitive view of phonetics, and all the changes perceived as possible concerned letters rather than sounds. Such changes were not recorded as diachronic, the differences in the shapes of words being registered theoretically from current forms of the languages. In spite of this, now and then there was some reference to changes following certain laws, in the vein of what would later be developed as systemic laws. However, the permutation method, when followed closely, could reveal enough of a system for the Germanic sound shift to be suggested at an early stage. This, as far as Swedish was involved, was a recurrent topic among Uppsala scholars.[10] For other Germanic languages it was also recorded in the 1660s in *Etymologicum Anglicanum* by Franciscus Junius, the above-mentioned editor of the Codex Argenteus.[11]

On the threshold of the eighteenth century a prominent figure of European scholarship was the philosopher and mathematician Gottfried Wilhelm von Leibniz (1646–1716) (Aarsleff 1975: 385–410). He exhibited an interest both in the structure of language and in its genesis, in a way that in fact pointed to later

9. Olof Rudbeck the Elder (1630–1702) and Olof Rudbeck the Younger (1660–1740), both professors at Uppsala University.
10. More on the issue below, in the next section. The subject was positively advanced by Benzelius in the 1720s and by Ihre in the 1760s.
11. The first edition of the dictionary was issued in Oxford in 1743 (Junius [1743]).

thoughts on language and linguistics. His most far-sighted work, dealing on a philosophical basis with all questions about language current at the time, was his *Nouveaux essais sur l'entendement humain*, written in 1704 but not printed until 1765. The main topic of the work was the knowledge and understanding of man. As far as language was concerned, he concentrated on the relationship between language and thought, a well-known philosophical question, as mentioned above, ever since Aristotle.

On the issue of the genesis of language, Leibniz was up to date in that he adhered firmly to the idea of Scythian as the original language, expressed by him in terms of a Celto-Scythian genesis. In this discussion he also showed a real interest in the linguistic evidence, and his interest in etymologies in particular was clearly guided by a desire to establish which was the original language. He expressly rejected Hebrew as that language in *De originibus gentium* (1710). Leibniz was very well read in authoritative literature, held firm views, and corresponded with many contemporary philologists, among them the German politician and Orientalist Hiob Ludolf,[12] with whom he exchanged thoughts on language affinities and not least on etymologies (their correspondence was published in 1755; Watermann 1978).

It must be pointed out that, in the analysis of language at the time, the notion of "etymology" was not used for methods of revealing the history of words. Instead its aim was to map forms of words, with or without an intention to compare. In 1698 the Uppsala Orientalist Gustaf Peringer Lillieblad published a new edition of a Latin grammar written by his uncle, the bishop Johannes Matthiae, teacher of Queen Christina (Resenär 2007: 16). In the foreword he expressed the common idea: *Partes grammaticae sunt quatur (sic): orthographia, prosodia, etymologia, et syntaxis.* From the content it is clear that *etymologia* refers to what in modern times came to be called morphology.

Leibniz, at all events, described some methodological requirements for etymologizing and stressed that kinship between words must be based on affinities in both sound and meaning. This was really anticipating the modern comparative linguistics that emerged in the nineteenth century, for which phonology and semantics were the two cornerstones. In fact, it was not until these requirements were generally accepted that it became possible to find recurring patterns and sound laws.

Furthermore, Leibniz and his scholarly correspondent Hiob Ludolf[13] were agreed that studies of words must be supported by languages of a similar nature,

12. Hiob Ludolf (1624–1704), German scholar and politician, a specialist in Oriental and classical languages.
13. Cf. above, footnote 12.

if a genetic relationship was to be proven, which was another rather far-sighted idea, not generally accepted at the time. Ludolf for his part also declared that it was not enough to establish a genetic relationship by studying words alone; grammar and its categories also had to be studied.

Thus the direction towards a comparative method for systematizing languages was staked out as early as the seventeenth century, and Leibniz was read by scholars in the eighteenth century. But the rudimentary comparative method suggested did not actually develop much further during that century.

6. Academic philology at Nordic universities

At the beginning of the eighteenth century, academies all over Europe were in any case, as has been noted, still dominated by theology and studies of the Bible. Traditional studies of texts in Greek and Hebrew, together with Chaldean (= Aramaic) and also Arabic, were represented by special professorships at any university. In the time of orthodoxy imposed by the state, the duties of the professor of Oriental languages were strictly confined to Christian texts (UUK 1890: 41).

Chairs in these subjects were first founded on papal initiative in 1311/12, in Rome, Paris, Oxford, Bologna and Salamanca. They were all intended to provide students with such knowledge that they would be able to pursue Christian mission in internal and external encounters with other religions. Such was the standard, which was repeated in Uppsala when the first professorship of languages in the Faculty of Philosophy was established in 1605, directed to studies of Hebrew and Greek.

Scholarship was not confined merely to interpreting the Biblical world; the classical world was also a living part of the whole culture, and the professor of Greek was responsible not only for the New Testament, but also for the Greek Auctores. The Latin Auctores were also taken care of, however, by professors of rhetoric. In the seventeenth century and at the start of the eighteenth, the Renaissance ideal of the *homo trilinguis* was still alive. Thus the languages of the Bible were not only to be studied for the sake of the texts, but also mastered as a means of expressing oneself. Latin especially, the language of the Church and of classical rhetoric and poetry, had to be well known and, in particular, put to practical use.

The Swedish universities (Uppsala, Lund, Åbo and Greifswald) did not breed many innovators of language studies, and scholars seeking to develop language theories, such as Leibniz, received little obvious attention. From our perspective, interest in language was in fact rather limited, mainly addressing two issues, as we see from the dissertations of the time. One was the recording of material, as it had developed after the Middle Ages; the second was textual interpretation, confined to religious, and to a certain extent philosophical and classical texts of different kinds, which were mainly studied in order to collect material information. In the professors' teaching, reading knowledge was thus the main concern.

Only occasionally were more conscious ambitions regarding the significance of the task of teaching expressed. A declaration in 1693 by the professor of Oriental languages at Uppsala University on what was to be read stated that he intended to strive for a philological interpretation that considered not only the meaning of the expressions, but also the syntax and the semantics and furthermore the deviations in the text in different versions thereof. (*Philologiae curam maxime habiturus, sic ut de dictionum significatu tum in se, tum in vario nexu & indole, tum de versionum discrepantia, sollicitus sit*; Resenär 2007: 37.)

To the general low profile of scholars of languages at Uppsala there is one outstanding exception, a person who even enjoyed an international reputation, Johan Ihre, born in 1707, the same year as Carl Linnaeus. He died in 1780 – to wit, some years before the renowned speech by William Jones to the Asiatic Society at Calcutta. He was also a colleague of Linnaeus as a professor, acting in that capacity from 1737 to 1780.

Johan Ihre has deservedly gained a name as the first critical scholar of language in Uppsala and a praiseworthy model in the use of philological methods. His extensive scholarly activities in many fields are also apparent from the more than 400 dissertations he presided over, probably being the author of most of them. Ihre's scholarly work can very well be seen as a bridge from the knowledge and ideas about language of the past to the comparative method of the future. Ihre worked extensively on comparisons of languages, and in his great Dictionary, *Glossarium Suiogothicum* (Parts 1–2 printed in Uppsala in 1769; Agrell 1955: 143–150),[14] he referred to material from Hebrew, Greek, Latin, Persian, the Scandinavian and also other Germanic languages (Alemanian and Anglo-Saxon), as well as Celtic languages. His view of the relations between different languages had its basis both in earlier ideas and in current trends, and in the Prooemium to the glossary he expresses the following on the topic, here summarized:

> The Greek and the Gothic languages are similar, both with Scythian as their origin. The Scyths called themselves Goths or Getes. The Celts are also Scyths by origin and they founded Rome, and therefore Latin is akin to their language. The

14. An annotated translation of the Latin introduction to the *Glossarium* is being prepared by Krister Östlund, Uppsala.

> Slavic languages, to which Finnish, Hungarian and Lappish belong, also directly originate from Scythian. That Persian is similar to Gothic is explained by the fact that Scythians once dominated Persia. Hebrew is different, though very ancient, but could not be proved to be the origin of other languages.

Although the ideas expressed by Ihre were built only to a certain extent on a deductive method, his intelligent and sensitive discussions, using all possible knowledge, might be seen as anticipating the methods of comparative linguistics. He even presented comparative tables of sounds, which make it clear that he had an advanced awareness of the Germanic sound shift. Ihre's most outstanding contribution as regards comparison was a more overall description and an ambition to sketch some sort of system for the relations of the consonants between languages.

He was to a certain extent building on a tradition, since a series of earlier Swedish scholars had noticed such features, although they established correspondences through the method of classifying by permutations. There were scholars like Johannes Bureus (1568–1652), Erik Schroderus (1575–1647), the above-mentioned Georg Stiernhielm (1598–1672) and Olof Rudbeck the Elder (1630–1702). Later on, some of Ihre's colleagues, whom he surely met, also enjoyed a scholarly reputation: Olof Rudbeck the Younger (1660–1740) and Erik Benzelius the Younger (1675–1743), both mentioned above, along with Olof Celsius the Elder (1670–1756) (Agrell 1955: 83–172).

Johan Ihre was also unique in his time and at his university in the sense that he was aware of the necessity to include a theoretical side of scholarship. In a speech, he expressed severe criticism of his contemporaries, arguing that it was a threat to the sciences if Leibniz, Newton and Wolff were neglected.

In all senses, Ihre was a rather special scholar of languages in Sweden. His position was that of Skytteanus professor, a professorship devoted to the study of eloquence and politics; in practice, however, he was very devoted to studies of languages in general, while his fellow professors of languages were rather traditional philologists of classical and Oriental languages and – it has to be said – not very outstanding as such either. The few Swedish scholars of languages who really made a contribution during the eighteenth century were, in the first half of the century, to be found among representatives of certain well-known learned families, such as Olof Celsius the Elder and Erik Benzelius the Younger (Benzelstierna), mentioned above. Later, there was also Carl Aurivillius (1717–1786; professor of Oriental languages 1772–1786), about whom it has even been held that he represented the new European humanism of Winckelmann, with reference to his views on antiquity (Lindroth 1978: 589–592). The next great personality in language studies, who was at work at the end of the century, Jacob Fredrik Neikter (1744–1803), librarian and Skytteanus professor with a special interest in philosophy of language, seems really to have taken a step backwards as far as comparative methods are concerned, claiming that individual languages resulted from the blending of other languages, which was the only reason for similarities.

7. On practical knowledge of languages

Some reflections on the practical side of language studies in general might be added here. Language studies should not merely be seen as an offshoot of Biblical and classical studies. Competence in the practical use of the classical and Oriental languages was never a primary concern of university education, but was often aimed for and came with the general academic training provided. Such ambitions, however, declined during the eighteenth century.

As for contemporary European languages, during the centuries of extensive peregrinations it must have been quite natural for a student abroad also to learn to use the language of the country in which he was studying; there is also contemporary testimony about acquiring a speaking knowledge of languages. A travelling student had reason to keep quite well informed about the political situation in Europe. In that respect, a knowledge of the indigenous languages must have been of importance, even though Latin as an academic medium was not commonly challenged. That, without any doubt, was still the situation in the eighteenth century.

In fact, it seems clear that the professors of Uppsala University did master European languages – some more so and some less – and this was mainly due to their international contacts and early studies abroad. During the seventeenth century the German universities were of importance, along with the Dutch ones. Universities in England and France, and even in Italy and Spain, were also visited. The tradition of peregrination was upheld in the eighteenth century as well, though to a lesser degree. It is quite possible, though disputed, that the famous Swedish scholar Linnaeus, for example, bears witness to the learning of languages abroad; that his travelling pupils were versed in languages is well documented.

The question of learning languages was obviously an issue on which views differed. Johan Ihre, the eminent but also controversial scholar at Uppsala University, gave a speech in 1754 in which he was very concerned about the use of vernaculars, strongly criticizing his university and the (lack of) erudition of his colleagues, and saying that it was already necessary to know French, English, German, and even Italian and Spanish. But if it became necessary to learn, say

Polish, Russian and "other languages of uneducated Sarmatians" as well, he claimed, one's whole lifetime would be wasted and the republic of academia would be divided. The question took a long time to resolve, and not until the nineteenth century were the activities of the so called "language masters" at Uppsala University placed on a regular academic footing and professorships of modern European languages established.

8. The following period: the late eighteenth and early nineteenth century

During the eighteenth century there was – as we have seen – no break in traditional ideas of language and language studies until the very end of the century. Few new answers were provided, and as for the question of the genesis of language little apart from the Hebrew and Scythian models was put forward, though it might be said that the preponderance of the latter slowly grew. Possibly, that model seemed to better satisfy an outlook on the world in which Asia was becoming better known at a time of wider colonization and growing imperial interests. The increasing interest in natural sciences also led to a certain scholarly questioning of the Bible.

Hints about languages changing over time were encountered as early as the seventeenth century, but it was not until the emergence of historicism during the nineteenth century, the century of Darwin and other scholarly pioneers, that this was clearly expressed in genuinely diachronic studies. However, the breeding ground for such ideas is to be found during the Enlightenment and in its *Dictionnaire raisonné* (Encyclopédie 1751–1780), which to a certain extent at least advanced a theoretical idea of a genetic and historical affinity among different languages, at the same time as the text of the Bible was challenged as the primary source for explaining the world. On the other hand, though, in no way do the texts of the Enlightenment show any break with tradition. Even many figures renowned as "enlightened" adhered to rather unreflected and traditional views; this could be said not only of the very traditional philologists at the universities. Even in the great Encyclopedia we find quite traditional descriptions of the genesis of languages, and also a kind of Gaulish Goropianism. Thus, in the article on languages in general, the contention is that French is similar to the language of the Druids and, in connection with this, that it is close to Celtic, a language which in turn shares similarities with Hebrew. It seems that the authority of the Bible is still not totally obliterated, only theoretically somewhat moderated.

To sum up, it might be claimed that language studies went through a gradual process over the centuries and that few bold leaps are in fact discernible. Even the new ideas of Wilhelm von Humboldt in the nineteenth century about the inner essence of languages – ideas that led to structuralism – were founded on tradition, though expressed in an innovative form. Most influential for the study of languages, and indeed of all the humanities, was of course the shift towards positioning all instances of human behaviour in a historical frame, bound up with evolutionary ideas.

William Jones, at the end of the eighteenth century, of course had an impact on language studies by opening up a world of other far-away languages, but in view of all that had already been discussed by scholars down the centuries, he did not establish the starting point; his contribution, rather, should be seen as part of an ongoing development of language studies. The decisive and technical turn, which created the methods of comparative linguistics, was provided by Franz Bopp (1791–1867), the German pioneer scholar, some decades after Jones, at the beginning of the nineteenth century (Bopp 1816). But he, in turn, also had his forerunners and forms an integral part of a process of development of scholarly thinking. Everyone, at any time, will inevitably be dependent on the tradition of scholarship.

At the end of the nineteenth century, interest in the history of languages shifted towards a deeper interest in the structure of language, but historical linguistics also survived. This fact clearly indicates that the two questions about language highlighted here, i.e. its structure and its genesis, seem to be of eternal interest to man: where does language originate, and how does it express thought?

References

Aarsleff, Hans
1975 The eighteenth century, including Leibniz. In: *Current Trends in Linguistics. Vol. 13. Historiography of Linguistics*, 383–479. The Hague & Paris: Mouton.

Agrell, Jan
1955 *Studier i den äldre språkjämförelsens allmänna och svenska historia fram till 1827*. (Uppsala universitets årsskrift 1955:13.) Uppsala.

Arens, Hans
1969 *Sprachwissenschaft. Der Gang ihrer Entwicklung von der Antike bis zur Gegenwart*. (Orbis Academicus. Problemgeschichten der Wissenschaft in Dokumenten und Darstellungen. Ed. by F. Wagner & R. Brodführer.) München.

Bopp, Franz
1816 *Über das Konjugationssystem der Sanskritsprache in Vergleichung mit jenem der griechischen, lateinischen, persischen und germanischen Sprache.* Frankfurt am Main.
Brekle, Herbert E.
1975 The Seventeenth Century. In: *Current Trends in Linguistics. Vol. 13. Historiography of Linguistics*, 277–382. The Hague & Paris: Mouton.
Droixhe, D., K. D. Dutz and H. Haarmann
2000 Ideas on the origin of languages from the sixteenth to the nineteenth centuries (in French and German). In: *Handbücher zur Sprach- und Kommunikationswissenschaft. An International Handbook on the Evolution of the Study of Language from the Beginnings to the Present*. 18:1, 1057–1094. Berlin.
Encyclopédie
1751–1780 *Encyclopédie ou Dictionnaire raisonné des sciences, des arts et des métiers*. Ed. by D. Diderot et M. d'Alembert. Paris.
Gesner, Konrad von
1555 *Mithridates de differentiis linguis*. Zürich.
[Junius, Franciscus]
[1743] Etymologicum Anglicanum. Ex autographo descripsit & accessionibus permultis auctum edidit E. Lye. Oxford.
Lindroth, Sten
1976 *Uppsala universitet 1477–1977*. Uppsala.
Lindroth, Sten
1978 *Svensk lärdomshistoria. 3. Frihetstiden*. Stockholm.
Mithridates
1806–1817 *Mithridates, oder allgemeine Sprachenkunde mit dem Vater Unser als Sprachprobe in bey nahe fünfhundert Sprachen und Mundarten*. 4 vols. ed. by J. C. Adelung & J. S. Vater. Berlin.
Postellus, Wilhelmus
1538 *Linguarum duodecim characteribus differentium alphabetum introductio*. [Paris.]
Resenär
2007 *En resenär i svenska stormaktstidens språklandskap: Gustaf Peringer Lillieblad (1651–1710).* Ed. by Éva Á. Csató, Gunilla Gren-Eklund, Folke Sandgren. (Acta Universitatis Upsaliensis. Skrifter rörande Uppsala universitet. C. Organisation och historia, 81.) Uppsala.
Strahlenberg, Philip
1730 *Das Nord- und Ostliche Theil von Europa och Asien*.
UUK
1890 *Upsala universitets konstitutioner af år 1655*. Ed. by Claes Annerstedt. (Uppsala universitets årsskrift 1890, Program 1.) Uppsala 1890

Waterman, John T.
1978 *Leibniz and Ludolf on Things Linguistic. Excerpts from Their Correspondence (1688–1703).* (University of California Publications. Linguistics 88). Berkeley & Los Angeles.

Zedler, J. H.
1732–1754 *Grosses vollständiges Universallexicon aller Wissenschaften und Künste*, (www.zedler-lexikon.de). Halle & Leipzig.

The Swedish Academy of Sciences: Language policy and language practice

Ulf Teleman

The Royal Swedish Academy of Sciences was founded in 1739. This was a great event in the history of learning in Sweden. Sciences had been taught before at the Swedish universities, but their prominent disciplines had been theology, law and the humanities. With the foundation of the extramural Academy, the study of sciences was now strongly promoted, and Swedish was established as a normal language of scientific writing.

Carl Linnaeus (1707–1778) was one of its founding members, but the real initiator was Mårten Triewald (1691–1747), a fortification officer with wide interests, especially in mine engineering and gardening. He had lived for some years in England and was even a member of the Royal Society in London.

Triewald and his friends planned an academy or society that would devote itself exclusively to sciences promoting the national economy of the country, i.e. to sciences and arts which were of real use in public life, as the statutes put it. The members and their associates would study nature through observations and experiments. Their utilitarian ideas were characteristic of the time. The first president of the Academy, Baron Anders Johan von Höpken (1712–1789), a prominent politician, was a member of the mercantilist party, which came to power in the year the Academy was founded (cf. Hildebrand 1939).

Some of the founders and the earliest members were also fellows of the Royal Society in London and of the French Academy in Paris, and these organizations most likely served as models for the Swedish enterprise. (On the corresponding members from foreign countries, see Lindroth 1967 vol. 1: 179. Sörlin 1994: 103 provides interesting statistics on the countries of origin of the references in the transactions. For a discussion of other academies, see also Gunnarsson's 'Introduction' and Bazerman's chapter in this volume.)

There was already a similar society at the University of Uppsala (Kungl. Vetenskapssocieteten in Uppsala; see Hildebrand 1939, 99–135), but it included scholars in the arts as well as scientists. It had been suggested that this society's transactions should be written in Swedish instead of Latin, but the proposal was turned down. The majority of the members wanted their texts also to be read by colleagues abroad, and consequently the language had to be Latin, the common language of trade in the international republic of learning.

The Academy in Stockholm decided otherwise, that is, to use the vernacular for its transactions (*Kungl. Svenska Vetenskapsakademiens handlingar*, VAH). The purpose of this chapter is to discuss some aspects of its linguistic ideology and practice. One important issue will be the successful transition from Latin to Swedish in writing.

1. Language policy

According to the statutes of the Royal Swedish Academy of Sciences, its language in speech and writing was to be Swedish, "*good and pure Swedish*". Carl Linnaeus said that "vi äro Svänske män, som skrifva våre handlingar på Svenska [...] enkannerl:n för Svenskar, våre landsman" [we are Swedish men, writing our transactions in Swedish for Swedes, our fellow contrymen]. The choice of language was obviously regarded as non-negotiable by the members, and it was never questioned afterwards.

The Academy declared now and then that the cultivation of the Swedish language was one of its primary objectives. Nevertheless, it was decided in 1740 that no one could be accepted as a member of the Academy for the sake of the Swedish language alone, i.e. if he was not qualified in "other sciences and arts" which were enumerated in the statutes. (This was the result of an intrigue engineered by two dominant fellows who wanted to get rid of a certain person whom they did not like.)

So the Academy did not come to include any linguistic expertise. In language matters every member was an amateur, at best a competent writer or speaker. Neither the famous professor Johan Ihre (1707–1780) in Uppsala nor the highly knowledgeable and clever lector Sven Hof (1703–1786) in Skara was invited to become a member.

This was a problem in many ways. The members were agreed that a manuscript should be revised before being delivered to the printer, but they were not able to lay down reasonably specific guidelines for this. Neither did they know what to do with the various memoranda on language that were sent to them as a consequence of their solemn declarations that the Academy cared about the fate of the mother tongue. No more were they able – or perhaps willing – to see to it that somebody wrote a competent and authoritative Swedish grammar or a manual on style and rhetoric for scientific writers.

Nonetheless, it may be possible to get a picture of the general opinions the members had on language, from the records of their meetings or from their letters and speeches.

Not unexpectedly, the choice between German and Roman letters was the problem that attracted the greatest attention and gave rise to the best discussion. The majority of the members preferred Roman types, while two of the most respected fellows wanted to stay with the German (or as they were normally called: Swedish) types. They believed that Roman types would repel the common man and make the transactions more difficult to sell. Nevertheless a transition took place in 1743. This was comparatively early. As late as in 1769 nearly all political pamphlets – and they were numerous – were still printed with German letters (Gustafsson 2009: 77).

Orthography was a more difficult issue. Everybody wanted a uniform orthography but there was no agreement on the norm, and no norm was sufficiently codified. A typical entry in the meeting records reads:

> Hr. Salvius upläste sina påminnelser vid orthographien, och Akademien lemnade frihet at härmed göra som han prövar vara aldra bäst. (Dahlgren (ed.) 1918: 75)
>
> [Mr. Salvius [i.e. the secretary] presented his views concerning the orthography and the Academy left it to him to do as he thought best.]

Equally complete and at the same time ineffectual was the unanimity of the members concerning the use of foreign words: of course, they should be avoided. Anders von Höpken wrote in his introduction to the transactions (VAH 1739):

> At ränsa våra Fäders Språk, ifrån alla främmande tilsatzer, och återkalla det til sin första renhet, är en anständig omsorg, för ett fritt Folck, som under de mångfaldiga swåra öden det undergådt, likwäl aldrig förlorat sig sielft, och tyckes fördenskuld böra i tal och skrifter wara äfwen så Swenskt som det warit, är och bör wara, i Dygd, Wett och Mandom.
>
> [To purge the language of our fathers of all foreign additions and to bring it back to its original purity is a proper concern for a free people, which has never lost itself under the many severe destinies it has undergone and which seems therefore to be obliged to be as Swedish in speech and writing as it has been in virtue, learning and manhood.]

This was the politically correct opinion of the day. But most of the members probably accepted Latin technical terms, and only one of them was against German words. (Probably no one would have been able to recognize a German loanword anyway when he saw one, since they looked very much like regular Swedish words.) So we are left with imports from French. It turned out that even in that regard one member suggested a moderate policy: foreign words should be accepted, according to him, provided that they were adapted to Swedish pronunciation.

So, the honourable gentlemen of the Academy may have had strict opinions on standardization and purism, but it did not prevent some of them from spelling irregularly or adorning their texts with Latin and French imports.

A programmatic statement on language was made in a memorandum by Count Carl Gustaf Tessin (1695–1770) on "language in general" (1744; printed in Hernlund 1888). It was anonymous, but everybody knew he was the author. The memorandum was positively received at the meeting, but all the same it was put aside afterwards, without further comment or action.

Tessin argued that the Academy should compile a comprehensive Swedish dictionary. He disliked synonyms: each word ought to have its own meaning, different from the meanings of other words. Words like *springa* 'run' and *löpa* 'run' should be semantically differentiated, so that *löpa* meant 'run on even ground' and *springa* 'run on uneven ground'. These ideas are akin to other utopian contemporary projects to create unambiguous artificial languages (e.g. as proposed by Leibniz and Wilkens).

What views did the Academy entertain about the ideal manner of writing? This issue was even more complicated than the others, and it comes as no surprise that the Academy took a defeatist stance on rhetorical or stylistic revision of manuscripts:

> Hvad I gemen ändringar uti handlingar vidkommer, så fann Akademien att stilen och skrifarten borde lemnas orubbad, medan den gifver vid handen hvar och en auktors geni, men stafningen kunde rättas [...] (minutes of an Academy meeting, quoted in Hernlund 1888)
>
> [When it comes to revisions of transactions, the Academy found that the style or manner of writing ought to be left unchanged, because it shows the genius of each author, while the spelling could be corrected ...]

It is obvious, though, that the ideal of the Academy was a simple, short and clear style. Linnaeus said in a letter to the secretary of the Academy that he could not speak if he had nothing to say, and he was certainly of the opinion that others should be like him in this respect. He expressed his views very clearly in his comments on a manuscript (Linnaeus's letters: 260):

> Men author har en stylum grallatorum, som jag ej kan fördraga och är oanständig för en academie, och liknar en markskriares exklamationer. Han will wara eloquent, men stannar I Turkisk wältalighet.

> [But the author has a bombastic style that I cannot stand and which does not befit an academy. It is like a street-monger's exclamations. He wants to be eloquent, but is stuck in Turkish oratory.]

According to Linnaeus, he and his contemporaries wrote much better than their colleagues a century before (Linnaeus's letters: 272)

> Beskrifningarne blefwo då insvepte uti långa och widlöftiga orationer, som nu lämnas så nakne, som då de äro födde, med så många ord, som äro betydande, utan snömos.
>
> [The descriptions were wrapped up in long and verbose orations, which are now left as naked as when they were born, with as many words as have a meaning, without empty talk.]

Tessin stressed that every author should understand exactly what he wanted to say and then express it so that any reader could understand it, too. He maintained that Swedish was a language which did not tolerate hogwash and wordy expressions

> wid hwilka man måste förlora andan, innan man hinner till ändan (after Hernlund 1888)
>
> [where the speaker runs out of breath before he reaches the end]

The spirit of Tessin's memorandum is very much related to Thomas Sprat's famous words in his History of the Royal Society from 1702:

> They [i.e. the members of the society] have therefore been most rigorous in putting in execution, the only Remedy that can be found for this *extravagance*: and that has been, a constant Resolution, to reject all the amplifications, digressions, and swellings of style: to return back to the primitive purity, and shortness, when men deliver'd so many *things*, almost in an equal number of *words*. They have extracted from all their members, a close, naked, neutral way of speaking: positive expressions; clear senses; a native easiness: bringing all things as near the Mathematical plainness, as they can: and preferring the language of Artizans, Countrymen, and Merchants, before that, of Wits, or Scholars. (p. 113)

2. The transactions and what was so special about them

So much for the language policy of the Academy. What about its practice? I shall turn directly to its most important work, the transactions. Five hundred copies were printed in quarterly instalments and in the early years additional copies had to be distributed. The prestige of the transactions was considerable.

Some of the members of the Academy were famous scientists in their fields: Carl Linnaeus in botany and zoology, Anders Celsius (1701–1744) in astronomy, Samuel Klingenstierna (1698–1765) in mathematics, Carl de Geer (1720–1778) in entomology, Torbern Bergman (1735–1784) and Scheele (1742–1786) in chemistry, among others. Many transactions contained information useful to their readers in their daily work. They were a great success from the start.

The transactions were efficiently distributed, especially after a few years when the entrepreneur Lars Salvius began to print and sell them. He got permission from the government to combine editing, printing, distribution and selling. He had bookshops in Stockholm and other towns around the country and was even able to export the transactions, as well as other works by prominent members of the Academy, to various European countries (Schück 1929, Lindroth 1967 and 1978).

The Academy's transactions were revolutionary in many respects. In spite of their contents they were written in Swedish, and the authors addressed the general public. (At the time nearly all Swedes could read.) The ambition was that every citizen should understand them. This may have been an unrealistic goal, though, since some articles require a basic knowledge of relevant disciplines (for instance, more mathematics than I myself can muster). In fact, after some decades, the secretary Pehr Wilhelm Wargentin (1717–1783) tried to reduce the number of difficult contributions, since he thought that they drove potential buyers away.

As the audience was partly non-professional, it may be wrong to regard the transactions as truly scientific; perhaps they should be characterized as popular science instead.

It was understood that authors were allowed to use Latin for their hardcore scientific writing in other publications, which were intended to be read by colleagues outside the country. But in Sweden, too, the position of Latin was still strong. This is indicated by Abraham Sahlstedt's new Swedish dictionary (1773), where the meanings of the Swedish lexemes were explained in Latin. It was assumed – realistically or not – that the users of the dictionary knew Latin better than their own mother tongue!

Not only the readers but also the authors of the transactions could be unlearned men or women, i.e. persons who did not write or read Latin. Many very advanced articles were written by civil servants, farmers, shipbuilders, engineers, mining experts, mechanics, inventors, artillery officers, pharmacists etc. (just like many *Philosophical Transactions* of the Royal Society, but unlike the *Mémoires* of the French Academy).

The articles often concerned new subjects, outside the disciplines studied at the universities. Their main source of information was not man-made books (like the ones studied by scholars), but the book of Nature, written by God himself. It is interesting to learn how the authors legitimate the information they offer. One way was to point to its usefulness. Another was to show how complex God's creation was and consequently how infinitely wise its creator. But the author could also call attention simply to how fascinating, surprising or even lovely the knowledge was.

To read the book of nature, the scientist had to make systematic observations and experiments. The Academy seems to have agreed that good science was descriptive and inductive. According to the records of one meeting, the transactions were not to be allowed to contain reasoning or judgements. Such elements could only be tolerated in mathematics, not in "physical matters". The members had noticed that the scientific speculations of a century ago were often considered ridiculous one hundred years later. That risk should not be taken by the Academy, they concluded. Only incontestable truth should be printed... Fortunately the authors did not abide by this decision.

The transactions brought hot information. The statutes of the Academy prescribed careful review procedures, but also that the handling of manuscripts should be as speedy as possible:

> as soon as a piece of work has reached full maturity, it should appear in Swedish without delay

Unlike the French *Mémoires*, the Swedish transactions did not say when the article was received, but sometimes the date may be evident anyhow. For instance, an article written in South Africa early in 1775 was published in the last quarter of the same year's transactions. Some references in the articles are made to other publications only one or two years old. The transactions were rather short, as a rule not more than 30 pages and sometimes much shorter than that, down to a single page. This should be compared to regular academic volumes, which often required a long time for writing, editing, printing and distribution (an exception, however, being the dissertations).

(On the Academy and its transactions, see especially the monographs by Hildebrand (1939) and Lindroth (1967). Oscarsson (forthcoming) has analysed the types of articles, i.e. applied vs. theoretical, in relation to the types of authors, i.e. professional scientists vs. amateurs. The ideas of the Academy concerning language cultivation are presented and commented on by Hernlund (1883 and 1888) and Wellander (1959).)

All these circumstances influence the language and rhetoric of the transactions. I have read cursorily a few volumes from the early years and a few from around 1775, and also studied some articles in detail for a closer look at their language and rhetoric. What interested me in particular was to find out what happened – if anything – in the transition between Latin and Swedish, in the field of force between traditional academic writing on the one hand and, on the other, speech and writing outside the learned world. I therefore picked out three authors from the first years of the transactions who had written on the same or a similar subject in Latin in the *Acta* of the Uppsala society:

Nils Wallerius:
De ascensu vaporum in vacuo [On the ascension of gases in a vacuum] (Acta 1738)
Försök, om Dunsternas upstigande uti lufttoma rum, giorde, ock til Academien insände [Experiments on the rising of gases in vacua] (VAH 1740)

Anders Celsius:
Observationes Meteorologicæ habitæ Upsaliæ Anno 1730 [Meteorological observations made in Uppsala in 1730] (Acta 1730)
Meteorologiske Observationer hålne i Upsala Åhr 1740 [Meteorological observations made in Uppsala in 1740] (VAH 1741)

H. D. Spöring:
Abscessus Ventriculi & Hepatis [Abscesses of the ventricle and the liver] (Acta 1733)
Beskrifning på en i Finland mycket gängse Barna-siukdom *Borst* kallad [Description of a childhood disease, very common in Finland, called *Bristle*] (VAH 1743)

I also compared three transactions on similar subjects from around 1740 and three others from around 1775 to see if anything had changed in the meantime:

Carl de Geer:
Beskrifning på en märkwärdig Fluga, kallad Ichneumon ater, antennis ramosis [Description of a strange fly, called Ichneumon ater, with branchy antennae] (VAH 1740)

Carl Peter Thunberg:
Pneumora, Ett nytt Genus ibland Insecterne [Pneumora, a new genus among the insects] (VAH 1775)

Pehr Elvius:
Sätt at mäta watnets hastighet [Ways of measuring the speed of running water] (VAH 1741)
Erik Prosperin:
Om Cometernes minsta afstånd ifrån Jordens Orbita [On the minimal distance of comets from the orbit of the earth] (VAH 1775)

Anders Celsius:
Observationer om twänne beständiga grader på en Thermometer [Observations concerning two constant degrees of a thermometer] (VAH 1742)

Nils Landerbeck:
Beskrifning På en förbättrad Luft-Pump [Description of an improved air-pump]
(VAH 1774)

Possible linguistic or rhetorical influences from the English or French transactions have not been included in this study, however.

3. Rhetoric and language of the transactions

Understandably, the transactions differ from each other in many ways: some are short, others longer, some theoretical, others more practical (or "economical", as it was called at the time), some come from older, more mature disciplines, others represent new fields still looking for their identity. Some articles or parts of them are narrative, others descriptive, and still others are more discursive and analytical. Not unexpectedly the proportion of theoretical articles increased from the 1740s to the 1770s, and the relative number of non-academic authors also decreased somewhat (cf. Oscarsson, forthcoming). We must keep these circumstances in mind if we are to understand the varying language and style of the transactions. The following aspects have been focused upon:

Textual organization
Objectivity
Rhetorical devices
Lexical apparatus
Syntactic machinery
Techniques of presentation
Standardization

(Inspiring studies of scientific prose in other languages and its development in modern times are Bazerman 1988, Gross, Harmon and Reidy 2002, Latour 1987, Swales 1990. In Sweden, Britt-Louise Gunnarsson and her group have been pioneers in the field of medical writing; see especially Gunnarsson 1988, 1990, 2005a, 2005b.)

3.1. Textual organization

On the whole the articles are well organized. This is so from the very start in 1739, although one might have the impression that the first generation of writers in Swedish are somewhat freer than their successors. Prosperin's article on astronomy (1775) is a very long, brilliantly written and planned text. De Geer's transaction (1740) is also well written, but its disposition is not as tight as Prosperin's. It is about an insect laying its egg in another insect's egg which is placed in the interior of an oak leaf. The transaction starts with a description of the host insect, and it takes the reader two pages to realize that the article is about the other insect. But the story is fascinating, and a non-academic reader may never notice the detour taken by the author. So the author's arrangement may have something to do with the fact that his audience is not exclusively academic but consists of laymen, too.

Another early article is about the rising of vapours in a vacuum. It is available in two versions, one in Latin, and the other a transposition of it in Swedish for the Academy's transactions. The Latin version uses a straightforward academic introduction, but when it is revised and transposed into Swedish the author adds an extra prelude on various kinds of invisible evaporation happening all the time in nature. It makes interesting reading, but as an introduction to the main contents of the article it is a little deceptive.

It is surprising that so many transactions lack an introduction and that even more of them lack any kind of final conclusion, in which the principal results are summarized and discussed.

Some authors of short articles might have thought that the title of the transaction was enough to introduce the text. When there is an introduction it is normally used for legitimation: the knowledge to be presented is characterized as either useful or "delightful". Only a minority of authors, even around 1775, offer theoretical justification, proceeding from present knowledge which is maintained to be incomplete or incorrect in a way that the article is going to remedy.

A genuine step forward during the period is the schematic model for the description and taxonomy of botanical and zoological species. This seems to have been an invention of Linnaeus and his pupils. It included a subtext schema for a so-called "half-list" (Bo-A. Wendt, personal communication), i.e. something between a regular running text and a numerical table, like this:

name of the referent
non-finite predication 1
non-finite predication 2
etc.

The name of the referent is a kind of heading, followed by a sequence of non-finite predications. These are expressed by verb phrases, adjective phrases, noun phrases etc. Another conventionalized arrangement was the one for autopsy reports (a popular genre in the transactions), but that was used in Latin, too.

3.2. Objectivity

A characteristic aspect of academic texts today is their apparent lack of subjectivity and personal style. When I was a young student I was told never to use the pronoun *jag* 'I' in academic writing. The style should be objective and impersonal, and the author should not be mentioned within the text proper. This was an idiotic rule, but I think most of us still carry it with us. Sprat's rhetorical ideal was "mathematical plainness", i.e. absolute objectivity, but also the language of artisans. These ideals, however, may contradict each other in practice, since the spoken language of artisans and other common people is full of personal remarks, i.e. not at all objective.

There are three cases where the author may refer to himself:

(1) He says he has observed something
(2) He says he has performed an action
(3) He says he has thought or thinks something

To avoid "I" in (1) is easy. "I observed that the water was boiling" can be replaced by a bare "The water was boiling". To avoid "I" in (2) the passive is the most straightforward solution: "I filled the vessel with air" is replaced by "The vessel was filled with air". (3) presents more intricate difficulties, although often epistemic auxiliaries and adverbs come in handy.

My comparisons between Latin and Swedish transactions indicate no difference in the use of the passive. We can look at some figures from Wallerius's articles on evaporation for sentences of the types (1) to (3) above:

	Sw. active 1st pers. sing.	Sw. passive
Lat. active 1st pers. sing.	10	3
Lat. passive	3	11

Both versions have approximately as many passives as actives in the relevant cases. There are differences, but they go in both directions. The relative share of passives in these cases is the same in both versions.

In other respects, however, the authors of the first generation may have been seduced by their mother tongue into writing a more personal prose. The astronomer Anders Celsius uses 1st pers. plur. about himself in Latin, but 1st pers. sing. in Swedish. (Prosperin, an astronomer of the following generation, though, also preferred 1st pers. plur. in Swedish.)

Some narrative and descriptive passages are quite entertaining, and the authors do not hesitate to make evaluative statements or to give extra, unnecessary information to create a more vivid and tangible impression. Celsius in his meteorological report says, for example, "there is an oak-tree in my garden" or refers to "the little river here in Uppsala". The entomologist de Geer is really captivated by the beauty of his fly, and the professor of medicine Spöring displays genuine sympathy for the children suffering from the illness he reports on.

Another kind of subjectivity is the one where authors like Spöring and Prosperin fake an impersonal style superficially, while managing to annihilate colleagues through irony and sarcasm. Very amusing, I can assure you, but I would not say that the objectivity goes very deep there.

The Swedish transactions give a more objective impression than the transactions of the Royal Society in London, since they do not reproduce the letters which accompany the manuscripts from the writer to the academy. In the English transactions the letters are printed as a kind of subjective framework story in which the objective article is embedded.

3.3. Rhetorical devices

That brings us to the rhetoric of the transactions. Their primary objective was to present useful truths, obtained by observations and experiments. Such texts are normally narrative or descriptive, and they can follow more or less universal patterns, familiar to the author from his native language, where they are used for telling stories and describing things.

Another kind of transaction, which is perhaps more interesting, is the one that the Academy said authors should avoid, i.e. texts characterized by analysis, interpretation, argumentation and even speculation. Such passages are found in many transactions, though. We can expect the author's rhetorical capacity to be put to a particularly tough test in this type of text, where his mother tongue experience may have provided no obvious models.

It is all the more impressive, then, that this transition seems to have taken place so smoothly. Of course the general speaker of Swedish had at his disposal – since the Middle Ages – the full range of logical and other relational markers: causal, conclusive, consecutive, conditional, concessive etc. subjunctions, conjunctions, verbs, adverbs and prepositions. What is surprising is that the authors also make use of a rich apparatus of formulas for argumentation and discussion, such as:

> it is well-known that
> somebody may say that
> somebody may doubt that
> nobody could reasonably accept that
> anybody can easily verify that
> hereby X's unfounded assertion fails
> etc.

Some authors can piece together long arguments and display the structure of the text nicely by various discourse markers like:

> sålunda 'thus, then'
> vidare 'further'
> nu 'now'
> åter 'again', 'on the other hand'
> etc.

How can it be that the authors have a command of this kind of rhetoric not only in Latin, but also when they write in their mother tongue? I shall return to this question in the concluding discussion below.

The rhetorical excesses of eloquence that Tessin had warned of are, as far as I can judge, totally non-existent in the transactions, but this also goes for their articles in Latin. Certainly, some authors love metaphors or similes, but these devices are always pedagogically motivated and often very illustrative, as for example in the text by de Geer:

> its shape [is] like a slice, cut lengthwise out of a pear

Perhaps de Geer is sometimes a little more anthropomorphic than would be accepted today in strict scientific prose, but to the general reader expressions like the following are quite attractive:

> the attack from hostile flies
> their horns are proud

3.4. Lexical apparatus

Some lexical elements have already been mentioned as useful devices in the analytic discourse of the transactions. Let us look a little more into the lexical apparatus available to the authors or created by them.

I have found few new technical terms. One possible reason for this may be that they are not introduced as such in the text, and perhaps not even recognized as such by the author. No formal definitions are provided and the lexemes are used to describe a phenomenon rather than to denote it. One example of this is the two basic terms in Anders Celsius's transaction on the graduation of thermometers:

> vattenfrysningspunkten 'the water-freezing-point'
> kokande vattenspunkten 'the-boiling-water-point'

Another reason may be that there are indeed few *new* technical terms to be found in the transactions. I have checked many candidates against the comprehensive historical dictionary of Swedish, SAOB, only to find out that most of the technical words – and there are many of them – had been introduced earlier than 1740. Obviously many terms had been coined for use in other environments than the universities: in pharmacy, mechanics, shipbuilding, surgery, navigation etc., long before these professions were scientified and academicized. These non-academic activities had needed technical terms and they were coined in Swedish or were somehow borrowed from other modern languages and adapted to Swedish. Sometimes they were taken from Latin as it was used in academic disciplines like mathematics, medicine and astronomy at the universities.

There are two kinds of technical terms: words for specific phenomena and words for abstract or general categories. I find the latter type of words particularly interesting. It is fundamental in advanced scientific thinking, but it is nevertheless often overlooked when you are looking for technical terms in an old text. I have encountered large numbers of them in the transactions. Most of them – like the more specific terms – were already available in the Swedish vocabulary when the Academy needed them. They include words such as:

art 'kind', beskaffenhet 'quality', väsende 'phenomenon',
substans, materia 'matter', anseende 'appearance';
skillnad/differens, proportion, kraft 'force', motstånd/resistens 'resistance', täthet 'density', massa, kapacitet, rum 'space', superficie/yta 'surface';
längd 'length', bredd 'breadth', vidd 'width', köld/kyla 'cold', värme 'warmth', tyngd 'weight', höjd 'height', medelhöjd 'average height', djuplek 'depth'
etc.

It is interesting to note that Anders Celsius in his article on the thermometer did not use a word for 'temperature'. He had to manage with the nouns for cold and warmth (or heat). The word *temperatur* appears a few years later according to SAOB. Probably, the invention of the standardized thermometer was the prerequisite for the concept of 'temperature' to emerge, but Celsius, who died in 1744, may not have been the one who introduced it into Swedish.

Some authors combine Latin and Swedish terms, as a precaution:

lönhål – crypta 'secret hole'
nedersta randen – basis 'lower edge'
sågtandad – crenata 'serrated'
de yttre delarna – integumenta 'exterior parts'

mellangården – diaphragma
starkt kokar – vehementer bullit 'is boiling vigorously'

The new words are found in particular in the descriptive articles. The authors of these – not unexpectedly – seem to have been more creative in Swedish than they were in Latin (even if Latin, too, had productive rules for lexical derivation). Some examples from Spöring (medicine) and Thunberg (entomology), who coined new words to describe what they saw (with literal translations):

svetthål 'sweat-hole'
honungsdeg 'honey dough'
talgaktig 'tallowish'
löpfötter 'running-feet'
hoppfötter 'hopping-feet'
hinnelik 'membrane-like'
vackert-grönt-nät-lik 'beautiful-green-net-like'

3.5. Syntactic machinery

Content words are to some extent chosen consciously, but the syntax produced by writers is normally an effect of choices they are not aware of. Now, what were the consequences syntactically when scientists began to write in Swedish instead of Latin?

Let us assume that the authors put the same amount of content into their sentences in Swedish as they did in Latin. We would then expect longer sentences and a flatter syntax, since some compact constructions, frequent in Latin, were not available in Swedish but had to be replaced by full finite clauses.

On the other hand, it might happen that the authors switched to a more paratactic way of writing, i.e. exchanged autonomous sentences for syntactically dependent clauses or non-finite constructions. They would do that if their mother tongue inspired them to use a more spontaneous style, with a more domestic ring to it. The outcome would then be more and shorter sentences – provided that the Swedish text contained the same amount of information as a corresponding text in Latin.

Before offering some figures, I have to make two reservations. Since Swedish is typologically more analytic than Latin, word-per-sentence counts are inevitably biased. The figures will be a little too high in Swedish, because Latin lacks, for instance, definite and indefinite articles, pronominal subjects and a few other things. It should also be kept in mind that the basis for my data is very small.

Let us look at the Latin and Swedish transactions by Spöring and Wallerius. First Spöring, where the figures are as expected, i.e. the number of long sentences is relatively higher in the Swedish text.

	Number of sentences:	
	Latin	Swedish
< 21 words	46	18
21 – 40 words	11	12
41 – 60 words	0	3
> 60 words	0	7

In the articles by Wallerius, the results are divided: the descriptive/narrative paragraphs show the same tendency as in Spöring's article, but in the analytical paragraphs the difference between the languages has disappeared. Or rather, it is reversed.

	Number of words per sentence		
	Latin		Swedish
Experiment			
7 sentences			
1	16	<	30
2	16	<	24
3	20	<	51
4	17	<	26
5–6	23	<	42
7	15	<	20
Discussion (one paragraph)			
5 sentences			
1	32	>	31
2	33	>	23
3	27	>	25
4	24	<	34
5	26	>	18

As these figures show, the sentences of the experiment section are longer in Swedish than in Latin. In the discussion section on the other hand they are, with one exception, shorter (or have the same length). A possible interpretation is that the syntactic influence of Latin is stronger in the discursive section than in the narrative account of the experiments.

Most Latin participial constructions (participium coniunctum, ablativus absolutus) have disappeared in the Swedish transactions, and finite clauses have been used instead. Here are the figures for finite verbs vs. participial constructions in the transactions of Wallerius, Celsius and Spöring:

	Latin	Swedish
Wallerius		
finite verbs	8	21
part. constr.	11	1
Celsius		
finite verbs	19	34
part. constr.	7	0
Spöring		
finite verbs	91	123
part. constr.	5	0

My corpus is small and cannot prove anything by itself, but my general impression as a reader of these and other transactions indicates that they are indeed representative.

In other respects the syntax of the Swedish transactions is still very Latin-like, and the sentences are sometimes very long and complex, in both the first and the second generation of writers. The authors normally manage to construct even their long sentences correctly, i.e. in accordance with the general rules of Swedish syntax at the time. I could quote some beautiful examples, but it would require too much space to explain their complexity.

Sometimes, though, the authors cannot keep it all together, and as a result their sentences become anacoluthic. The examples of syntactic breakdowns I have found were produced by the writers of the first generation (Celsius and Elvius), and that might be significant. The second generation have had more opportunities to handle a complex Swedish syntax in scientific writing than their predecessors and have learnt from it. The generation who wrote the articles of the first volumes were the pioneers.

3.6. Techniques of presentation

Let us return to the conscious side of text production, and to a less linguistic aspect of it: techniques of presentation. The phenomena I have assigned to this section are things like

Typography
Abstracts
Headings
References, quotations
Footnotes

Tables
Indexes
Illustrations

An important step was taken when the Academy abandoned German letters and switched to Roman types. From 1743 onwards the pages looked as nice and neat as in the French and English transactions.

The Swedish transactions had no abstracts. The French Academy was more advanced in this respect: each annual volume had full abstracts of the important transactions in a comprehensive introductory section.

Some of the longer transactions have headings announcing the principal sections of the text, while others signal the transitions between sections verbally within the text. The headings are rather uninformative, apart from their function of signalling boundaries. Sometimes the paragraphs are numbered. Sometimes headings are used which say very little about the contents, normally metaexpressions like "the first experiment" or "remarks" or "various comments". The Swedish texts represent no advance in this respect from earlier writing in Latin.

Many transactions include references and quotations, but the degree of bibliographic specification is normally low, often with no information about the title or the pages referred to. The usual way of doing it in Latin or Swedish transactions is to mention the surname of the author preceded by a title (often only *herr* 'Mr') and quite often also an epithet of praise like "learned" or "famous".

Among the second generation of authors, footnotes had become popular for references or for minor remarks – as they are today.

Simple statistical data of different kinds were neatly presented in tables. The conventions for this were taken over from Latin writing. Tables of contents and indexes had been invented quite early in the history of printing, and the Swedish Academy of Sciences (like her sister academies in Paris and London) was very ambitious with these. Each quarterly instalment had a table of contents, and so too did the annual volume, in addition to a comprehensive subject index with a large number of keywords.

Naturalistic illustrations were indispensable for descriptive articles on plants, animals, machines, instruments, other inventions, diseases etc. They were created by draughtsmen and then copied by copperplate engravers. Some figures in the Swedish transactions were beautiful and informative, although they could not compete with their French and English counterparts, at least not to begin with. Illustrations were placed at the end of each quarterly instalment and the authors referred to them by means of letters and numbers. Linnaeus complained in his letters about the artists he had to put up with and often asserted that his output was seriously hampered by incompetent draughtsmen. He envied continental colleagues who were able to use drawings which were so outstanding "that the most insensitive Hottentot was moved by admiration and affection for God's creation".

Other more abstract graphic representations were very rare: just a few geometric figures or curves to show the orbits of planets or trajectories of bullets. Linear graphs were not yet invented (Gross, Harmon and Reidy 2002: 154).

3.7. Standardization

Academic teachers had some difficulty when they started to give lectures in Swedish, and some of them mixed Latin and Swedish freely, at least in the late seventeenth and the early eighteenth century. The Academy's transactions, however, were written in fluent Swedish with very few Latin elements in them. Even their orthography was quite uniform.

The records of the Academy's meetings show that the members were concerned to achieve a standardized language with a standardized orthography. Their first secretary Lars Salvius introduced the spelling rules of the sister society in Uppsala, but they were insufficient and not specific enough.

Salvius was soon dismissed as secretary, but came back as the leading printer of the Academy. To him a homogeneous orthography was as important as a clean and distinct typography for the quality of his products. The manuscripts were to some extent revised by him or his employees before going to print, but there were still no comprehensive spelling guidelines and still no authoritative dictionary to guarantee a uniform standard. We cannot rule out the possibility, though, that the spelling converged nevertheless, without codified rules, whether these things were discussed within the Academy or whether the orthography became more uniform "spontaneously" within the national community of public writers.

The best method to measure the relative degree of standardization between texts or text corpora is to compare the spelling of words occurring more than once to see if each instance of each specific word is spelled differently or not. I have not been able to carry out a systematic investigation, but have simply made notes of differing spellings within texts, as I read through them a few times. I noticed 15 pairs of differing spellings within the early texts and only 3 in the texts from the 1770s. Wallerius (1740) spells seven words in different ways:

watu – wattu 'water'
göra – giöra 'make', 'do'
toma – tomma 'empty'
rumet – rummet 'the room'

ymnig – ömnog – ömnig 'abundant'
ricktig – riktig 'correct'
talrik – tallrik 'plate'

Of course we come across words occurring only once in an article which might be orthographic variants, but we cannot know for sure if the author is consistent in his spelling or not, since phonetically irregular spellings may be lexically determined. One author, for example, always uses the spelling *här* 'here', but also always writes *der* 'there'. Another writer uses the spelling *där* for the separate word, but *der-* in compounds. So we cannot know whether the following spellings are meant to be standard or not:

Wallerius (1740): -achtig, wigt, fukt
Prosperin (1775): rickta, frukta, magt

My general impression is that the orthography of the Academy's transactions is fairly uniform even in the early texts and that it has become a little bit more standardized in the 1770s.

There is also still some variation in the use of capital letters in nouns. Capitals are more often used in nouns with a specific meaning than in frequent nouns, but sometimes the writing of a word may vary in the same transaction. Prosperin (1775) writes *Jorden* 'the earth', *Massa* 'mass' but also *jorden, massa.* Landerbeck (1774) writes *Conen, Recipienten* but also *conen, cylindern*.

Compounds (especially nouns) can be written in three ways: (a) as two separate words, (b) as one word with a hyphen between the constituents, (c) as one word without a hyphen. Alternative (a) is rare, but some authors, such as Prosperin, make frequent use of alternative (b):

medel-avstånd 'average distance', drag-kraft 'attractive force', lutnings-vinkel 'angle of inclination', cirkel-rund 'circular'

Landerbeck (1774) uses both (b) and (c):

luft-pump 'air-pump', järn-ten 'iron rod', con-formigt 'cone-shaped', tilbaka-dragande 'drawing back'
järnten 'iron rod', fyrkantig 'quadrangular', bomolja 'industrial olive oil', renshud 'reindeer hide'

The use of commas is at least as irregular as in contemporary Swedish. Grammatical punctuation prevails, but often other heavy constituents than clauses are demarcated, as in this sentence from Landerbeck (1774):

På ena sidan af rörets framkant, är en ögla, genom hvilken är en skruf D af Stål, för at fästa en Barometer.

[On one side of the tube, there is a loop, through which there is a screw D of steel, to attach a barometer.]

Morphology is also on its way to becoming standardized in eighteenth-century Swedish. In the transactions, as in other printed texts of the time, some suffixes have variant forms such as *-a/-e* (plural and/or definiteness in adjectives) and a few other inflectional morphemes (cf. Santesson 1986, Teleman 2002).

It is impossible on the basis of these informal observations to decide whether the Academy's transactions reflect the general degree of standardization in contemporary printed texts, or whether they are more standardized or less. It is obvious, however, that the transactions – like other forms of public writing in Swedish at the time – have still not reached a modern level of standardization.

4. To conclude

What, then, was the role of the Academy and its transactions for the development of Swedish scientific prose and of the Swedish written language?

During the eighteenth century an increasingly sharp distinction between the language of fact and the language of fiction and eloquence was emerging, and people were beginning to become aware of it. Factual prose was the dominant canonical form of written language, and the transactions of the Academy may have been regarded by many as the true manifestation of this kind of Swedish. This was a time when utilitarianism, economism and republican utopianism had taken over from ideologies like religious orthodoxy, antiquarian historicism and the glorification of autocracy. The new era needed a new prose for the purposes of information and discussion, a prose from which unnecessary verbosity was banished, and in which language was not allowed to obscure the naked truth.

For the establishment and development of this kind of prose the prestige of the Academy was important, but significant roles were also played by other expanding agents like the press, journals and non-fictional books or pamphlets for the general public. The outcome was a bifurcation between two main types of written Swedish. A balance of power was established, with one side, factual prose, dominating in the eighteenth century, while the other side, the prose of belles-lettres, was to have its golden age in the 19th.

This was also an era when Swedish as a written language was being standardized. The Academy may have influenced this process, too, but not by its transactions and not by any codified language norms. Its important contribution was to stimulate debate on the standardization of language at the time. As mentioned above, the Academy itself lacked linguistic expertise. But it supported essential language projects, financially and above all morally. Among these projects, Sven Hof's treatise on the Swedish written language (1753) and Abraham Sahlstedt's Swedish dictionary (1773) were the most prominent. These works may never have come into being, if they had not been backed by the Academy.

The Academy had neither the capacity nor the energy to formulate the norms of a uniform orthography, but it had prepared the ground for the new Swedish Academy founded by the king, Gustav III, in 1786. This body had as its sole purpose the cultivation of the Swedish language and its literature. The ideological and intellectual climate of the country was by then ready for the Swedish Academy's authoritative manual of orthography (Leopold 1801).

Another question concerns the contribution of the Academy of Sciences to the development of a scientific language and a rhetoric for scientific writing in Swedish. Of course, it was significant that the Academy had chosen Swedish as the language of its transactions. A language is developed above all through its use, and the Academy had demonstrated that complex scientific matters could be treated in Swedish. The language choice of the Academy was certainly one reason why the scientific departments of the Swedish universities soon abandoned Latin in their research and teaching.

The transactions may not have entailed any dramatic change in the practice of writing on scientific matters, though. The transition from Latin to Swedish seems to have been rather effortless. Why was that? It is possible that Swedish was more widely used at the universities before 1739 than we have had reason to believe. Some professors in Uppsala, Lund and Åbo may have given lectures and colloquia in Swedish from the end of the seventeenth century. One example of this is the private lectures given by Professor Petrus Lagerlöf in Uppsala on the Swedish language. (Cf. Teleman 2002, chapter 1.)

The narrative and descriptive parts of the transactions could follow general patterns, the same as in non-academic, non-scientific Swedish, and an impressive repertoire of technical terms already existed in Swedish.

The language of analysis, reasoning and argument, too, could obviously be transposed into Swedish from Latin. It is even possible that students had been trained in a similar process of transfer when they learnt how to preach. (On the education of future clergymen in the art of preaching, see Askmark 1943: 350.) Many science students studied theology, too, and became priests after leaving university. To write a sermon implied the same kind of transposition from Latin theology to preaching in the mother tongue as the transposition from academic science in Latin to the kind of popularizing scientific prose that was expected in the Academy's transactions. The only difference was that scientists read and interpreted the book God had written himself, the book of nature, instead of the Bible and other books written by men.

And abracadabra! The scientific community of Sweden had found a language and a rhetoric for scientific writing in Swedish. Perhaps it was no more difficult than that.

References

Acta = Acta literaria et scientiarum Sueciæ [Scholarly and scientific transactions of Sweden]. Uppsala.

Askmark, Ragnar
1943 *Svensk prästutbildning fram till år 1700* [Swedish education of clergymen up to 1700]. (Samlingar och studier till Svenska kyrkans historia 7.) Stockholm: Svenska kyrkans diakonistyrelse.

Bazerman, Charles
1988 *Shaping Written Knowledge: The Genre and Activity of the Experimental Article in Science*. Madison: University of Wisconsin Press.

Bazerman, Charles
2011 Church, state, university, and the printing press: Conditions for the emergence and maintenance of scientific publication in Europe. In: Britt-Louise Gunnarsson (ed.), *Languages of Science in the Eighteenth Century*, 25–44. Berlin and New York: De Gruyter Mouton.

Dahlgren, E. W. (ed.)
1918 *Svenska Vetenskapsakademiens protokoll 1739–1741* [The record of the Swedish Academy of Sciences]. 1–2. Stockholm.

Fries, Th. M.
1908 *Bref och skrifvelser af och till Carl von Linné* [Letters and communications by and to Carl von Linné]. Första afdelningen. Del II. Skrifvelser och bref till K. Svenska Vetenskaps-akademien och dess sekreterare. Stockholm: Ljus.

Gross, Alan G., Joseph E. Harmon and Michael Reidy
2002 *Communicating Science*. Oxford etc.: Oxford University Press.

Gunnarsson, Britt-Louise
1988 Textmönster i vår äldsta vetenskapliga tidskriftsprosa [Text patterns in our oldest prose of scientific journals]. *Nysvenska studier* 67 1987, 155–180.

Gunnarsson, Britt-Louise
2005a Medical Discourse: Sociohistorical construction. In: *Encyclopedia of Language and Linguistics* (2nd Ed.), Keith Brown (ed.), 709–716. Oxford: Elsevier.

Gunnarsson, Britt-Louise
2005b Icke-verbal representation i vetenskapliga artiklar [Non-verbal representation in scientific articles]. *Språk i tid. Studier tillägnade Mats Thelander på 60-årsdagen*, 303–314. Uppsala: Institutionen för nordiska språk, Uppsala universitet.

Gunnarsson, Britt-Louise
1990 Makrotematiska och pragmatiska mönster i medicinska artiklar [Macrothematic and pragmatic patterns of medical articles]. *Svenskans beskrivning 17*, 97–114. Åbo: Åbo akademis förlag.

Gunnarsson, Britt-Louise
2011 Introduction: Languages of science in the eighteenth century. In: Britt-Louise Gunnarsson (ed.), *Languages of Science in the Eighteenth Century*, 3–21. Berlin and New York: De Gruyter Mouton.

Gustafsson, Anna
2009 *Pamfletter! En diskursiv praktik och dess strategier i tidig svensk politisk offentlighet* [Pamphlets! A discursive practice and its strategies in the early Swedish political public sphere]. Lund.

Hernlund, Hugo
1883 *Förslag och åtgärder till svenska skriftspråkets reglerande 1691–1739* [Proposals and measures to regulate the Swedish written language 1691–1739]. Stockholm.

Hernlund, Hugo
1888 *Vetenskaps-Akademien och Lars Laurels rättskrifningsförslag* [The Academy of Sciences and Lars Laurel's proposal for a new orthography]. (Program för Nya Elementarskolan 1887–1888.)

Hildebrand, Bengt
1939 *Kungl. Svenska vetenskapsakademien: förhistoria, grundläggning och första organisation* [The Royal Swedish Academy of Sciences: previous history, foundation and original organization]. Uppsala: Almqvist and Wiksell.

Hof, Sven
1753 *Swänska språkets rätta skrifsätt* [The correct way of writing the Swedish language]. Ed. Mats Thelander. Uppsala 1985.

Leopold, Carl Gustaf af
1801 *Afhandling om svenska stafsättet* [Treatise on Swedish orthography]. (Svenska Akademiens handlingar ifrån år 1796. Första delen.) Stockholm.

Liljencrantz, Axel
1957 *Kungl. Vetenskapssocietetens protokoll 1732–1784* [The record of the Royal Society of Sciences 1732–1784].

Lindberg, Bo
1984 *De lärdes modersmål. Latin, humanism och vetenskap i 1700-talets Sverige* [The mother tongue of learning. Latin, humanism and science]. Acta Universitatis Gothoburgensis. Göteborg.

Lindroth, Sten
1967 *Kungl. Svenska Vetenskapsakademiens historia 1739–1818* [The history of the Royal Swedish Academy of Sciences]. 1–3. Uppsala: Almqvist and Wiksell.

Oscarsson, Ingemar
(forthcoming) Cultivating only those sciences and crafts that serve the common good. Scientific strategy and editorial policy in the Transactions of the Royal Swedish Academy of Sciences in mid-eighteenth century (1739–1755).

Sahlstedt, Abraham
1773 *Swensk ordbok. Med latinsk uttolkning* [Swedish dictionary. With Latin interpretation]. Stockholm.

Santesson, Lillemor
1986 *Tryckt hos Salvius. En undersökning om språkvården på ett 1700-talstryckeri med hänsyn till ortografi och morfologi* [Printed by Salvius. An investigation of the language cultivation and planning of an eighteenth-century printing house with regard to orthography and morphology]. (Lundastudier i nordisk språkvetenskap A 37.) Lund.

SAOB = *Ordbok över svenska språket utgiven av Svenska Akademien* 1893– [Dictionary of the Swedish language published by the Swedish Academy].

Schück, Henrik
1929 *Lars Salvius. Minnesteckning* [Lars Salvius. A memorial sketch]. Stockholm: Norstedts.

Sprat, Thomas
1958 *History of the Royal Society*. First published [1702]. Eds. Jackson I. Cope & Harold Whitmore Jones. Saint Louis Miss.: Washington University.

Swales, John M.
1990 *English in Academic and Research Settings*. Cambridge: Cambridge University Press.

Sörlin, Sverker
1994 *De lärdes republik. Om vetenskapens internationella tendenser* [The republic of learning. On the international tendencies of science]. Malmö: Liber-Hermods.

Teleman, Ulf
2002 *Ära, rikedom och reda* [Glory, wealth and order]. Stockholm: Norstedts ordbok.

VAH = *Kungl. Svenska Vetenskapsakademiens handlingar* [Transactions of the Royal Swedish Academy of Sciences]. Stockholm.

Wellander, Erik
1959 Svensk akademisk språkvård under 1700-talet [Language cultivation and planning by the Swedish academies in the eighteenth century]. *Septentrionalia et Orientalia. Studia Bernhardo Karlgren dedicata*. (Kungl. Vitterhetsakademiens handlingar 91.), 429–449. Stockholm: Almqvist and Wiksell.

Section 2.
The emergence of new languages of science

Scientific literacy in eighteenth-century Germany

Renata Schellenberg

Critics tend to qualify German intellectual life in the eighteenth century as a development that assumed a separate course from that of its French and English counterparts.[1] The German *Sonderweg* is attributed to the country's complex history of fractious politics that did not support the notion of centralized authority, but instead encouraged a state of outright instability, which presented itself as an ever-changing establishment and re-establishment of political allegiances to suit momentary rule. Germany's *Kleinstaaterei*, its existence as a loose alliance of small princely states, posed an obvious challenge to the formation of a uniform intellectual culture because its disparate nature undermined the very idea of commonality in favour of more regional tendencies. Yet, German intellectual culture did flourish in the eighteenth century, establishing a national presence and a strong tradition of academic thought, developments that are now seen as intertwined and inextricably linked. German *Aufklärung* was strongly felt at all levels of society, as it informed opinions, affected points of view and profoundly changed the manner in which one regarded the world. Arguably, its most significant effect was, however, on language itself, for writers of the German Enlightenment increasingly endorsed German as a language of learned discourse, deeming it not only capable of communicating abstract expression, but also entrusting its vernacular quality to be credible enough to encompass the depths of their own intellectual discoveries.

The purpose of this chapter is to highlight the development of this type of learned discourse as it occurred through scientific debate. The emergence of scientific literacy in Enlightenment Germany was a driving force that propelled overall public literacy as it set a standard, tone and objective for academic discourse which was later emulated in other fields of study. Propitious to the development of the scientific academic argument was the unprecedented expansion of print culture in Germany. The ready availability of printed matter afforded access to textual media and information for members of all levels of

1. There are a number of seminal texts that deal with this topic. For more information please see, for example, Gay (1969); Lepenies (2006); Eckhart (1990).

society, allowing the individual to appropriate knowledge in a solitary and personal manner. With this, the production and exchange of scientific knowledge also changed. This new method of appropriating knowledge created a pluralist communication system that challenged the very idea of a singular intellectual authority. Rather than being unconditionally respected, scientific treatises and theories were now openly questioned and authors had to defend their work in a much more public, and generally literate, manner than before. As this chapter demonstrates, this exchange of opinion elevated levels of literary expression by making everyone pay greater attention to the medium of language itself.

These contestations and rebuttals took place, for the most part, in periodical print media and as such were left available for open scrutiny long past the occurrence of the initial argument. In order to demonstrate the longevity and multifarious turns such discussions could take, this study examines the case of the preformation-epigenesis debate as it transpired at the height of the Enlightenment in eighteenth-century Germany. This debate was long, effusive and caustic. It originated in a scientific dispute between Caspar Friedrich Wolff and Albrecht von Haller, but then expanded to engage other members of the intellectual elite, polarizing opinions of some of the leading thinkers of the age. It spurred written reactions and counter-reactions from its participants and extended across disciplines to include both scientific observation and philosophical belief. The complexity and apparent insolubility of the argument left an unmistakable narrative trace that was documented in various journals, letters and essays of the day, providing therewith a veritable example of scientific literacy as it evolved and changed in eighteenth-century Germany.

This chapter isolates this singular, yet significant, debate by recounting some of the more important exchanges shared between the young scientist Wolff and his would-be mentor von Haller. It exposes the role print media played in the dissemination of knowledge and it highlights the use (and abuse) of that media by members of the establishment, while showing the overall importance of conviction and communication in formulation of new scientific theory. The involvement of other prominent thinkers of the age (such as Johann Friedrich Blumenbach and Immanuel Kant) is included as part of the chapter in order to illustrate the growing scope and depth the preformation-epigenesis argument assumed over time.

1. German language and the academic tradition

The absence of a clear and established protocol regarding a single authoritative language of scientific discourse becomes jarringly apparent when one compares the leading academies of science in the German speaking world in the eighteenth century: the academies in Berlin, Göttingen and Munich. The Berlin Academy, founded 1700, conducted its affairs initially in Latin, but by mid-century (1744), reflecting the Francophile nature of both its curriculum and staffing, it switched to French. *Die Akademie der Wissenschaften zur Göttingen* (1751) taught in Latin, while at the *Churfürstliche-Bayerische Akademie der Wissenschaften* in Munich, established 1759, the language of instruction had always been German.[2]

This lack of consensus regarding the status of German as a scientific language reflects not only an uncertainty over the merit of its institutional usage, but more importantly, an ambiguity over its credibility in matters of scientific discourse. Institutional skepticism vis-à-vis the validity of German as a viable academic language would prevail for most of the eighteenth century. For example, in the Berlin Academy French continued to be the official language of conduct as late as 1804, for it is only then that its journals *Mémoires* and *Histoire* were replaced with an in-house German version (*Abhandlungen*) to chronicle the work and publish the proceedings of the Academy. (For a discussion of other academies, see also Gunnarsson's 'Introduction' and Bazerman's chapter in this volume.)

Efforts to standardize German were made early in the century by academics themselves, who attempted to counter the neglect the language suffered in the official scholarly realm. At the University of Halle philosophers Christian Wolff and Christianus Thomasius famously adopted German (rather than Latin) as their language of instruction, thereby endorsing its usage in research and in the promulgation of academic findings.[3] Support to the cause of developing German academic literacy was also given, perhaps somewhat less radically, by men of letters such as Johann Christoph Gottsched, who despite deep personal Francophile leanings, promoted the usage of German through extra-curricular acts of sociability. In his *Deutsche Gesellschaft* (1727), a literary society at the University of Leipzig, he insisted that all communication be strictly conducted in German.[4] Gottsched's involvement in his literary society anticipates later developments of the era, for over the course of the century the emergence of such private organizations and gatherings became increasingly commonplace,

2. For more comprehensive information on the development of the academy in Germany please see Grau (1988); Hartkopf (1992); Vierhaus (2003).
3. For a brief contextual introduction to the intellectual climate in eighteenth-century Germany and a synopsis of Thomasius's and Wolff's work please see Martini (1965: 168–226).
4. The history of the *Deutsche Gesellschaft* is well chronicled in Döring (2002).

as people assembled to discuss intellectual and cultural issues in an increasingly casual setting. For Germany these types of gatherings bore additional significance, as they also constituted an important linguistic occurrence for the members; due to the relative informality of these meetings, affairs were quite naturally conducted in German.

The formation of such *Societäten* (Wittmann 1999: 191) – to mention but one of the many names by which they are known – was a common occurrence in eighteenth-century Germany. These societies had an invariably cohesive nature. They operated under the auspices of commonality, sharing a basis of interest and reference among their members, regardless of one's class or standing. The result of this policy was unprecedented. For the first time members of the academe, the bourgeoisie and nobility could congregate in a neutral setting with the objective of articulating and debating a particular issue, making the act of the intellectual pursuit the common thread of their confluence.

The regularity of these learned meetings furthermore assured a continuity of discussion and the establishment of networks of information, guaranteeing a productive interaction of their members. As the dissemination of knowledge mixed with an overall eighteenth-century tendency to socialize, an occurrence that was becoming an essential component of proper social conduct, the basic nature of information-exchange changed. Most significantly, the perception of conveying knowledge had altered. In these congenial social settings the presentation of knowledge was seen as an interactive public event, which, because it was no longer confined to predictable fora, welcomed the influx of capricious and entertaining ideas as a primary stimulus.

2. Periodicals and the communication of knowledge

This democratization in the exchange of ideas had an indelible effect on print culture, as members of these societies quickly moved beyond the immediacy of debate to create a legitimate body of publishable knowledge to be disseminated to the public at large. Members published the ideas and findings of their societies either through their own periodicals or through literary magazines, establishing a firm literary presence on the intellectual scene in Germany. Because the programs of these societies were determined by the interest of individual members, rather than by the expertise of a given group/school of thought, the mandate of these societies was not decidedly set, which meant that it often ran not only concurrent, but in deliberate contest, to that of the establishment. Moreover, members-turned-writers directed their attentions to a wide range of non-academic topics. They demonstrated through debate, and subsequent publication, that their inquisitiveness was to be understood as more than mere intellectual exercise, and they frequently sought practical solutions to day-to-day conundrums that had little to do with academic issues.

The act of publicizing predicaments and soliciting the general public for assistance became the standard means of intellectual enquiry in these polemical societies in the eighteenth century. Posing the "open question" in printed matter provided an opportunity for debate, allowing both the access and distribution of ideas, while also maintaining a fluidity of interaction - a principle that was at the very core of many of these societies. The results of these non-discriminatory discussions often had far-reaching consequences for the community as a whole, a development that attests to the benefits of open interaction between writer and public. One must only think of the impact the Berlin *Mittwochsgesellshaft* (the Wednesday Society) had on the development of Western intellectual thought with the publication of its *Berlinische Monatsschrift*. In 1783 Johan Friedrich Zoellner, pastor, theologian and member of the society, opened up debate on the practical consequence of the German Enlightenment. Among the responses received, there were missives from both Moses Mendelsohn and Immanuel Kant, the latter of whom voiced a clear call to intellectual courage (*Sapere aude!*) and maturity, which stands as the seminal definition of *Aufklärung* to this day.

In light of such occurrences, the significance and potential intellectual range of the periodical cannot be underestimated. As a popular and new genre on the literary scene it irrevocably changed the nature of reading culture in Germany. Much has already been written about the explosion of print matter in the eighteenth century, and specifically about the German fanaticism with reading at the time.[5] The statistics on periodicals are particularly astounding in this regard, for two-thirds of all scientific journals in Europe were published in Germany, making the German scientific periodical one of the biggest literary enterprises on the continent. The magnitude of influence it exerted and the popularity it garnered were eventually overshadowed by questions of authenticity and quality; however, in the strict sense of functioning as a means of communication with the public, it did prove to be extraordinarily efficient.

When studying the meteoric rise of the periodical in the eighteenth century, and its incredible success, it is easy to overlook other important aspects of this literary genre.[6] Setting the practical advantages aside, one must consider the laden and overlapping interaction that this literary form sustained. Publication of any periodical implied a constant collaborative effort between author, editor and publisher, working with an intensity of interaction which had not been seen

5. Guthke (1975); Giles (1981); Wittmann (1999).
6. See Johns (2003: 536–560).

previously, and whose joint work served both to accentuate and to validate the importance of the written medium itself. As the notion of a secular publishing culture and book trade took root in popular consciousness, it transformed the perception of German-speaking cities such as Leipzig, Halle and Berlin. These cities were no longer seen as mere geographical locales, for as they became known for their printing houses, book fairs and the publication of a particular periodical, they also acquired the reputation as a "reading centre", a designation that carried a clear and, more importantly, apolitical intellectual connotation.

As scholars and thinkers increasingly employed the medium of print to exchange and contest ideas, periodicals created a new dynamic network of communication between readers and writers, a contribution that was particularly significant in a country as geographically disparate as eighteenth-century Germany. Many differences of opinion were aired through the pages of the periodical as many established authorities took advantage of this open forum both to promote and to argue their case in the public sphere. By gravitating towards print and opting to reveal their queries and complaints in such an open (and permanent) way, writers themselves sanctioned debate and dispute as acceptable modes of intellectual discourse. It is important at this point to differentiate open periodical writing from other forms of written expression, such as, for example, correspondence. For although many grievances have been vented (and subsequently published) through personal correspondence, these grievances were initially envisioned as private. The disagreements declared within a periodical, on the other hand, became the immediate and inescapable property of the public sphere and therefore thoroughly out of the control of both writer and recipient.

Despite its potentially damaging effect on the reputation of the people involved, the polemical aspect of the periodical proved surprisingly helpful in the proliferation of scientific ideas. Through periodicals science became an integral part of public discourse. This development is due in part to the accessible and egalitarian format of the periodical, but more importantly, to the manner in which the periodical presented science to its readership. With its emphasis on discussion and its increasingly disputatious prose, periodicals presented science as a dynamic discipline intent on presenting knowledge as a work in progress, rather than as a static set of rules. This proved appealing to a general public with little real invested intellectual interest in the discipline of science and with limited abilities to participate in all tenets of a purely scientific discussion.

Public scientific debates could furthermore often be a source of entertainment for readers as the rhetorical rancour inherent in a scientific quarrel was often the element that transformed the debate into a type of spectacle for all parties involved. Some scientists, however, abhorred this sort of unnecessary and unprofessional exposure. Swedish taxonomist Carl Linnaeus, for example, was well-known to have abstained from professional disputes for this very reason. In his opinion, scientific disputes that played out in public were never helpful to an elaboration of subject matter and therefore clearly unworthy of the scientists themselves.[7]

Scientific disputes in eighteenth-century Germany were, however, common, and as they occurred they had an impact not only on the immediate members involved, but also on the evolving scientific literacy of the general public. An incident that is worthy of mention within this context is the famous epigenesis-preformation debate that dominated literature of the life sciences after the 1750s. This debate is significant, not only because of its protracted and rather complicated duration, but also because it clearly attested to the reality of an evolving textual presence of scientific discourse in German. It firmly took place in print and in the public domain and involved some of the leading thinkers of the era, transgressing in the process the boundaries of science to find proper expression for the predicament in question.

3. Disagreement and debate: preformation vs. epigenesis

At the centre of the original debate were Albrecht von Haller (1708–1777) and Caspar Friedrich Wolff (1734–1794), men of different age groups and of two very separate philosophical leanings. Haller was a recognized scientific authority who was well-known for his work on anatomy and physiology at the University of Göttingen. He was also an established figure in Germany's nascent vernacular print culture with extensive personal experience in writing and publishing. In addition to being the poet of a very popular collection of poems,[8] Haller exerted literary authority over institutional affairs by presiding over the academy's review journal, *Die Göttingischen gelehrten Anzeigen,* and assuming editorial control over its content. Scientifically, Haller was an ardent believer in preformation theory and he participated in the generation debate of the day by adopting a traditionalist view. Haller had struggled to find a conclu-

7. In a letter to Albrecht von Haller dated September 13 1748, Linnaeus counsels Haller how to respond to professional criticism. He writes: "If you will but listen to me as a friend, I would advise you to write no answers to Hamberger and such people. He is not on level with you; and the more he is your inferior, the more consequence you give a man who would otherwise remain in obscurity, known only to those immediately about him." Quoted from Smith (1821: 421).
8. Haller's nature poem "Die Alpen" (1729) was particularly popular with the reading public.

sive formulation for this theory for over a decade and in his early research he did briefly consider evolution as a possibility for form creation. However, in his mature work, he conclusively asserted the notion of the embryo as a preformed, pre-existing entity that was *not* subject to the laws of evolution, thereby refuting all other theories of evolution. As critics have noted, this viewpoint was consistent with his personal religious upbringing and therefore it was not surprising that it surfaced so obviously in his late work.[9]

Haller's approach clashed with the concurrent theory of epigenesis, an opposing school of thought that endorsed evolutionary treatment of development. Epigenesis was process-oriented and it sought to address the more creative dimensions of nature by actively analyzing change, a move that discredited the premise of preformation in favour of a dynamic understanding of natural phenomena (cf. Pahta, this volume). Haller's preformation study was published in a two-volume work entitled *Sur la formation du coeur dans le poulet* in 1758 and it marked the beginning of the epigenesis-preformation dispute. In the text Haller argued the manifest visibility of the heart, claiming that it existed from the very beginning of the embryo and that it was, to cite Müller-Sievers, "terrified into action [i.e. visibility] by the 'rotten odor' of the semen and which then pumped up… all other organs" (Müller-Sievers 1997: 36). What is interesting here is that Haller based his doctrine almost entirely on his personal ability to observe the experience, stating *de facto* that this testimony alone guaranteed the scientific validity of the argument. He supplied experiments to support the argument, but maintained that observability of structure was the best and final proof for the existence of structure. His preformation theory depended thus heavily upon the astute perception of the scientist, but also upon his/her philosophical convictions: for in order to endorse this preformation theory, one needed to believe that scientific investigation does need not to go further than the eye can see.

With this outlook, however, as Elke Witt has so correctly identified, the study of organism form no longer lies within "the scope of scientific research" (Witt 2008: 653), but becomes rather a matter of one's approach to science as a discipline. It becomes a matter of personal belief and principle and therefore impossible to prove in objective, empirical terms. By presenting his findings in

9. As Helmut Müller-Sievers asserts, preformation was very much in keeping with Haller's pietistic beliefs. He writes: "And this closed economy of the organism, into which life is "knocked" from the outside and which is terminated by external death, can serve as a religious and moral bastion against the openness and transitoriness of the Buffonian model where, at least since puberty, the organism is always dying and where the physiological (and sexual and political) exchange with the outside is of vital and mortal importance" (Müller-Sievers 1997: 36).

the manner in which he did, Haller conveyed his preformation study as scientific truth, rather than as personal opinion, and this precluded any future alternative investigations from his objective professional consideration. Haller's study was strongly challenged by Caspar Friedrich Wolff, who presented a divergent theory of generation in 1759 in his dissertation *Theoria Generationis.* Wolff claimed the opposite viewpoint from Haller, by stating that the emergence of organisms was in fact the production of something irreducibly new, and therefore not something that could exist within the scope of opinion of a scientist, but rather, needed to be understood as part of nature itself. He refused to reduce nature to a pre-formed mechanistic model and explained epigenesis as a progressively creative dimension existing within nature, terming the workings of this force a vital force, a *vis essentialis.*

The debate that ensued between Haller and Wolff continued for the next ten years and it happened at a very public level within the open domain of print. After an initial gracious acknowledgement, Haller responded to Wolff peremptorily, reviewing his dissertation negatively (and anonymously) in *Göttingische gelehrte Anzeigen*. Revealing little generosity of mind, and no consideration for Wolff's young career, Haller lambasted the theories of his junior colleague, announcing to the scientific community that Wolff's work was fundamentally flawed (Haller 1760: 1226–1231). Wolff refused to acquiesce to Haller's established scientific authority. Instead, he translated his theory into German and republished it in 1764 as *Theorie von der Generation,* introducing the work to a wider audience by rearticulating some of its more salient points in accessible language. It is rather poignant to read the passion with which Wolff wrote to express his commitment to the cause of epigenesis, for one sees how deeply, and personally, involved he was in the entire affair. Preformation seemed to have offended Wolff's very understanding of nature. Addressing Haller's interference, Wolff wrote:

> Until now it was a living Nature, which brought about infinite changes by its own forces. Now it is a creation, which only seems to bring about changes, but actually and by its nature stays unchanged in the way it was constructed, besides getting more and more worn-out. Before it was a Nature that destroyed itself and created itself anew, to thereby bring forward infinite changes and to present itself again and again in a new angle. Now it is a lifeless lump, losing one part after the other until the whole stuff finds its end. Such a miserable nature I cannot stand. (Cited in Witt 2008: 654)

Wolff's conflict with Haller lasted until Haller's death in 1777, with the unproductive result of damaging Wolff's career and not bringing any lasting solution to the debate itself.

As Shirley A. Roe has asserted within her own extended study of the matter, the Haller-Wolff debate was not merely an intergenerational disagreement on a particular theory; it touched on deeper philosophical issues that went beyond the scientific investigation at hand. She states:

> The debate between Haller and Wolff rested, however, on more fundamental philosophical grounds, which included not only the problem of the operation of forces in development but also such issues as the role of God in the world, the question of spontaneous generation, the relationship between observability and existence, and, more generally, the nature of scientific explanation as applied to the problem of generation (Roe 1975:189).

It could be successfully argued, however, that the difficulty of reconciling the divergent scientific approaches between Haller and Wolff was also a matter of deficient linguistic means on both sides, for, as they so spectacularly demonstrated, neither party was able to create or appropriate the necessary nuance of argument that would reach solution.

Investigations into generation theory continued through the concerted work of Johann Friedrich Blumenbach (1752–1840), the Göttingen anatomist and anthropologist who advocated epigenesis over preformation on the basis of his own research on green hydra and its regenerative powers. In 1780 he published his findings in an essay entitled "Über den Bildungstrieb" in *Göttingisches Magazin der Wissenschaften und Litteratur.* A year later he reiterated his findings more expansively in his monograph *Über den Bildungstrieb und das Zeugungsgeschäfte.* He revisited the same topic yet again in various editions of his seminal work *Handbuch der Naturgeschichte.* The difficulty of constructing a single conclusive theory on the subject matter of epigenesis becomes quite apparent through the shifts in formulation that occurred in successive editions of his work, which demonstrates that Blumenbach was actively working out tenets of the theory by constantly and diligently revising his research.

A significant trait of Blumenbach's efforts was his conscious use of language, which he modified throughout the various editions of his work in order to accommodate his growing research. He did not fixate on a particular terminology or on a given mode of expression; instead he adapted language to the purpose of his research, modifying his texts as he saw appropriate. Fully aware of the complicated nature of the subject matter he was presenting, Blumenbach acknowledged previous accomplishments in generation theory by accrediting his predecessors for the advances made. In his work he duly chronicled the names of scientists who had furthered the study of organism formation, noting his own contribution as a matter of continuing this historical study of a natural phenomenon. As the following passage demonstrates, he presented his research very cautiously, seeking to integrate it within existing scientific theories: "Without entering into a long and tedious detail of all the doubts and difficulties which rise up in opposition to so singular theory, I shall content myself with adding a few reflections..." (Blumenbach 1792: 9).

By framing his epigenesis argument in such a conciliatory manner, Blumenbach paid homage to the scientific establishment, while also detaching his work from its precedent. He also took the notable step of distancing himself from the nomenclature of his predecessors, as he understood full well the objectionable connotations Wolff's *vis essentialis* had for the scientific community. Rather than allowing his research to be interpreted through the lens of an esoteric philosophical vitalism, a trap that Wolff had fallen into with Haller, Blumenbach designated aspects of his research more astutely. He collegially acknowledged Wolff's work, but also distinguished it as separate from his own. Most importantly, he reformulated his approach with carefully selected vocabulary. He identified his epigenetic vital force as *Bildungstrieb*, a formative drive, cleverly binding its existence to that of the organism itself, thereby eliminating any suggestion of a speculative approach.

Despite its linguistic reformulation, Blumenbach's formative drive continued to share some of the problematic traits of the *vis essentialis*; most significantly an innately abstract existence that could not be verifiably shown or empirically maintained. The careful re-designation of the mysterious vital force as a *Bildungstrieb* clearly suggested a purposive formation (unlike Wolff's *vis essentialis*), yet this assertion still could not be demonstrated in positivistic, scientific terms. Rather than succumbing to scientific scrutiny and the inevitable criticism of philosophical abstraction, Blumenbach received unexpected encouragement for his work precisely from the field of philosophy, which helped not only maintain but also develop his epigenetic theory further. His work drew the attention and support of Immanuel Kant (1724–1804).

Kant was drawn to teleological principles found in Blumenbach's work and he agreed with the principle of self-organization of organic matter that permeated the epigenetic outlook. Blumenbach's theory overlapped in many ways with his own work at the time, as Philip R. Sloan explains: "...the *Bildungstrieb* theory was put forth by Blumenbach as a phenomenological force that acted in lawful ways, and in this way it was in agreement with Kant's interpretation of forces in the *Metaphysische Anfangsgründe* of 1786" (Sloan 2002: 249). Kant was thus in a good position to praise Blumenbach for the advances he made in the field of epigenesis and made reference to him in his *Critique of Judgment* (Kant 303–307). What ensued between the two men is interesting, for despite operating from divergent fields of research – epitomizing in many ways the 'Two Cultures' divide – they engaged in an active exchange of their work, mutually sustaining their intellectual efforts. As Timothy Lenoir asserts, Kant's intervention was particularly fruitful, for it provided the necessary language "for clarifying issues at the heart of Blumenbach's work" (Lenoir 1980: 88), allowing for its mature formulation in subsequent editions.

Not surprisingly, there are critics who argue the exact opposite scenario, namely that it was Blumenbach who influenced Kant, and that it was science, rather than philosophy, that provided primary intellectual stimulus to the epigenesis discussion (Sloan 2002: 248). In the end what is remarkable about this interaction between these two seemingly opposite fields of study is the reciprocity of exchange and the attention paid by both men to the language of each other's work. Both were searching for a more apposite mode of expression that could express the subtleties of their work; neither found it in their own discipline and therefore they were forced to look for it elsewhere. In their quest for a more deliberate and precise articulation of ideas, they not only consciously wrote, but in doing so they popularized the premises of their respective sciences, thereby contributing both to scientific literacy and the knowledge base of their readers.

4. The marketability of scientific discourse

The evolving sophistication of German as a language of precise scholarly discourse is unfortunately difficult to trace at the popular everyday level. Reading had become an incredibly fashionable pastime, with everyone embracing the availability of print culture and contributing to its growing social importance. The popularity of the *Lesewut* trend, as it is also known,[10] regrettably did not always correspond to the acquisition of knowledge. Critical understanding of the new and exciting book culture in Germany in the eighteenth century took time to develop, as the process of meaningful reading was frequently encumbered by issues of authenticity, quality and piracy that were prevalent on the literary market. The deluge of print material available to the general public was often of dubious quality and the information presented needed to be carefully scrutinized for accuracy. Texts were consequently often accredited as manifestations of knowledge with little questioning of the legitimacy of their contents.[11]

An additional threat to the intellectual autonomy of books was the economic success of the enterprise as a whole. As the book trade became a profitable business, the reading public assumed a powerful market presence, becoming a com-

10. See Erlin (2007: 145–164).

11. Woodmansee (1984: 425–448) offers an excellent overview of the emergence of book culture in eighteenth-century Europe.

peting source of authority in determining the popularity, and by implication, the validity of a particular text. The readers' favourable reception of a text often became the deciding factor in the overall merit of a book. Writers, and especially writers of science, in eighteenth-century Germany were keenly aware of the precarious situation they faced and they responded to the capricious patronage of the general marketplace not only by appropriating their style of expression, but also by constantly negotiating a delicate balance between universality and exclusiveness in their work.[12]

It is under these circumstance that we see authors like Johann Kaspar Lavater (1741–1801), the founder of the quasi-science of physiognomy, having enormous success on the literary market. His multi-volume *Physiognomische Fragmente zur Beförderung der Menschenerkenntnis und Menschenliebe* (1775–1778) was extraordinarily popular, despite its considerable cost and its many scientific inaccuracies and flaws.[13] Understanding the importance of display, Lavater formatted his book in a particularly reader-friendly way. Part of the attraction of the book rested on its appearance and handsome style of the publication, which actively incorporated word and image and had many elaborate illustrations accompanying the text. On the other hand, there were also legitimate scientists who at this time embraced the literary mode as a natural extension of their scientific activities. The name that most readily comes to mind in this context is that of Georg Christoph Lichtenberg (1742–1799), a writer whose ironic and playful literary exploits in his so-called "waste books", the *Sudelbücher*,[14] eclipsed in fame his more serious accomplishments in the field of physics.

It becomes thus evident that acts of reading and writing in eighteenth-century Germany were not always driven by the most erudite of causes, and that the great production and consumption of reading materials of the era can be indeed tied into a larger debate on the overall phenomenon of an emerging consumer culture.[15] If one, however, side-steps this sort of interpretation and focuses on the manifestation of this prolific literary production (rather than concentrating only on its underlying motivation), it is impossible to disregard the achievement and advancement created by the appearance of textuality in everyday German life. The presence of textuality on a day-to-day basis, regardless of whether it was manifested through journals, periodicals, newspapers or books, offered the public the opportunity to participate in the acquisition of knowledge, render-

---

12. For an interesting point of comparison for the German market see Fissell (2007).
13. See Graham (1979); Perceval and Tytler (2005).
14. For more information on the impact of Lichtenberg's aphoristic work, see Knauf (1977).
15. An excellent case study of this issue can be found in Wurst (2005).

ing epistemological goals as attainable to anyone with the sense to read. Within learned circles this increased literacy served as an impetus to explore new discourses of difference, establishing both dialogues and disputes between disciplines, and forcing everyone to take a little more note of what was expressed on the written page.

References

Bazerman, Charles
2011 Church, state, university, and the printing press: Conditions for the emergence and maintenance of scientific publication in Europe. In: Britt-Louise Gunnarsson (ed.), *Languages of Science in the Eighteenth Century*, 25–44. Berlin, New York: De Gruyter Mouton.

Blumenbach, Johann Friedrich
1792 *An Essay on Generation*. Translated by A[xx]. Crichton. Edinburgh: Cadell.

Döring, Detlef
2002 *Die Geschichte der Deutschen Gesellschaft in Leipzig: Von der Gründung bis in die ersten Jahre des Seniorats Johann Christoph Gottscheds.* Tübingen: Niemayer.

Eckhart, Hellmuth (ed.)
1990 *The Transformation of Political Culture: England and Germany in the Late Eighteenth Century.* London and Oxford: German Historical Institute.

Erlin, Matt
2007 Useless Subjects: Reading and Consumer Culture in Eighteenth-Century Germany. *The German Quarterly* 80: 145–164.

Fissell, Mary
2007 The Marketplace of Print. In: Mark Jenner and Patrick Wallis (eds.) *Medicine and the Market in England and its Colonies, c. 1450–1850,* 108–132. London: Palgrave.

Gay, Peter
1969 *The Enlightenment: An Interpretation*. New York: Harper and Row.

Giles, Barber (ed.)
1981 *Buch und Buchhandel im achtzehntem Jahrhundert.* Hamburg: Hauswedell.

Graham, John
1979 *Lavater's Essays on Physiognomy: A Study in the History of Ideas.* Bern: Lang.

Grau, Conrad
1988 *Berühmte Wissenschaftsakademien: Von ihrem Enstehen und ihrem weltweiten Erfolg*. Leipzig: Edition Leipzig.

Gunnarsson, Britt-Louise
2011 Introduction: Languages of science in the eighteenth century. In: Britt-Louise Gunnarsson (ed.), *Languages of Science in the Eighteenth Century*, 3–21. Berlin, New York: De Gruyter Mouton.

Guthke, Karl
1975 *Literarisches Leben im achtzehnten Jahrhundert in Deutschland und in der Schweiz*. Bern: Francke.

Haller, Albrecht von
1760 *Göttingische gelehrte Anzeigen* 143: 1226–1231.

Hartkopf, Werner
1992 *Die Berliner Akademie der Wissenschaften: ihre Mitglieder und Preisträger 1700–1990.* Berlin: Akademie Verlag.

Johns, Adam
2003 Print and Public Science. In: Roy Porter (ed.) *The Cambridge History of Science,* 536–560. Volume 4. Cambridge: University Press.

Kant, Immanuel
1987 *Critique of Judgment.* Translated by Werner S. Pluhar. Indianapolis: Hackett.

Knauf, Manfred
1977 *Lichtenbergs Sudelbücher: Versuch einer Typologie der Aphorismen.* Dreieich: Rau.

Lenoir, Timothy
1980 Kant, Blumenbach and Vital Materialism in German Biology. *Isis* 71.1: 77–108.

Lepenies, Wolf
2006 *The Seduction of Culture in German History.* Princeton: University Press.

Martini, Fritz
1965 *Deutsche Literaturgeschichte.* Stuttgart: Kröner.

Müller-Sievers, Helmut
1997 *Self-Generation: Biology, Philosophy and Literature around 1800.* Stanford: University Press.

Pahta, Päivi
2011 Eighteenth-century English medical texts and discourses on reproduction. In: Britt-Louise Gunnarsson (ed.), *Languages of Science in the Eighteenth Century*, 333–355. Berlin, New York: De Gruyter Mouton.

Perceval Melissa and Graeme Tytler (eds.)
2005 *Physiognomy in Profile: Lavater's Impact on European Culture*. Newark: University of Delaware Press.

Roe, Shirley A.
1975 The Development of von Haller's Views on Embryology. *Journal of the History of Biology* 8.2: 167–190.

Sloan, Philip R.
2002 Reforming the Categories: Eighteenth-Century Generation Theory and the Biological Roots of Kant's A Priori. *Journal of the History of Philosophy* 40.2: 229–253.

Smith, James Edward (ed.)
1821 *A Selection of the Correspondences of Linnaeus and other Naturalists from the Original Manuscript*. London: Longman.

Schmidt, James
1989 The Question of Enlightenment: Kant, Mendelsohn, and the *Mittwochsgesellschaft*. *Journal of the History of Ideas* 50: 269–291.

Vierhaus, Rudolf
2003 *Etappen der Göttinger Akademiegeschichte*. Göttingen: Vandenhoeck & Ruprecht.

Witt, Elke
2008 Form – A Matter of Generation: The Relation of Generation, Form, and Function in the Epigenetic Theory of Caspar F. Wolff. *Science in Context* 21.4: 649 –664.

Wittmann, Reinhard
1999 *Geschichte des deutschen Buchhandels*. Munich: Beck.

Woodmansee, Martha
1984 The Genius and the Copyright: Economic and Legal Conditions of the Emergence of the 'Author'. *Eighteenth-Century Studies* 17.4: 425–448.

Wurst, Karin
2005 *Fabricating Pleasure: Fashion, Entertainment, and Cultural Consumption in Germany 1780–1830*. Detroit: Wayne State University Press.

From vernacular to national language: Language planning and the discourse of science in eighteenth-century Sweden

Anna Helga Hannesdóttir

1. Introduction

In the eighteenth century, the linguistic climate in Sweden was permeated with constructing the language we now call Swedish.

The Swedish language has undergone an evolution from the poorly codified vernacular that appears in medieval texts to the standardized, developed and thoroughly described official language it is today. This progression can be described within the framework of language planning. In the eighteenth century, the language is in the midst of this process, and a noticeable feature of the linguistic climate is the movement towards set norms and standardization. In the present study, some characteristic aspects of the linguistic climate of the time are related to relevant phases in the continuous process of linguistic development which Einar Haugen has called "language planning" (LP).

I will here present the LP model as developed by Haugen. I argue that it should be regarded not simply as a model of standardization, but rather as a way of analysing and explaining language change from a sociolinguistic point of view. The adequacy of Haugen's model is demonstrated with reference to various processes in the development of the Swedish language.

One of the individuals involved in the project of constructing the Swedish language was the eminent botanist and scientist Carl Linnaeus (1707–1778). Linnaeus' own contribution to the process is discussed briefly hereinafter in the perspective of language planning as defined by Einar Haugen.

2. Language change, standardization and language planning

When Haugen introduced the concept of language planning, he defined it as "the activity of preparing a normative orthography, grammar, and dictionary for the guidance of writers and speakers in a non-homogeneous speech community" (Haugen [1959, 1961] 1972: 133). The concept was later developed further into a four-step model (Haugen 1966: 18–26). At the same time, however, he came to regard the specific activities of preparing an orthography, a grammar and a dictionary as the outcome of the LP process, rather than constituting the process itself. The LP process he then redefined as "*the evaluation of linguistic change*" (Haugen [1966] 1972: 162, his emphasis). The concept of LP has since then been discussed, modified and supplemented, by Haugen himself as well as by other scholars (see Deumert and Vandenbussche 2003 for an overview). In 1987, Haugen presents a revised version of the LP model. He refers to two dimensions of language planning: one regarding *form* (*policy planning*) and the other concerning the *function* of the language (the *cultivation* aspect). Both dimensions include the language *society* on the one hand (*status planning*) and the linguistic material or "the *language* itself" on the other (*corpus planning*).

The four-step model is organized as a matrix, in which the four components of the process are identified as phases of (1) Selection, (2) Codification, (3) Implementation and (4) Elaboration (figure 1).

The four phases of the process can be successive, but they can also be simultaneous and cyclical. According to the model, the standardization procedures affecting the writing system and the orthography, the grammar and the lexicon are significant to only one of the four phases that constitute the entire process, namely the phase of codification (figure 1).

Haugen himself used various kinds of language change, historical as well as recent or ongoing, to illustrate the concept of language planning. Now, how-

	Form (policy planning)	***Function (cultivation)***
Society (Status planning)	1) **Selection** (decision procedures) a) Identification of problems b) Allocation of norms	3) **Implementation** (educational spread) a) Correction procedures b) Evaluation
Language (Corpus planning)	2) **Codification** (standardization procedures) a) Graphization b) Grammatication c) Lexication	4) **Elaboration** (functional development) a) Terminological modernization b) Stylistic development

Figure 1. Haugen's model of language planning (Haugen 1987: 64).

ever, the concept is mainly associated with more specific kinds of activities. One is the process that often takes place in new, post-colonial states, involving the establishment of an official language, chosen from a number of languages of varying status spoken in the country, and, as an extension of that, the settling of relations between the majority language and various minority languages within the speech community. Another kind is the continuous effort that occurs within the field of what Haugen calls language elaboration or functional development, i.e. activities led by official or semi-official language authorities. Both activities require an elaborated notion of a norm, as well as an incentive to comply with it. The distinction between LP and standardization has become blurred, and now language planning is often more or less synonymous with standardization. Thus, Haugen's four-step model of language planning is referred to as "Haugen's four-step model of language standardization" and the four dimensions in the model are presented as "norm selection", "norm codification", "norm implementation", and "norm elaboration" (Deumert and Vandenbussche 2003).

As standardization research has developed and grown as a field of sociolinguistics, the perspective has been widened with respect to the socio-political aspects of the process of establishing one authorized standard version of a national language by evaluating or reducing the number of linguistic alternatives available to the speaker. These socio-political aspects, however, highlight non-linguistic notions such as explicit motivation and goals, status and power, as well as a deliberate striving for established, adopted and ubiquitous norms of linguistic behaviour. To the two dimensions of language form and language function, a third has thus been added: *language ideology* (Deumert 2003). Such ideological notions, however, generate certain restraints as to what kind of language change can be evaluated within the framework of LP.

There is good reason to maintain the initial distinction between standardization and LP. If standardization is regarded as but one aspect of the process of LP, a vast number of different kinds of language change are left to be evaluated in relation to the LP process. LP can then be a useful explanatory tool when applied to selection and codification procedures taking place at remote stages in the history of language, stages previous to variation reduction first and foremost being a deliberate means of constructing a standard language. There is also another reason to keep standardization and LP apart. When studying changes in the language structure, it is crucially important to consider whether the process of codification has affected the feature in question. Once the variants involved in the change have been evaluated and rated in relation to a norm, i.e. what is right and what is wrong, the speakers cannot be expected to be unaffected by the "ideology of standardization" (Milroy and Milroy 1997).

3. Language planning as an explanatory model

Haugen states that the discipline of LP is "largely descriptive and hypothetical, not having reached a stage of 'explanatory adequacy"' (Haugen 1987: 63). Recently, however, Haugen's schema has been made to serve as a frame of reference for the description of standardization processes in the Germanic languages, among them Swedish (Deumert and Vandenbussche (eds.)). With reservations concerning the adequacy of Haugen's model to account for the "motivations and non-linguistic goals of the 'standardizers"' (Deumert and Vandenbussche 2003), the model is nonetheless considered to be "broad as well as detailed enough to function as a frame of reference for the description of highly varied standardization histories" (Deumert and Vandenbussche 2003). The standardization history of Swedish is provided by Teleman. In a lucid survey, the standardization undertaken by authorities such as kings, chanceries and other official and semi-official actors is reviewed, and the socio-political and ideological aspects are accounted for to some extent (Teleman 2003).

The unrestricted four-step model of Haugen can indeed serve as an explanatory model. Thus, some of the changes that can be observed in the history of Swedish can be assigned to distinct phases defined in the LP process. Haugen's model is thereby quite appropriate for assessing the different diachronic processes constituting the transition of a language from a domestic vernacular, still not properly codified, to a standardized national language, suitable for use in all domains of a modern society. The efforts observed, made by individuals as well as institutions in the language society, successful or otherwise, can then be evaluated as contributions to the ongoing construction of the language in question. In the study of the history of Swedish lexicography (Hannesdóttir 1998), the LP model has been introduced as a descriptive framework to illustrate the efforts of the lexicographers of the eighteenth century. Later, different aspects of the early lexical description of Swedish have been analysed in relation to the various phases of the LP process as described by Haugen (Hannesdóttir 2000a, 2000b, 2008).

In this study, the linguistic landscape as it appears in the eighteenth century is described with the LP process as a point of departure. Some of the prominent features of that time form part of long-term processes, while others were innovative and highly debated at the time.

3.1. The selection phase

One significant feature of eighteenth-century Swedish is the socio-political status of the Swedish language in relation to Latin. The choice of language has not yet been settled once and for all. Indeed, there had already been a long process, which can be traced back to the introduction of Christianity.

According to Haugen (1987), the prerequisite for the selection phase is the presence of conflicting languages or a conflict between different norms concerning language. Here, the concept of "norm" is fundamental, but Haugen allows quite a broad definition. He uses the replacement of English with Irish in Eire, as well as the transition from Norman French to English, to exemplify conflicts of this kind. The latter process we really do not know much about – except for its outcome. The main issue is, he says, that the selection in question becomes normative for the whole language community. The actors involved in the selection process may be the leaders of the society, but they may also be influential individuals or institutions.

We know that in medieval Swedish society, different languages had been established as means of communication in different linguistic domains. The language of Swedish legislation seems never to have been anything but Swedish, the language of trade and craft was strongly influenced by German, while, when it comes to religion, the situation is somewhat more complex. Not only had the old local religion been discarded, the language used in performing religious acts had also been abandoned in favour of Latin. Christianity brought with it both a new language and a new writing system. While the Christian religion and the Latin alphabet replaced earlier religions and the runic writing system, respectively, the linguistic side of the Reformation in the sixteenth century promoted a reversed language selection, whereby Swedish regained the domain of religion.

Within one linguistic domain, however, the choice of language was still at issue and much debated in the eighteenth century, namely academic language and the language of science. Latin was the academic lingua franca at the time, and in academic circles its status was – by and large – taken for granted. Swedish was not a conceivable alternative to Latin. The balance of power between the two languages was a dynamic one and it was deeply influenced by the ideological climate in Europe. The ideas of the continental Renaissance had reached Sweden during the seventeenth century, ideas that brought the local culture and language into focus. The Renaissance gave rise to calls for the elaboration of the vernacular languages so that they could meet all the communicative needs of society – including those of academic discourse.

3.2. The codification phase

Throughout the eighteenth century, much effort was devoted to codifying the Swedish language. In the codification phase Haugen includes the standardization procedures of graphization, grammatication and lexication (figure 1). An early step in the graphization process was establishing a local version of the Latin alphabet. The alphabet had indeed been introduced and accepted for writing, not only in Latin but also in the vernacular, as early as the Middle Ages. At the beginning of the eighteenth century, however, it had not yet been fully adjusted to the Swedish language. As I have discussed elsewhere, the number of letters, their relative position, their names and their status were not yet established (Hannesdóttir 2008). In 1696, the grammarian Nils Tiällmann states that there are 26 – or 28 – letters in the Swedish alphabet, depending on whether *j* and *v* are regarded as consonants. He includes the specific letters <å>, <ä> and <ö> and places them at the end of the alphabet (Tiällmann 1696: 48). Some twenty years later, in 1722, another grammarian reckons with 32 letters (Swedberg 1722: 1). He too includes the specific letters in the same position and relative order as are familiar to us today.

The eighteenth century was a vital period when it comes to the standardization of the orthography, and much has been said and written about this aspect of the graphization process (Teleman 2002, 2003; for a comprehensive study see Santesson 1986). This subject will not be dealt with here, except to draw attention to the effects of orthographic standardization. Once the orthographic norms have been settled and are being implemented in the speech community, they interfere with spontaneous phonological development. Thus, when studying phonological change in a longer diachronic perspective, it is necessary to consider whether and how the features in question have been affected by standardization (Hannesdóttir 2000a).

The grammatication of the Swedish language was also a fundamental issue at the time. Grammatication concerns the structure of the language, and the structure of Swedish was now being scrutinized. The early descriptions of the grammar of the vernaculars quite naturally took the structure of the thoroughly studied and well-known Latin language as a point of departure. It is not until quite late in the eighteenth century that this process takes place on the basis of the Swedish language itself (see also Watts 1999 concerning the English grammar tradition).

The grammatication process advanced throughout the eighteenth century. Not only grammarians were involved; lexicographers, too, played an active part in the description of the structure of Swedish, thereby providing basic data as to the actual variation occurring. The efforts of Swedish lexicographers in this phase of the language planning process have been discussed both in general (Hannesdóttir 1998: 11–13) and in more detail with respect to specific structural features (Hannesdóttir 2000b).

That brings us to lexication. The lexicon, too, had to be settled, elaborated, described and standardized. One of the dominant arguments for the superiority of Latin over Swedish as the language of science was the underdeveloped vocabulary of Swedish. Enlargement and development of that vocabulary were sorely needed and had been at stake from the middle of the seventeenth century. The continental Renaissance brought with it different approaches to this problem. In the prevailing ideological climate in Sweden at this time, the alternatives favouring indigenous linguistic material were preferred. One way of doing this was to promote dialectal and common popular words of the native language, another was to use archaic words, and the third was to make up new words. Adopting words from Latin and Greek was not really an option for the time being.

The lexicographers of the eighteenth century struggled with the organization of the Swedish vocabulary. The lexicon was not only to be enriched, it was also to be studied and described and a generally accepted, representative and effective lexicon established. What should be accepted as good Swedish words and thus included in their dictionaries, and which words should be regarded as "foreign" or "un-Swedish" and therefore omitted, was a crucial question that kept the lexicographers occupied (Hannesdóttir 2002). Foreign words, in the sense of strange, unknown words, could be Swedish as well as not Swedish. Archaic, dialectal words or words for special purposes could very well be domestic in origin and at the same time unfamiliar to the vast majority of the general public, while many Latin and German words were well known and frequently used, even if they were easily recognized as being of foreign origin. Preferences vary through the century. During the early decades, archaic and dialectal words were favoured, rather than Latin and Romance ones. Later on it was the other way around, the archaic and "Gothic" items becoming more or less obsolete. By the turn of the century, the general opinion had yielded to a less black-and-white point of view.

3.3. The implementation phase

The eighteenth century is the time of Enlightenment and utilitarianism. Learned societies and academies were established and they played an important part in the implementation of ongoing linguistic efforts (see Teleman in this volume).

A logical and rational perspective thus dominated the prominent ideology of the eighteenth century. The natural resources available in the country were to be put to the best possible use for the benefit of the state. This required extensive efforts to explore local resources and conditions in general in various parts of the country. The findings of these expeditions were to be reported and the knowledge spread. This activity – as well as academic progress in the sciences – called for a suitable language and for efficient channels of communication.

Phase 3 in Haugen's model, the implementation phase, is concerned with implementing and spreading the results of the selection and codification processes (figure 1). How efficient the implementation process can be depends on the existence of appropriate communication channels. In this respect, however, the emerging picture of the situation in the eighteenth century is somewhat ambiguous. While one process involving a major issue – like the language shift in the discourse of science – is quite effective, other processes, like the establishment of a Swedish version of the alphabet and the relative order between the letters, are slower. Although the alphabetical order seems to have been settled early in the eighteenth century, the lexicographers of the time are clearly reluctant to adopt it in their alphabetically arranged dictionaries (Hannesdóttir 2008). It is mainly the three Swedish letters, <å>, <ä> and <ö>, in word initial position, that cause them problems. It is, of course, difficult to know whether their reluctance is due to ignorance or to scepticism about the efficiency of communication channels at that time. Not until the end of the eighteenth century was this process completed. The final chapter in the graphization process of Swedish was perhaps written during the middle of the present decade, when the Swedish alphabet was augmented to include the letter <w>. The question has been the subject of a thorough discussion at the Language Council of Sweden, the decisions have been made, and we are now in the midst of the implementation process. So far, the latest edition of *The Swedish Academy Glossary* (2006), which is regarded as the unofficial norm for the spelling and inflection of Swedish words, has implemented the new letter in its alphabetically ordered presentation of the words. How quick and efficient the process of implementation will be in the Swedish language community remains to be seen. The result, however, will probably be shaped more by attitudes than by a lack of efficient communication channels.

3.4. The elaboration phase

Phase 4, the elaboration phase, refers to the functional aspects of the language: an elaborated terminology and stylistic development.

The sixteenth, seventeenth and eighteenth centuries were periods of important conquests by the Swedish language. The religious domain was recaptured in the sixteenth century, which involved an enormous task of translation and a major effort to consolidate an appropriate and efficient Swedish vocabulary relevant to the domain. The seventeenth century brought with it stylistic development, as poets chose the Swedish language as a means of expression for their artistic ambitions. At the end of that century, Latin still dominated one field of application, viz. the academic one. The eighteenth century was a time not only of enriching and standardizing the vocabulary in general, but also of constructing terminology and a language for special purposes. The spread of knowledge in wider circles also demanded a straightforward and effective prose – common as well as academic. Here the eminent botanist Linnaeus enters the scene.

4. Linnaeus' contribution to the elaboration of a discourse of science

Carl Linnaeus was one of the founders of the Royal Swedish Academy of Sciences in 1739. He in fact played an active part in promoting the Swedish language by suggesting that research findings within the realm of the Academy should be transmitted to the general public in Swedish, both by way of public lectures and in the regularly published transactions (Hildebrand 1939: 253; Wollin in this volume). It is well known, however, that in his own writings, Linnaeus was somewhat ambivalent about using Swedish in a strictly scientific context. The Swedish language could not meet the demands of a scientific description of his findings and, for obvious reasons, it could not function as a means of communication with the international community of scholars. In his travelogues, in which he gave an account of his expeditions across the country, he thus combined the two languages, Latin for the scientific content and Swedish for the description of more everyday events.

4.1. Linnaeus' findings and namings

The enlargement of the vocabulary of the sciences was indeed an urgent task, in which Linnaeus himself also played an active part. Hence, according to the *SAOB* (*Dictionary of the Swedish Academy* (*a–trivsel*)), there are approximately 1,900 Swedish words that appear for the first time in print in a text written by Linnaeus. A quick look at these 1,900 words reveals that most of them are more or less transparent compounds. Albeit trivial, some of them are significant for the linguistic situation at the time. Already established Latin or Greek names of scientific disciplines make their debut in a Swedified form in Linnaeus' writings: *entomologi*, *iktyologi*, *metallurgi* and *semiologi* (for *entomology*, *ichthyology* etc.). He also introduces *naturvetenskap* [*science of nature*] to designate the field of *natural sciences*. Not all of the words introduced by Linnaeus were successful, however; e.g. *petrifikationskonst* [*science of petrification*] was later on replaced by the classical *paleontologi*. Other words attributed to him, relating to academic education in general, are for example *examinand* [*examinee*] and *licentiatexamen* [*licentiate degree*].

Naturally, the words attributed to Linnaeus reflect his realm of science. As one of the modern scholars studying his style has put it: "allt är viktigt eller allt lika viktigt" [everything is important or equally important] (Abenius 1971: 56

f.). Nothing was too small or trivial to attract Linnaeus' interest. As one of his fields of study was entomology, he has distinguished, described and named a number of insects. Thus, words for different kinds of insects – a tangible element in people's everyday lives, but, in pre-Linnaean times, hardly worth wasting ink and paper on – are abundantly represented. In Linnaeus' texts, we find the first appearances of several compound nouns with *myra* [ant], *mask* [worm] or *mygg/mygga* [mosquito] as the second element. The different kinds of ants he names are *arbetsmyra* [worker ant], *flogmyra* [flying ant] and *stackmyra* [ant hill-ant, i.e. wood ant], the worms are *binnekemask* [binnike worm, i.e. tapeworm], *bokmask* [bookworm], *knutmask* [knot worm, i.e. tadpole], *kormmask* [i.e. the larva of the reindeer warble fly], *nosmask* [nose worm, i.e. the larva of Oestridae], *ormmask* [snake worm, i.e. *Arenicola marina*], *skeppsmask* [shipworm, i.e. *Teredo navalis*], *syrmask* [acid worm] and *tagelmask* [horsehair worm, i.e. nematomorpha]). He also distinguishes and names three specific kinds of mosquitoes, *aftonmygga* [evening mosquito], *braxenmygg* [bream mosquito, i.e. *Gerris lacustris*] and *havsmygga* [sea mosquito]. On top of that, he distinguishes and names 24 different kinds of snakes as well as 14 kinds of flies:

> aftonfluga [evening fly], dagfluga [day fly], husfluga [house fly], kofluga [cow fly], ljusfluga [light fly], lysfluga [light fly], nattfluga [night fly], ostfluga [cheese fly], pussfluga [puddle fly], rovfluga [prey fly], siktfluga [sieve fly], skeppsvarvsfluga [shipyard fly], stinkfluga [stink fly], stugfluga [cottage fly].

None of these words can be said to have become central or even common in the Swedish vocabulary, except maybe *husfluga* and one of the words for ants, *arbetsmyra*, the latter also being used metaphorically of a person who is "as busy as a bee". Some Linnaean words that actually were incorporated into the common vocabulary are *bladlus* [leaf louse, i.e. 'aphid'], *fästing* 'hard tick' and, from a different field of the fauna, *guldfisk* [goldfish].

Most of the words that are attributed to Linnaeus in the *SAOB* designate stones, plants and animals. In the report from his expedition to Westrogothia in 1746 (Linnaeus 1747), we learn about Linnaeus' own view on the subject of his travelogues. He declares that even if many of his readers expect to learn about the most spectacular things he has come across, he must admit that he has not seen anything but some stones, plants and animals. He adds that he has consulted other explorers, alive as well as dead, who have travelled across the world, on what spectacular things they have seen. They all testify that all there is to be seen is stones, plants and animals.

> Mången lärer wänta uti denna Resa ganska sällsamma ting; men jag måste tilstå, at jag på hela wägen ej något annat sedt, än några STENAR, ÖRTER och DJUR,

dem jag upteknat, där jag dem igenfunnit. Jag har frågat lefwande och döda, som anstält widlöftiga resor omkring hela Jordklotet, hwad märkwärdiga saker de sedt uti fjärran länder; men alla hafwa instämt, at de icke eller sedt annat än Stenar, Wäxter och Djur [...]. (Linnaeus 1747, Företal s. 3)

[Reading this travelogue, many will expect to learn about quite spectacular things, but I must admit that during the entire journey I have seen nothing but some Stones, Plants and Animals, which I have recorded where they were observed. I have consulted those, alive and dead, who have performed grand journeys around the entire Earth, as to what remarkable things they have seen in remote countries; but they have all agreed that they have not seen anything but stones, plants and animals [...].]

What certainly will cause problems for his readers, he says, is that they might not recognize the names for the different stones, plants and animals, as these differ in the various parts of the country. Readers will therefore not be able to make use of his findings.

Swårigheterna, som wid denna Resas genomseende möta de mäste mine Läsare, lära bestå uti Namnen på Stenar, Wäxter och Djur, hwilka ofta torde blifwa så obekante, som sjelfwa tingen; men jag kan icke häfwa deße hinder, ty den som wil tala med någon, måste nödsakeligen förstå språket: Om jag nämner Öga, Björk, Abborre, eller Orre, och Läsaren ej förstår hwad som förstås med deßa namnen, kommer han icke fort med Texten. (Linnaeus 1747, Företal p. 3)

[The problems that reading this travelogue will cause the majority of my readers will probably be the names of stones, plants and animals, which may well often remain as unknown as the objects themselves; but I cannot remove these obstacles, because he who wants to talk to someone must necessarily understand the language; if I mention Eye, Birch, Perch or Black grouse, and the reader does not know what these names refer to, he will not understand the text.]

Another problem is that not all the stones, plants and animals Linnaeus observes have a name at all. He therefore constructs a system of references to his major works: *Systema naturæ*, the *Flora* and the *Fauna*. There he gives all the stones, plants and animals designations in Latin, but not in Swedish.

Another aspect of the terminological problem that occupies Linnaeus' mind is the shortcomings of the Latin nomenclature when it comes to describing different findings in the Swedish countryside. The Latin vocabulary is outdated, and the language lacks words for basic, ordinary objects of everyday life (see also Wollin in this volume). Linnaeus discusses this obvious conflict between the Latin and the Swedish language as an instrument in the discourse of science. He regards the development of Swedish husbandry, and primarily the large number of reports on the subject written in Swedish, as problematic. It would be difficult, he says, to do justice to the state of the art in the "Roman and the common language", as there are no modern terms in Latin. Therefore, he states, it would be a good thing if someone could translate the old terms of rural life and husbandry so that they can be incorporated into Swedish.

> wår *Oeconomie* har på några år stigit til ansenlig högd med oändeliga rön, som ock mästadels äro skrefne på modersmålet, hwilka ej så lätt skulle kunna skrifwas på det *Romer*ska och allmänna målet, förnämligen derföre at oß fattas *termer* i *latinen*; hwaraf händer at då något *Oeconomi*skt skal skrifwas uti en *disputation*, *termerne* ofta antingen updiktas eller bråkas, ty woro önskeligt, at någon wille förswänska de gamlas *Oeconomi*ska termer, at wi må kunna få dem rena *applicerade* til wårt språk. (Linnaeus 1751: 426)
>
> [in a few years, our husbandry has risen to considerable heights, with countless findings, mainly reported in the vernacular, which could not easily be reported in the Roman and common language, principally due to a lack of terms in Latin; with the result that when a matter of husbandry is treated in a dissertation, the terms are often either made up or deformed; therefore it is important that someone should update and Swedify the old husbandry terms so that we might add them pure to our language.]

The established Swedish words were not precise enough for terminological use and many of the Latin terms were obsolete. The obvious need for elaboration in the Latin language, especially in terms of terminological modernization, will be left aside here.

4.2. Linnaeus' contribution to Swedish style

Linnaeus himself is not very outspoken about his idiom in the travelogues. His comments are mostly of a kind quite representative of his time. Thus, in the report from his expedition to the islands of Öland and Gotland, he modestly focuses on the simplicity of his style: "Skrifarten är mycket enfaldig, hwarföre jag torde blifwa hårt ansedd af många *Plinii* Nächtergahler. Språket pryder en wetenskap som kläderne kroppen, den som intet sielf kan hedra kläderne, måste låta dem hedra sig. om hos andra folkslag allenast *Eloqventiæ Doctores* fått skrifwit, torde werlden i dag wetat mindre." [The style is very simple, and therefore I will probably be severely criticized by many of Pliny's nightingales. Language adorns a science like clothes adorn the body; he who can't himself honour the clothes, must let them honour him. If, among other nations, only eloquent doctors had been allowed to write, the world would have known less today.] (Linnaeus 1745: Företal [p. 7].)

Linnaeus did possess a quality that is very clearly reflected in his writings, i.e. "ögats genialitet" [the genius of the eye] (Hildebrand 1939: 295). Although

Linnaeus was no novelist, the literary qualities of his texts have indeed attracted the attention of 20th-century scholars. First and foremost, it is his vivid and descriptive scenes which, owing to their simplicity and literary qualities, have been appreciated by scholars – and most likely – successful novelists as well. Thus, it has been argued that Swedish novelists of the 19th century, among them the internationally recognized Swedish author and playwright August Strindberg, have benefited from his style (Ralph 2007; Ralph in this volume). Ralph also argues that it is within scientific prose that the impact of Linnaeus should be most apparent. Linnaeus' stylistic ideals influenced his disciples in their reports from their own expeditions to distant and exotic places. The style crystallized in Linnaeus' travelogues from his expeditions around Sweden thus became a foundation for the elaboration of factual prose in Swedish.

5. Summary

In this article, some of the crucial lines of development which the Swedish language has followed have been briefly discussed. By relating these processes to language planning as defined by Einar Haugen and to the model he constructed, I have illustrated how the notion of LP can serve as an instrument for the evaluation of linguistic change in a broader perspective than standardization proper.

The process of elaboration which the Swedish discourse of science underwent during the eighteenth century can easily be accounted for and explained in terms of language planning. One eminent individual who in several ways left his mark on the academic discourse is the Swedish botanist Linnaeus. He supported the selection and implementation phases of the LP process by promoting the Royal Swedish Academy of Sciences' use of the Swedish language rather than Latin. He also contributed directly to the elaboration phase by creating a Swedish vocabulary appropriate to his scientific findings. This terminological elaboration was, in addition, instrumental in the lexication of Swedish, i.e. the codification phase of the LP process. In this particular respect, the progress of the discourse of science is a simultaneous as well as a cyclical LP process.

Linnaeus was not, however, one of the official or semi-official actors who dominated the scene, debating the linguistic issues of the time. In spite of that, he too is now included among those who renewed and refined Swedish prose in general, and his academic style is acknowledged as a kind of ideal model for Swedish academic prose. Contrary to his own assumptions, Linnaeus has – one way or another – become one of Pliny's nightingales in Swedish factual prose. He certainly contributed in a significant way to the development and sophistication of the Swedish language.

References

Abenius, Margit
1971 Allt är viktigt. En stiltendens hos Linné [Everything is important. A stylistic tendency in Linnaeus' writings]. In: Sigurd Fries (ed.), *Linnés språk och stil*, 56–73. Stockholm: Prisma.

Deumert, Ana
2003 Describing language standardization – models and methods. In: Kristján Árnason (ed.), *Útnorður. West Nordic standardization and variation*, 9–32. Reykjavík: University of Iceland Press.

Deumert, Ana and Wim Vandenbussche (eds.)
2003 *Germanic standardizations. Past to present*. Philadelphia: John Benjamins Publishing Company.

Deumert, Ana and Wim Vandenbussche
2003 Standard languages. Taxonomies and histories. In: Ana Deumert and Wim Vandenbussche (eds.), 1–14.

Dil, Anwar (ed.)
1972 *The ecology of Language. Essays by Einar Haugen*. Stanford: Stanford University Press.

Hannesdóttir, Anna Helga
1998 *Lexikografihistorisk spegel. Den enspråkiga svenska lexikografins utveckling ur den tvåspråkiga* [History of lexicography reflected. The development of monolingual Swedish lexicography from the bilingual]. Ph.D. dissertation, Department of Swedish, University of Gothenburg. (Meijerbergs arkiv för svensk ordforskning. 23.) Göteborg.

Hannesdóttir, Anna Helga
2000a Ordböckernas roll i svenskans grammatiska standardisering [The role of dictionaries in the grammatication of Swedish]. In: *LexicoNordica* 7, 23–40. Oslo.

Hannesdóttir, Anna Helga
2000b Tvåspråkig lexikografi och språkplanering i 1700-talets Sverige [Bilingual lexicography and language planning in Sweden in the eighteenth century]. In: Lars-Erik Edlund (ed.), *Studier i svensk språkhistoria* 5, 224–235. (Nordsvenska 5.) Umeå: Institutionen för litteraturvetenskap och nordiska språk.

Hannesdóttir, Anna Helga
2002 Svenskt och osvenskt i 1700-talets svenska ordböcker [Swedish and non-Swedish in Swedish dictionaries of the eighteenth century]. In: Åsa Karlsson and Bo Lindberg (ed.), *Nationalism och nationell identitet i 1700-talets Sverige,* 87–99. (Opuscula Historica Upsaliensia 27.) Uppsala: Uppsala universitet.

Hannesdóttir, Anna Helga
2008 Hur det blir skillnad på rätt och fel i språket [The rise of the notion of right and wrong in language]. In: Folkmålsstudier 46, 11–26. Helsinki.

Haugen, Einar
1966 *Language conflict and language planning. The case of modern Norwegian*. Cambridge, Massachusetts: Harvard university press.

Haugen, Einar
[1959, 1961] 1972 Language planning in modern Norway. In: Anwar Dil (ed.), 1972, 133–147.

Haugen, Einar
[1966] 1972 Linguistics and language planning. In: Anwar Dil (ed.), 159–190.

Haugen, Einar
1987 *Blessings of Babel. Bilingualism and language planning. Problems and pleasures*. (Contributions to the sociology of language. 46.) Berlin, New York & Amsterdam: Mouton de Gruyter.

Hildebrand, Bengt
1939 *Kungl. Svenska Vetenskapsakademien. Förhistoria, grundläggning och första organisation* [Royal Swedish Academy of Sciences. Prehistory, foundation and initial organization]. Stockholm: Kungl. Vetenskapsakademien.

Linnaeus
1745 = *Carl Linnæi Öländska och Gothländska resa på Riksens högloflige ständers befallning förrättad år 1741* [Linnaeus' Öland and Gotland journey 1741]. Stockholm and Upsala.

Linnaeus
1747 = *Carl Linnæi* [...] *Wästgötaresa* [...] *förrättad år 1746* [Linnaeus' journey to Westrogothia 1746]. Stockholm.

Linnaeus
1751 = *Carl Linnæi Skånska resa* [...] *förrättad år 1749* [Linnaeus' journey to Scania 1749]. Stockholm.

Milroy, James and Lesley Milroy
1997 Exploring the social constraints on language change. In: Stig Eliasson and Ernst Håkon Jahr (eds.), *Language and its ecology*, 75–101. Berlin and New York & New York.

Ralph, Bo
2007 August Strindberg – unik genombrottsman med föregångare [August Strindberg – a unique pioneer with predecessors]. In: Lars Wollin, Anna Saarukka and Ulla Stroh-Wollin (eds.), *Det moderna genombrottet – också en språkfråga?* Studier i svensk språkhistoria 9, 229–239. (Skrifter från Svenska institutionen vid Åbo Akademi 5/2007.) Åbo.

Ralph, Bo
2011 Linnaeus as a connecting link in Swedish language history. In: Britt-Louise Gunnarsson (ed.), *Languages of Science in the Eighteenth Century*, 247–261. Berlin, New York: De Gruyter Mouton.

Santesson, Lillemor
1986 *Tryckt hos Salvius. En undersökning om språkvården på ett 1700-talstryckeri med särskild hänsyn till ortografi och morfologi* [Printed by Salvius. A study of the linguistic usage of an eighteenth-century printing house, with special reference to orthography and morphology]. (Lundastudier i nordisk språkvetenskap. A 37.) Lund: Lund University Press.

SAOB = *Ordbok över svenska språket, utgiven av Svenska Akademien* 1893 – [Dictionary of the Swedish Academy]. Lund.

Swedberg, Jesper
1722 *En kortt Swensk Grammatica* [A brief Swedish grammar]. Stockholm.

Teleman, Ulf
2002 *Ära, rikedom och reda. Svensk språkvård och språkpolitik under äldre nyare tid* [Glory, wealth and order. Swedish language cultivation and politics in early modern times]. Stockholm: Norstedts ordbok.

Teleman, Ulf
2003 Swedish. In: Ana Deumert and Wim Vandenbussche (eds.), 411–429.

Teleman, Ulf
2011 The Swedish Academy of Sciences: Language policy and language practice. In: Britt-Louise Gunnarsson (ed.), *Languages of Science in the Eighteenth Century*, 63–87. Berlin, New York: De Gruyter Mouton.

Tiällmann, Nils
1696 *Grammatica suecana*. Stockholm.

Watts, Richard J.
1999 The social construction of standard English: Grammar writers as a 'discourse community'. In: Tony Bex and Richard J. Watts (eds.), *Standard English. The widening debate*, 40–68. London: Routledge.

Wollin, Lars
2011 From Latin *and* Swedish to Latin *in* Swedish: On the early modern emergence of professional vernacular in Sweden. In: Britt-Louise Gunnarsson (ed.), *Languages of Science in the Eighteenth Century*, 123–139. Berlin, New York: De Gruyter Mouton.

From Latin *and* Swedish to Latin *in* Swedish. On the early modern emergence of a professional vernacular variety in Sweden

Lars Wollin

An important aspect of our heritage from the eighteenth century is the interplay between the vernacular (e.g. Swedish, Danish or German) and Latin in scientific discourse. Very significant is the shift of roles experienced by Latin as the traditional language of learning: from a general medium of thinking on abstract as well as technical levels, to a supplier of lexical material for the vernacular variety that gradually took over the function of Latin in the same domain. In this historical development, the main emphasis shifted from Latin *and* the vernacular to Latin *in* the vernacular. Both aspects raise intriguing questions concerning the refinement and international adaptation of national language in a gradually widening spectrum of functional varieties.

In the present paper, this complex will be studied at a lexical level, in a quantitative perspective supported by statistical data. In the light of the overall proportions of Swedish and Latin in older New Swedish book printing (1), the stock of Latin loan words in written Swedish from the sixteenth century to modern times is described and discussed with a focus on elements and patterns of word formation (2). In the latter section, the frequencies of loanwords with eight particular Latin suffixes and prefixes are shown, as distributed between common language and scientific professional language; this distribution is related partly to domains of usage, partly to specific suffixes and prefixes (2.1). Next, the same lexical material, likewise divided between common and professional language, is examined with regard to the distribution of single words and compounds (2.2). In both perspectives, the relationships observed result in specific chronologies, the general conclusion leading to a hypothesis concerning the character of the gradual integration of loan words into the receiving language.

1. Latin and Swedish

I set out from a broad outline of linguistic conditions during Sweden's era as a great power (roughly 1650–1720), i.e. the period preceding the time when scientists like Carl Linnaeus, Anders Celsius and Nils Rosén von Rosenstein were productive. I will be looking at the distribution of languages in literary writing, as reflected in a statistical account based on the production of printed books in seventeenth-century Sweden (Figure 1). This production is documented in the Swedish National Bibliography;[1] the sum total of printed titles during the period amounts to approximately 4,000.

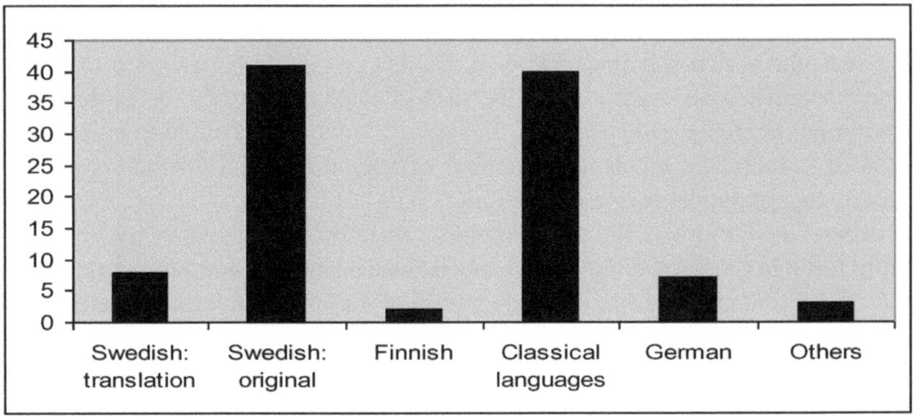

Figure 1. Languages in seventeenth-century Sweden. Printed books. Percentage distribution. Source: Hansson 1982.

Two languages obviously stand out as uncontestedly predominant: Swedish and Latin. At the time, the former had emerged quite recently as a relatively cultivated written language, partly on the basis of a fairly substantial medieval, Old Swedish tradition, partly as a result of the Reformers' modern, historically crucial biblical translations. Swedish appears in the printed literature of this period mainly as an original language. The significance of translation is fairly marginal: only one book out of seven printed in Swedish in the seventeenth century was translated.

The share of the total accounted for by original Swedish texts is approximately 40 %, which is roughly the same as that of Latin. Among other languages, the largest is German, with about 7 %, while third position is held by Finnish, with 3 %. Other languages are quite marginal in this period; in English,

1. Collijn 1943–44. For the statistical account, see Hansson 1982: 57.

for example, hardly a single book was printed in Sweden in the seventeenth century.

The distribution of Swedish and Latin was functional and complementary, of course. Latin was still in a very wide sense the language of science and of the learned world in general. Almost all printing in the Latin language consisted of the publications of the universities at Uppsala, Lund and Åbo/Turku: dissertations and other products of scholarly and scientific endeavour, normally published in very small editions. Edifying and educating books, written in the spirit of Lutheran orthodoxy and historical patriotism, were those most extensively printed in Swedish.

This picture of a complementary relationship between Latin and the vernacular is the one that is generally maintained. Nevertheless, it is slightly simplified. First, the use of the vernacular in communicating professional knowledge has a long history, going back at least to the High Middle Ages. The extensive vernacular *medical* literature of the late Middle English and Early Modern English periods, for example, which has been recorded and investigated, linguistically and in other respects, in modern research (most recently by Norri 2004), has a parallel in Sweden and Denmark in the medieval *läkeböcker* ("medical books"), written in Old Swedish and Old Danish within similar patterns of Latin influence. Second, a few stray works in other more or less "learned" genres were in fact printed in Swedish as early as the seventeenth and the early eighteenth century. Third, even edifying religious or patriotic literature in the vernacular could very well exhibit significant features of a scientific discourse, though perhaps somewhat speculative features.

At any rate, the editors of the comprehensive historical dictionary of Swedish, *Dictionary of the Swedish Academy* (*Ordbok över svenska språket* 1898–2007, referred to below under the common designation *SAOB*, 'Svenska Akademiens ordbok'), attach labels referring to "professional language" (the Swedish term is *fackspråk*) to thousands of words even from periods before 1700. This rudimentary professional Swedish was developed, though, under a very tangible influence from the general European language of learning. Humanist and Renaissance Latin was a mental property of contemporary scholars. It was deeply rooted in their thinking and thoroughly cultivated in practically all their learned writings. Scientists laboriously writing in Swedish do not even attempt to conceal their dependence on established Latin modes of expression, a fact that manifests itself in several ways. One of the more striking of these is a widespread inclination to *code-switch* with Latin, a procedure typical of course of scientific writers who do not feel entirely at home with their mother tongue in contexts where it is still unusual (cf. Hannesdóttir, this volume; interesting surveys of code switching in medieval and Early Modern English medical writing are given by Pahta 2004 and Schendl 2009). An illustrative example is the following passage in an early phonetic description of the dialect of the island of Gotland, discussing the historical origin of the diphthong *au* (which is very typical of this dialect):

> Alratydeligast talar Kongl. Secreteraren Peringskiöld: The twå vocaler, säger han, **a** och **u**, hafwa hwarthera sitt liud, thet the gamle hafwa låtit höras wid uttalandet, *sed seqvioris ævi hominibus solenne est, duarum harum vocalium* ***a*** *et* ***u*** *sonos in unum* ***ö*** *contrahere* [emphasis mine]. Af exempel skal nu thetta bli klarare …
>
> 'Most clearly of all speaks the Royal Secretary Peringskiöld: The two vowels, says he, **a** and **u**, have each their respective sounds, which our forefathers allowed to be heard in pronunciation, *but to people of succeeding periods it is decent to contract the sounds of these two vowels* ***a*** *and* ***u*** *in a single* ***ö***. Examples will now make this clearer …'
>
> (L. Neogard: *Gautau-Minning* 1732)

When Carl Linnaeus, along with his learned contemporaries, set about developing a Swedish scientific prose in the printed transactions of the Royal Swedish Academy of Sciences in the mid-eighteenth century, it was a matter of something radically new.[2] Uncontrolled code switching, for example, is now far away: in the Academy's transactions, strict linguistic discipline prevails, and the language is genuine Swedish throughout. This does not entail any contradiction with a well-developed Latin terminology, which is obviously used without restraint. No consistent purist endeavour to replace Latin and Greek morphemes with domestic material is to be seen.

This did not, of course, end the pivotal position of Latin in the Swedish linguistic community. In the strictly scientific sphere its dominance was total, and remained so for at least a century. Well into the nineteenth century, Latin was the natural, and normally the sole, means of communication at the universities, in speech and in writing, at all levels of academic life.

It was only in the capital of Sweden, Stockholm – at that time a city where a relatively open-minded cultural climate prevailed – that a change in favour of the national language was within reach. The Royal Swedish Academy of Sciences (cf. Teleman, this volume) had recently been founded and, as indicated above, was pursuing a radical linguistic policy, as stable as it was enlightened. Under its protection, energetic efforts were possible. Linnaeus and his collaborators, who had the strength and the courage to take such a step, struck the first heavy blow to the state of the linguistic culture in Sweden, which appeared to

2. Gunnarsson 1987, Fries 1996, Teleman in this volume.

many as a Latin tyranny, ruling in science and learning almost up to modern times.

This widening domain for the more advanced use of Swedish is one side of the matter.[3] Another is the character of the vernacular actually used to replace the universal language of learning. As initially suggested, this modern professional variety of Swedish not only took over a Latin heritage in its terminology, but even elaborated it further, constantly refining it in accordance with international patterns. To exaggerate slightly, it can be argued that Latin never actually left the arena of learning and science, but was simply transformed. It appears today, not as an overall medium of human thinking (that state of affairs is lost for ever), but as the predominant supplier of lexical material for national languages, which readily interweave it with their own domestic material, in accordance with their own structure, thus making Latin part of national language history as well. This role may, certainly, be a more limited one, but it is hardly less significant. Latin itself has changed to another, more modest level of discourse – where, on the other hand, its position as yet seems unchallenged.

2. Latin in Swedish

The pattern of usage will be studied here in closer detail as it appears during the New Swedish period (from 1520 to the present time), and as reflected in a few significant Latin elements, in distinct varieties of written Swedish. We will focus on three dimensions: language type, i.e., common (non-professional) v. professional; word formation, i.e., single (simple) words v. compounds; and chronology, i.e., Older New Swedish period (before 1750) v. Younger New Swedish period (after 1750).[4]

The overall question considered is the emergence and development to a high level of a professional variety, or professional varieties, of written Swedish, in its constant interplay with influences from the second dominant language, i.e., Latin. Hardly surprisingly, the principal level of analysis is the lexicon.

---

3. Important contributions to research on the historical development of Swedish non-literary prose have been made in recent decades, particularly within the framework of the project *Svensk sakprosa* [Swedish non-literary prose]; see Melander and Olsson 2001 and Englund and Ledin 2003. In a Gothenburg project (ORDAT; Malmgren 2000, Wollin 2004), a lexical perspective has been adopted. Very informative, too, are the studies of Gunnarsson 1997 and 2005 on economic and medical language.
4. This topic has been treated more comprehensively, though with another focus, in Wollin 2004.

My starting point is the quantity and the quality of the Latin element in the vocabulary of these linguistic varieties. The source material immediately accessible for such purposes has been found in the *Dictionary of the Swedish Academy* (SAOB), mentioned above. The editors of this work attach labels referring to "professional language" ("fackspråk") to thousands of words from the entire period covered. This lexicographic marking is, quite simply, my principal criterion for distinguishing a professional variety of written Swedish from a non-professional, or "common", variety. But of course it is only reasonable to present some kind of definition, to fix the concept we are talking about. My own definition goes as follows:

> (Professional language, Fachsprache, fackspråk):
> a linguistic variety used in a particular field of art, science or technology and employing, within that field, a specific set of linguistic devices, including an exactly defined terminology.

The editors of the SAOB dictionary have performed a comprehensive inventory of the lexicon of written Swedish since the Lutheran Reformation in the sixteenth century. Today the results are also available in an electronic version,[5] significantly facilitating sorting of the material by several crucial variables. They have provided a solid base for my selection of material for the present study. Another efficient instrument has been the pattern of word formation in the Latin language itself, and its derivational morphology in particular.[6] Accordingly, a half-dozen Latin suffixes and two prefixes – or rather the various modern manifestations of these Latin morphemes in the endings and initial syllables of New Swedish loanwords – have been selected from the electronic version of the dictionary. Subsequently, an inventory has been made of the total stock of lemmas with these word endings and initial syllables. The years of the "first instances" have been noted and the instances have been counted. The term employed here for this kind of "first instance" is *word usage*.

What is actually measured here, then, is the varying *inclination* of users to adopt lexical material of Latin origin – here understood as morpheme material – to meet new needs of expression. This may be carried out either by borrowing new words or by semantic reinterpretation of words already borrowed.

---

5. http://g3.spraakdata.gu.se/saob/ . For a more detailed presentation, see Malmgren 2000.
6. For a survey, see Hofmann and Szantyr 1965 (allgemeiner Teil): 67–74; for a Swedish application, see Helander 2005.

I have chosen for closer study the following Latin suffixes and prefixes, together with the Swedish word endings and initial syllables corresponding to them:

Latin	Swedish	Examples
-or	*-or*	*doktor* ('doctor'), *lektor* ('lecturer')
-or	*-ör*	*chaufför* ('driver'), *lektör* ('reader')
-bil-	*-bel*	*möbel* ('piece of furniture'), *ämabel* ('amiable')
-iv-	*-iv*	*motiv* ('motive'), *provokativ* ('provocative')
-(if)ic-	*-(if)icera*	*domesticera* ('domesticate'), *pacificera* ('pacify')
-iz-	*-isera*	*civilisera* ('civilize'), *urbanisera* ('urbanize')
per-	*per-*	*perforera* ('perforate'), *permission* ('permission')
prae-	*pre-*	*predika* ('preach'), *prefix* ('prefix')

2.1. Common and professional language

As mentioned, the editors of the SAOB dictionary attach labels denoting professional language to words from all periods of history from the Reformation onwards (the oldest instances dating from the year 1520). The majority of the "first instances" found lack a marker of professional language: 3,386 or 69 % of the total represent "common language", whereas 1,523 or 31 % are marked as some kind of "professional language".

There is a sharp increase in the overall frequency of instances over time, as shown in Figure 2.

However, as we can also observe, the balance between the two varieties remains constant in all three periods of New Swedish. The increase in frequency seems to be exclusively a question of a general expansion of vernacular writing.

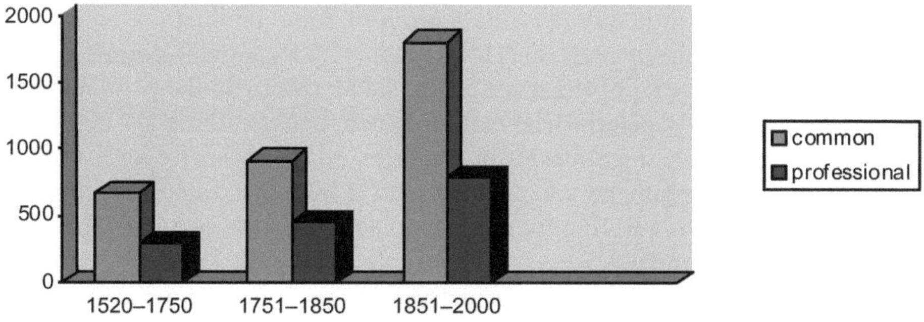

Figure 2. First instances in the SAOB material: unmarked ("common") v. marked as "professional" language, in three periods after the Reformation. Absolute frequencies.

A crucial dimension of this pattern is the variation of *domains of usage* within the professional variety. Four domains are distinguished here: theology and humanities; law, economics and social science; science and medicine; and technology. A fifth, unspecified domain includes a few dubious instances. The distribution is illustrated in Figure 3 (confined here to the prefix derivations).

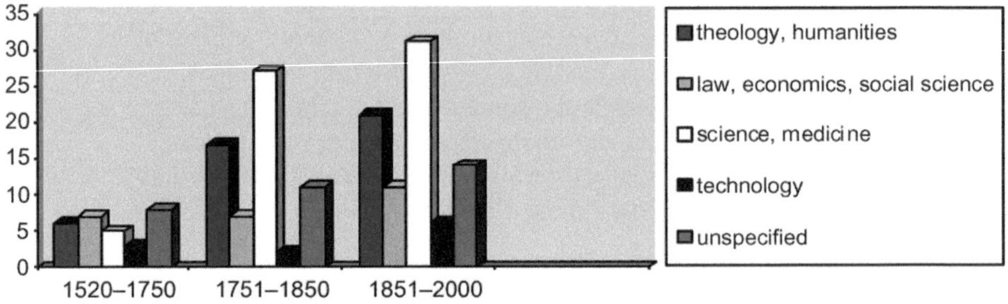

Figure 3. First instances in professional language in the SAOB material: marked as four particular domains of "profession" and one unspecified domain, for three periods after the Reformation. Absolute frequencies (confined to prefix derivations in *per-* and *pre-*).

The increase over time seems basically common to all the domains (the unspecified one included), with a certain reservation for law and technology in the middle period; a general uncertainty may be attached to the minute differences in some cases. It is clear, though, that some significant changes in the relations between the domains do occur. Particularly striking is the triumph in the middle period of the scientific and medical domain, which rises from a rather insignificant position before the mid-eighteenth century to undisputed dominance in the two later periods.

We will look a little closer at the pattern of some particular elements of Latinizing word usage in Swedish. The development from the Reformation onwards of the frequencies of the word endings under study in the two language varieties in the SAOB material has been examined for common and professional language respectively. Frequencies of instances are distributed over fifty-year intervals during the entire period of New Swedish (except the sixteenth century, which is measured as a whole; for the second half of the twentieth century, the information is incomplete). This is demonstrated in Figure 4a–c.

The first two groups of words considered are those involving the suffixes *-iv*, as in *arkiv* ('archive') and *massiv* ('massive'), and *-bel*, as in *nobel* ('noble') and *variabel* ('variable'), illustrated in Figure 4a.

The *-iv* ending presents roughly the same pattern in both varieties, with a very marked peak in the professional variety in the first half of the nineteenth

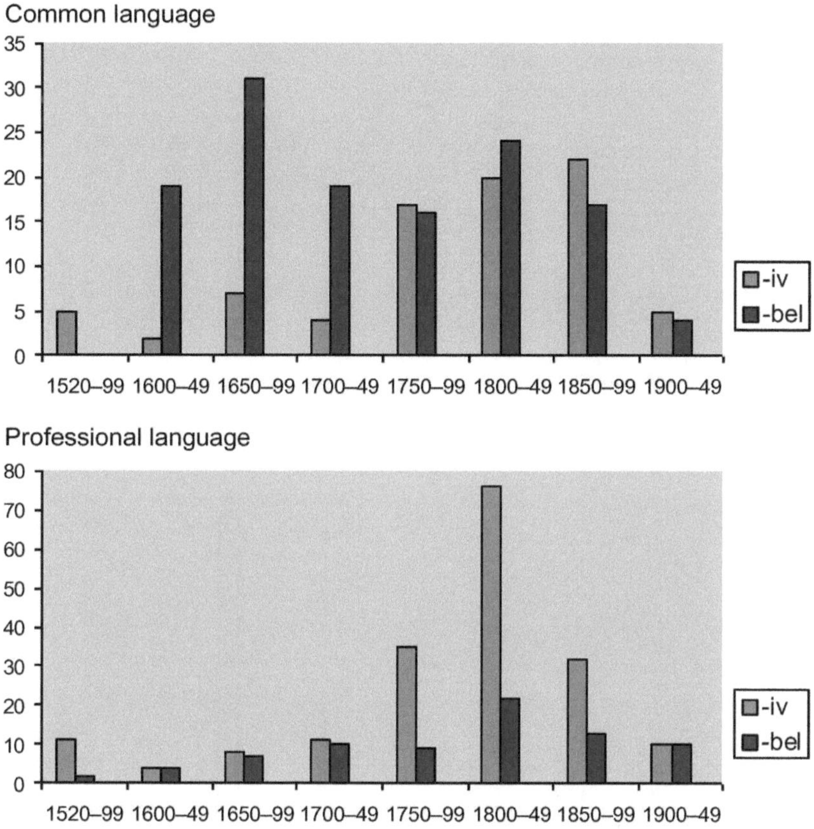

Figure 4a. Words ending in *-iv* as in *arkiv* and *massiv* and *-bel* as in *nobel* and *variabel* in the SAOB material in the sixteenth century and the seven subsequent half-centuries.

century, whereas *-bel* seems to be substantially more typical of common than of professional language, and particularly so in the later seventeenth century.

Another group of words actually involves a single Latin suffix, that of the nomen agentis *-or*, and two Swedish word endings: either preserved Latin *-or*, as in *doktor* ('doctor') and *lektor* ('lecturer'), or the substituted French syllable *-ör*, as in *direktör* ('director') and *lektör* ('reader'). Figure 4b illustrates this.

Like *-bel* in the previous pair of diagrams, but with quite a different pattern, the two Swedish word endings shown here exhibit a fairly tangible contrast between the professional and the non-professional variety. Both increase their frequencies in later periods, the Latin ending, however, much more obviously so in professional language, the French in non-professional.

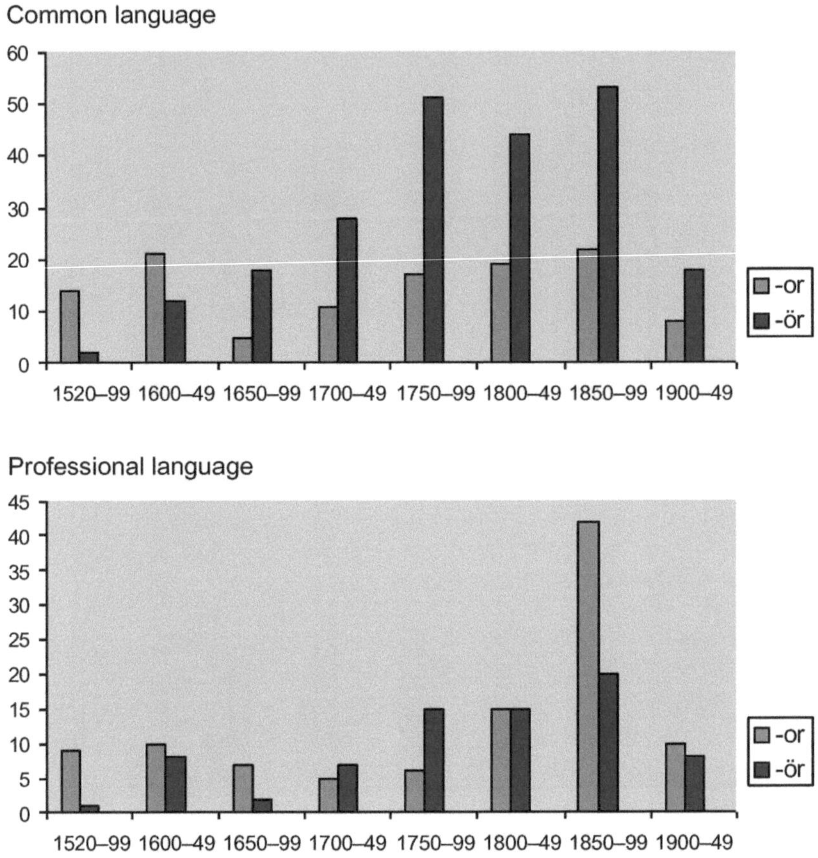

Figure 4b. Words ending in *-or* as in *doktor* and *lektor* and *-ör* as in *direktör* and *lektör* in the SAOB material in the sixteenth century and the seven subsequent half-centuries.

Finally, the development of the two verb suffixes -*(if)icera*, as in *domesticera* ('to domesticate') and *kvalificera* ('to qualify'), and *-isera*, as in *civilisera* ('to civilize') and *realisera* ('to realize'), is shown in Figure 4c.

The frequencies are quite different for these two suffixes: after the mid-eighteenth century, *-isera* seems to be almost the only one of them that is productive. However, in that later period the pattern is strikingly similar in both language varieties.

In the absence of more penetrating studies, the causes of these very different chronological variations in particular Swedish word elements remain beyond our reach.

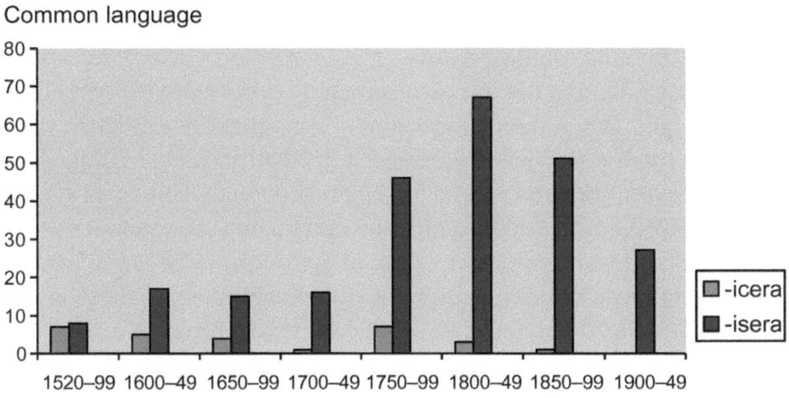

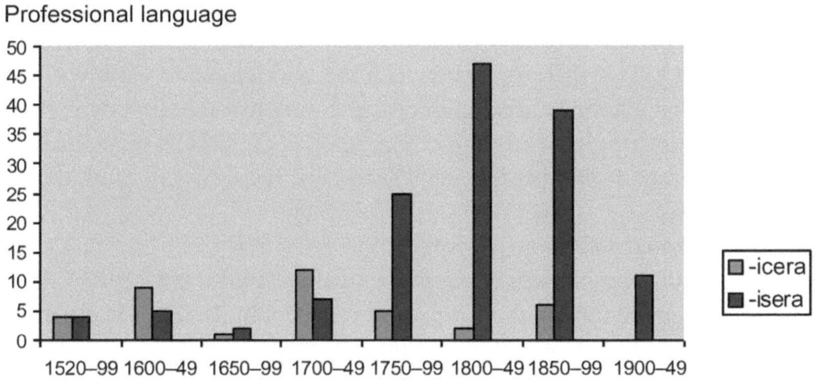

Figure 4c. Words ending in *-(if)icera* as in *domesticera* and *kvalificera* and *-isera* as in *civilisera* and *realisera* in the SAOB material in the sixteenth century and the seven subsequent half-centuries.

2.2. Word formation and language type

To the dichotomy of common and professional language we can fruitfully relate the second dimension of this study, that of word formation. Typical examples of compounds with Latin-suffixed words as the second element are:

engelsklektor ('senior master in English')
förlagslektör ('publisher's reader')
kontorsmöbel ('piece of office furniture')
mordmotiv ('motive for murder')

The compounds selected here follow this pattern, i.e. those with one of the Latin suffixes or prefixes under study in their second element (irrespective of the character of the first element). Their total number in the corpus is 2,794, as compared to 2,115 non-compound (single) words.

It seems clear, first, that the use of compounds is not a device typical of professional language. The general frequency is considerably higher in the common variety. In total, compounds are used in 64% of instances in the common language corpus, as compared to 41% in the professional. This hints at a pattern where authors working in strictly matter-of-fact genres (i.e., authors of professional texts), when borrowing new words of Latin origin or using established Latin loanwords in a new sense, are obviously quicker to adopt these new items as simple words than to use them in productive word formation, viz. as compounds. Writers of common language, then, show the opposite preference. This is no doubt intuitively reasonable, in a way predictable: the frequent, more "domesticated" loan words, most apt for use as compounds, are more likely to appear in ordinary than in professional language. This considered, one might even have expected a bigger difference between the two language varieties.

As regards this tendency, too, an interesting dimension of time may be noted. The distribution of single words and compounds in the SAOB material has been examined in common and professional language respectively from the Reformation onwards.

The professional language variety is the most clear-cut one in the period before 1750, with compounds accounting for less than a tenth (9%) of usage. In common language during the same period compounds also form a minority, but a considerably larger one: they are used here in more than a third (36%) of instances. After 1750, the frequency of compounds in loan word usage increases very markedly in both varieties: in professional language to half the number of instances (49%), in the non-professional variety to more than two thirds (71%).

As suggested above, the frequency of compounds in loan word usage presumably reflects quite another dimension of the users' linguistic competence than the one that manifests itself in the borrowing of Latin words or in semantic shifting. The inclination to use Latin derivatives as the final elements of compounds – often with a domestic first element – does not necessarily correspond to the Latin, or even "foreign", origin or identity of these words: the Latin character may have faded more or less entirely during the long usage of a well-established and well-assimilated loan word.[7] One might possibly test the hypothesis here of a statistically provable positive correlation at any given moment in language history between, on the one hand, the degree of "acclimatization" of a loan word (measured most simply as the time that has elapsed from its first occurrence) and, on the other hand, the productivity of central pat-

7. cf. Helander 2005:2095.

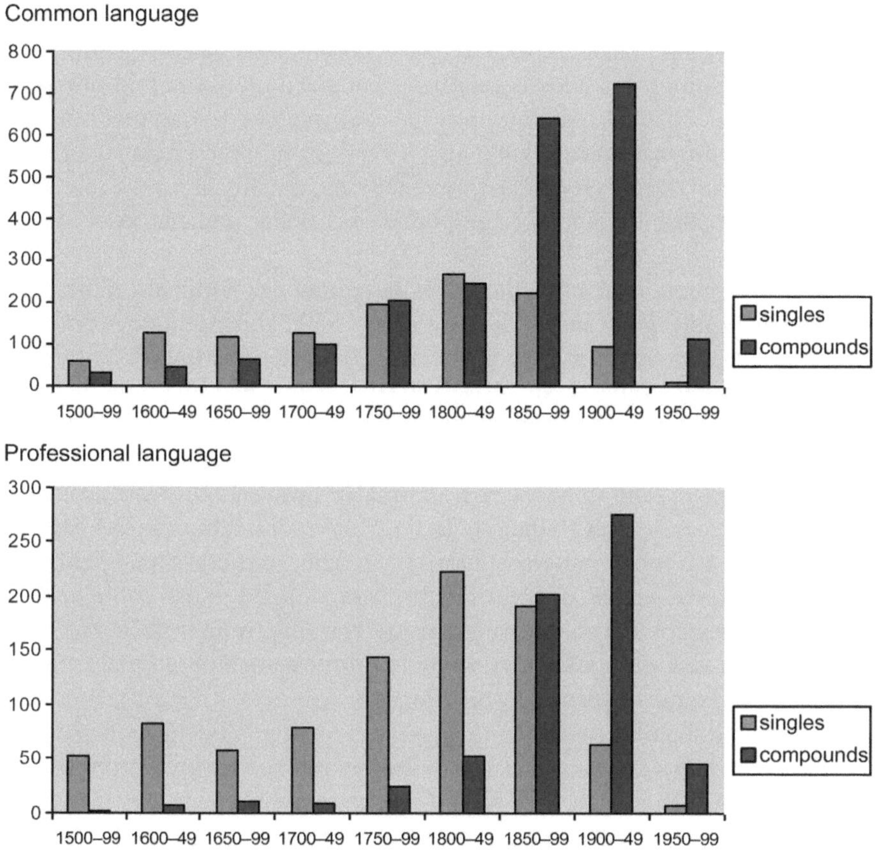

Figure 5. Distribution of single words and compounds in the SAOB material in the sixteenth century and the eight subsequent half-centuries.

terns of word formation, like compounding, related to the same loan word. The proportions indicated here no doubt offer some foundations for this hypothesis.

Again, the dimension of time may be fruitfully specified in some detail. This is demonstrated in Figure 5. The frequencies are distributed over the same fifty-year intervals during the entire New Swedish period (except the sixteenth century).

The use of compounds increases in both language varieties throughout the period. Unlike the trend for single words, it also shows a constant increase (apart from a minor decrease in professional language in the first half of the eighteenth century).The breakdown over time is of most interest, however, in the middle of the later New Swedish period: a marked "jump" upwards can be noted for the use of compounds in the late nineteenth century. The difference between the language varieties is considerable, though: in common language, the proportion of single words to compounds is relatively balanced up to the mid-nineteenth century and then changes quite abruptly into a strong and lasting predominance of compounds. In professional language, a very strong predominance of single words prevails up to the same time, before passing, equally abruptly, into a state of balance, the predominance of compounds not being realized until the 20th century.

The development over time, then, is not continuous and linear; rather, there is a distinct turning point in the decades after 1850. This point does not, however, coincide with the boundary between the two periods traditionally designated as Older and Younger New Swedish (the established line of demarcation being the first edition of Olof von Dalin's *Then Swänska Argus*, 1732), but a full century later. Up to 1850, writers in non-professional genres, i.e., users of common language, tend to adopt new, normally simple Latin words, or to use old ones in a new sense, as frequently as they incorporate them as last elements of compounds in a living pattern of word formation. In professional genres, the writers of the same period are considerably less inclined to use compounds. In other words: newcomers to the language are certainly numerous before 1850, but they are treated, particularly by writers of professional or learned language, with a certain degree of formal reservation.

After the middle of the nineteenth century, a relatively open-minded attitude to Latin-based innovations in the vocabulary is replaced with a more creative use of loan words that have long since become established. In this pattern, the writers of professional and nonprofessional texts react in basically the same way. The former though, who are constantly more inclined to borrow words, react in quite different proportions.

3. Concluding remarks

The interplay between Latin and the vernacular in the historical development of a national linguistic variety of scientific discourse, in Swedish as in other comparable languages, certainly takes place on several levels, not just the lexical. Particularly in syntax, the imitation of Latin patterns makes itself strongly felt – whether the work of translators or of stylistically creative authors. This is also true of the shape of the argumentation itself and the structure of the text. The writers' dependence on models provided by Latin discourse is an essential element in establishing a particular scientific variety of text, formed in good Swedish. As suggested above, this particular variety assumes a more definite shape in the middle of the eighteenth century and is brought to maturity in the course of the succeeding hundred years. A cultural product like this is, in the course of time, laboriously worked out by writers and carefully polished in official guidance on modern Swedish usage. In the present-day linguistic community it is a valuable asset in its own right. Its future survival may not necessarily be more threatened by English today than it was by German a hundred years ago: the basic terms of existence of scientific Swedish have always included its use in close cooperation – often fruitful rather than detrimental – with other and larger civilized languages.

In the particular lexical dimension applied in this paper, an interpretation along the lines of the facts observed so far suggests the general idea – perhaps somewhat strongly worded – that a comprehensive body of linguistic material of Latin and Greek origin was fixed and stabilized in written Swedish in the mid-nineteenth century, within a given quantitative framework and with a distribution over the varieties of language, based on tradition. At that time, it ceased to grow. Instead, it made itself at home.

And, I would like to add, it is equally at home today.

References

Collijn, Isak
1943–44 *Sveriges bibliografi. 1600-talet* [Sweden's bibliography. The seventeenth century]. 1–2. Uppsala: Svenska litteratursällskapet.

Englund, Boel and Per Ledin (eds.)
2003 *Teoretiska perspektiv på sakprosa* [Theoretical approaches to nonliterary prose]. Lund: Studentlitteratur.

Fries, Sigurd
1996 Lärdomsspråket under frihetstiden [The language of learning in the Age of Liberty]. In: Moberg, Lena and Margareta Westman (eds.): *Svenskan i tusen år. Glimtar ur svenska språkets utveckling* [Swedish over a thousand years. Glimpses from the development of the Swedish language], 88–103. Stockholm: Norstedts.

Gunnarsson, Britt-Louise
1987 Textmönster i vår äldsta vetenskapliga tidskriftsprosa. En analys av medicinska rön i Vetenskapsakademiens Handlingar 1750–1769 [Textual patterns in the prose of our oldest scientific journals. An analysis of medical findings in the Proceedings of the Academy of Sciences 1750–1769]. In: *Nysvenska studier* 67: 155–180.

Gunnarsson, Britt-Louise
1997 On the Sociohistorical Construction of Scientific Discourse. In: Gunnarsson, B.-L., Linell, P. and Nordberg, B. (eds.), *The Construction of Professional Discourse*, 99–126. Longman. London and New York.

Gunnarsson, Britt-Louise
2005 Medical Discourse: Sociohistorical Construction. In: *Encyclopedia of Language and Linguistics*, 2nd ed., edited by Keith Brown, Vol. 7, Article 2360, 709–716. Elsevier: Oxford.

Hannesdóttir, Anna Helga
2011 From vernacular to national language: Language planning and the discourse of science in eighteenth-century Sweden. In: Britt-Louise Gunnarsson (ed.), *Languages of Science in the Eighteenth Century*, 107–122. Berlin, New York: De Gruyter Mouton.

Hansson, Stina
1982 *Afsatt på Swensko. 1600-talets tryckta översättningslitteratur* [Translated into Swedish. Printed translated literature of the seventeenth century]. (Skrifter utgivna av Litteraturvetenskapliga institutionen vid Göteborgs universitet 5). Gothenburg: Department of Literature, University of Gothenburg.

Helander, Hans
2005 Language contact outside Scandinavia V: Loans from Latin and Greek. In: Bandle, Oskar, Kurt Braunmüller, Ernst Håkon Jahr, Allan Karker, Hans-Peter Naumann, and Ulf Teleman (eds): *The Nordic Languages. An International Handbook of the History of the North Germanic Languages*. Vol. 2: 2086–2095. Berlin, New York: de Gruyter.

Hofmann, J. B. and A. Szantyr
1965 *Lateinische Syntax und Stilistik*. (Handbuch der Altertumswissenschaft, begr. von I. von Müller et al., 2:2.) München: Beck'sche Verlagsbuchhandlung.

Malmgren, Sven-Göran
2000 *Projektet Det svenska ordförrådets utveckling 1800–2000. Utgångspunkter* [The project "The development of the Swedish lexicon 1800–2000". Starting points]. (ORDAT 1.) Gothenburg: Department of Swedish, University of Gothenburg.

Melander, Björn and Björn Olsson (eds.)
2001 *Verklighetens texter. Sjutton fallstudier i svensk sakprosa* [Texts of reality. Seventeen case studies in Swedish non-literary prose]. Lund: Studentlitteratur.

Norri, Juhani
2004 *Entrances and exits in English medical vocabulary, 1400–1550*. In: Taavitsainen & Pahta 2004: 100–143.

Ordbok över svenska språket [The Dictionary of the Swedish Academy] 1–35
1898–2007 Lund: Svenska Akademien. (Electronic version: http://g3.spraakdata.gu.se/saob/)

Pahta, Päivi
2004 *Codeswitching in medieval medical writing*, In: Taavitsainen & Pahta 2004: 73–99.

Schendl, Herbert
2009 *William Harvey's Prelectiones anatomie universalis (1616): Code-switching in Early Modern English lecture notes* [Brno Studies in English 35 (2): 185–198]. Brno.

Taavitsainen, Irma and Päivi Pahta (eds)
2004 *Medical and Scientific Writing in Late Medieval English*. Cambridge: Cambridge University Press,

Teleman, Ulf
2011 The Swedish Academy of Sciences: Language policy and language practice. In: Britt-Louise Gunnarsson (ed.), *Languages of Science in the Eighteenth Century*, 63–87. Berlin, New York: De Gruyter Mouton.

Wollin, Lars
2004 *Inlåning och användning av latinska ord under (yngre) nysvensk tid* [The borrowing and use of Latin words in (Younger) New Swedish time]. (ORDAT 23.) Gothenburg: Department of Swedish, University of Gothenburg.

Science and natural language in the eighteenth century: Buffon and Linnaeus

Richard Sörman

What form of knowledge is accurate knowledge when it comes to delineating the principles of nature? What form of language should be used to describe and to classify the immeasurable variety of plants and animals? Is it necessary and always fruitful to elaborate abstract, conceptual systems of explanation in order to attain knowledge about nature? Or is it sometimes preferable to describe reality as it actually reveals itself and somehow makes sense to us? These are some of the most fundamental questions appearing in the preliminary discourse to the *Histoire naturelle* by Georges Louis Leclerc de Buffon (1708–1788), from 1749, in which the author severely criticizes his colleague and rival Linnaeus for doing natural science in a most unnatural way.

Buffon's critique of Linnaeus can very well be related to the general issue of language and science in the European eighteenth century. Buffon is the most famous eighteenth-century French naturalist, and he was, just like Linnaeus, a great scientific writer, but what is interesting is that he was also explicitly opposed to the way Linnaeus used language in his scientific work to describe objects of nature. There was indeed a scientific controversy between Linnaeus and Buffon; it is quite famous and has been studied previously[1], but as far as I know this controversy has not been examined in the light of the precise question of how language should be used in order to describe and explain nature. In brief, Buffon rejects Linnaeus' system for relating the kingdoms of plants and animals, saying that it is a system of meaningless words and advocating instead the use of a common sense-based language enabling us to describe nature in the way it actually appears to us.[2] I will show in my study that Buffon's opinions on writing largely relate to the aesthetics of French seventeenth-century classi-

1. See, for example, Larson 1967, Sloan 1976 and Le Guyader 1992.
2. Scott Atran has already used this concept of "common sense" to characterize Buffon's scientific method. Claiming for example that common sense was to Buffon a "method of understanding" (Atran 1992: 231), he writes: "Buffon was to become the most persistent and influential opponent of the idea of system. Like Linnaeus, how-

cism and that he expresses, in the midst of the French Enlightenment, what we may very well consider to be an anti-modern view of scientific research. The main conclusion, however, is that Buffon's critique of modern scientists' use of language is far from out of date and that it raises the general and ahistorical question of the effective value of abstract language as a tool for understanding and describing reality.

1. Definitions and descriptions

Buffon published his major work *Histoire naturelle* in 36 volumes between 1749 and 1789. Together with the famous *Encyclopédie* (published from 1751), it is one of the great French scientific monuments of the whole eighteenth century. We will focus here on the way Buffon characterizes his own scientific method in the preliminary discourse published in the first volume from 1749: "Premier discours. De la manière d'étudier et de traiter l'Histoire Naturelle" [First discourse. How to study and to deal with Natural History].

It is striking how much Buffon speaks about Linnaeus in this text and how disdainful he is towards his Swedish colleague. In order to briefly situate Buffon's relationship to Linnaeus in the intellectual and scientific context of eighteenth-century France, we can quote Phillip R. Sloan who writes about the importance of Linnaean botanies in French scientific circles at the time:

> The opening of the controversy between Linnaeus and Buffon dates from early in 1744 when Buffon delivered to a *séance* of the Académie a preliminary version of what was to be the *Premier discours* opening the first volume of the *Histoire naturelle*. By the time of the actual publication of his arguments in 1749 Buffon was taking on a formidable adversary. By this date Linnean systematics were sweeping European and English biological circles, with deep inroads even into French biology, in spite of the powerful influence of Pitton de Tournefort. All Linnaeus' main taxonomic treatises, with the exception of the *Species plantarum*, had been published or were in press by this date, with the *Systema naturae* in its seventh edition. Under the guidance of Bernard de Jussieu, the second edition of Linnaeus' *Genera plantarum* and the fourth edition of the *Systema* had been published at Paris within the decade, giving French equivalents of Linnaeus' Latin names and supplying powerful impetus to the growth of French Linneanism. (Sloan 1976: 358)

---

ever, Buffon did accept certain basic folk biological assumptions of common sense as conditions on any adequate reflection about natural history" (Atran 1992: 231).

Outlining his own scientific methods and his own approach to describing nature and designating its components, Buffon could hardly avoid, in 1749, marking his position on Linnaeus' new taxonomies and on the new tendencies in natural science in general. What is remarkable is that the position of this important and influential eighteenth-century scientific writer essentially consists in saying that the tendency of modern scientists to describe nature on the basis of an abstract system is an inexcusable *mistake*.

The problem, Buffon claims, is that abstract systems have nothing to do with nature since they can refer only to themselves. Taking the example of contemporary botanies, Buffon states more precisely that if modern-day naturalists are unable to give an accurate image of nature this is essentially because they permit themselves to classify plants on the basis of a limited number of criteria. Buffon reminds his readers that a certain botanist (he has not yet named Linnaeus) categorizes plants on the basis of the number of their stamens, and says that this implies the obvious blunder of designating the whole on the basis of the part, that is, letting one single aspect of the object represent its entirety:

> [Les botanistes contemporains font l'erreur de] vouloir juger d'un tout, & de la combinaison de plusieurs touts, par une seule partie : car vouloir juger de la différence des plantes uniquement par celle de leurs feuilles ou de leurs fleurs, c'est comme si on voulait connoître la différence des animaux par la différence de leurs peaux ou par celle des parties de la génération ; & qui ne voit que cette façon de connoître n'est pas une science, & que ce n'est au plus qu'une convention, une langue arbitraire, un moyen de s'entendre, mais dont il ne peut résulter aucune connaissance réelle. (Buffon 1749, 1: 15–16)
>
> [Contemporary botanists make the mistake of] wanting to judge a whole or a combination of several wholes on the basis of a single part. For to desire to discern the differences of plants only by the difference of their leaves or their flowers is as if one set out to discern the differences of animals by means of the variations in their skins or generative organs; and who does not see that this way of knowing is not a science, and that it is at the very most only a convention, an arbitrary language, a means of mutual understanding, and that no real knowledge of things can result from it.

The use of a limited number of criteria to describe nature, or the principles of nature, thus implies for Buffon a disjunction between representation and reality. We must use language in order to represent nature, he says, and it actually seems to be a matter for Buffon of how to present nature to his readers, that is, in concrete terms, how to make it *present* to his readers. Recent botanical theories cannot be regarded as botanies since they are nothing but an arbitrary language of botanies, a language enabling specialists to communicate with each other but not to actually refer to the realities of nature. And the most absurd of all botanical systems is the one elaborated by Linnaeus.

According to this system, Buffon argues ironically, a human being must be considered unable to recognize a plant or a tree unless he is armed with a microscope. He must likewise be considered unable to understand for example what flower he is looking at unless he can perceive the number of its stamens:

> […] il faut aller le microscope à la main, pour reconnoître un arbre ou une plante ; la grandeur, la figure, le port extérieur, les feuilles, toutes les parties apparentes ne servent plus à rien, il n'y a que les étamines, & si l'on ne peut pas voir les étamines, on ne sçait rien, on n'a rien vû. Ce grand arbre que vous apercevez, n'est peut-être qu'une pimprenelle, il faut compter les étamines pour sçavoir ce que c'est, & comme ces étamines sont souvent si petites qu'elles échappent à l'œil simple ou à la loupe, il faut un microscope ; mais malheureusement encore pour le système, il y a des plantes qui n'ont point d'étamines, il y a des plantes dont le nombre des étamines varie & voilà la méthode en défaut comme les autres, malgré la loupe & le microscope[3]. (Buffon 1749, 1: 19–20)

> […] it becomes necessary to go with a microscope in one's hand to recognize a tree or a plant. The size, the form, the external appearance, the leaves, all the obvious features are useless. Nothing is important except the stamens, and if one cannot see the stamens, one does not know anything, one has not seen anything. This large tree which you perceive is perhaps only a bloodwort, it is necessary to count its stamens in order to know what it is, and since its stamens are often so small that they escape the naked eye or the magnifying glass, one must have a microscope. But unfortunately for the system there are plants which do not have stamens; there are plants in which the number of stamens varies, and therein lies the shortcoming of this method as in the others, in spite of the magnifying glass and the microscope.

Buffon consequently claims that Linnaeus' system is counterproductive as a scientific tool since it does not help us to understand nature, but only makes it more difficult for us to know what we are really looking at. Linnaeus, Buffon says implicitly, uses language to *represent* nature in the sense of "standing in

3. Buffon gives a note at the end of this passage in which he quotes Johann Georg Siegesbeck, one of the most active opponents of Linnaeus, who had written in 1741: "Hoc vero systema, Linnaei scilicet, jam cognitis plantarum methodis longè viliùs & inferiùs non solum, sed & insuper nimis coactum, lubricum & fallax, imò lusorium deprehenderim" [I have found this system, Linnaeus's that is, to be not only most contemptible and inferior to previous methods of describing plants but also too strained, uncertain and fallacious and I would say even insignificant]. Buffon refers to "Vaniloq. Botan. Specimen refutatum a Siegelbeck. Petropoli 1741". This is the only explicit reference to Linnaeus in Buffon's text.

the place of nature" or "supplying nature's place", when he should use it in the sense of "bringing nature into presence".

It is important to understand that the starting point of Buffon's discussion on nature and scientific language is a profound conviction that nature cannot be an object of absolute knowledge. The very first phrase of the work stresses the point that every naturalist has an unlimited body of material to consider:

> L'Histoire Naturelle prise dans toute son étendue, est une Histoire immense, elle embrasse tous les objets que nous présente l'Univers. Cette multitude prodigieuse de Quadrupèdes, d'Oiseaux, de Poissons, d'Insectes, de Plantes, de Minéraux, &c offre à la curiosité de l'esprit humain un vaste spectacle, dont l'ensemble est si grand, qu'il paroît & qu'il est en effet inépuisable dans les détails. (Buffon, 1749, 1: 3)

> Natural history, taken in its fullest extent, is an immense history. It embraces all objects which the universe displays to us. This prodigious multitude of quadrupeds, birds, fishes, insects, plants, minerals, etc., offers to the curiosity of the human mind a vast spectacle, the totality of which is so grand that it appears, and indeed is, inexhaustible in its details.

This is why no scientific system in the world can thoroughly describe and explain all of nature's manifestations. What we can do is to give a picture of nature as it appears to us and to draw reasonable conclusions about the way it is organized. Buffon writes subsequently that scientists should devote themselves to characterizing the objects of nature and not to defining or classifying them:

> Les choses par rapport à nous ne sont rien en elles-mêmes, elles ne sont encore rien lorsqu'elles ont un nom, mais elles commencent à exister pour nous lorsque nous leur connoissons des rapports, des propriétés ; ce n'est même que par ces rapports que nous pouvons leur donner une définition : or la définition telle qu'on peut la faire par une phrase, n'est encore qu'une représentation très-imparfaite de la chose, & nous ne pouvons jamais bien définir une chose sans la décrire exactement. C'est cette difficulté de faire une bonne définition, que l'on retrouve à tout moment dans les méthodes, dans les abrégés qu'on a tâché de faire pour soulager la mémoire ; aussi doit-on dire que dans les choses naturelles il n'y a rien de bien défini que ce qui est exactement décrit : or pour décrire exactement, il faut avoir vû, revû, examiné, comparé la chose qu'on veut décrire, & tout cela sans préjugé, sans idée de système, sans quoi la description n'a plus le caractère de la vérité, qui est le seul qu'elle puisse comporter. (Buffon 1749, 1: 25)

> Things in relation to us are nothing in themselves; they are still nothing when they have a name, but they begin to exist for us when we become acquainted with their relations to each other and their properties; it is even only by these relations that we can give them a definition. Now, a definition such as we can construct by a phrase is still no more than a very imperfect representation of the thing, and

> we are never able adequately to define a thing without describing it exactly. It is this difficulty of forming an adequate definition that is found constantly in all systems; in all the epitomes which have been attempted in order to relieve the burden of the memory. It must also be said that in natural things nothing is well-defined but that which is exactly described. Now, in order to describe exactly, it is necessary to have seen, reviewed, examined, and compared the thing which one wishes to describe; and it is necessary to do all this without prejudging things and without an eye to systematization. Otherwise the description would not have the character of truth, which is the only characteristic it can contain.

In order to attain and to communicate knowledge about nature we must be able to describe nature in the way we actually perceive it. As we can see from the quote above, a complete description of an object (as opposed to an abstract definition) is according to Buffon a description of its relations ("rapports") and its properties ("propriétés"). That is, we have to compare objects as they appear in their similarities and differences, as well as depicting them as they appear in themselves. Buffon thus says that we can never re-present, or render present, objects of nature by using language otherwise than to thoroughly describe those objects. For Buffon it essentially seems to be a question of being truthful to nature and truthful to the way nature reveals itself to us. Definitions are nothing but words and do not procure real knowledge.

2. Classical French aesthetics

In the paragraph we have just read, Buffon also comments on the *style* of description:

> Le style même de la description doit être simple, net & mesuré, il n'est pas susceptible d'élévation, d'agrémens, encore moins d'écarts, de plaisanterie ou d'équivoque ; le seul ornement qu'on puisse lui donner, c'est de la noblesse dans l'expression, du choix & de la propriété dans les termes. (Buffon 1749, 1: 25)
>
> The very style of the description should be simple, clear, and measured. The nature of the enterprise does not allow of grandeur of style, of charm, even less of digressions, pleasantries, or equivocation. The only adornment one can give it is nobility of expression, of choice, and of propriety in the use of terms.

Correctness and simplicity are thus important to Buffon, and this because scientific language should never be anything more than a medium for truthful and instructive communication of knowledge and truth. Let us not forget that when Buffon was received into the French Academy in 1753, mainly thanks to the beautiful French he had written in the first volumes of his *Histoire naturelle*, he delivered a speech known as *Discours sur le style* [Discourse on style], in which he set out his conception of style and artistic beauty.

Having previously worked a great deal on seventeenth-century French literature I find it obvious that Buffon's opinions on writing relate to a great extent to the aesthetics of French seventeenth-century classicism. Let us take for example the great importance Buffon attaches to premeditation and control. A real writer, he claims, must perfectly know his subject; he must spend a great deal of time considering its different features, and he must necessarily advance in his work according to a rational plan. Buffon asks his readers why the creations of nature are always so perfect. The answer is that nature in all its creations works according to a plan from which it never deviates. It is likewise a good plan, as well as ambitious preparation, that liberates the creativity of a writer and ensures the quality of his style:

> C'est faute de plan, c'est pour n'avoir pas assez réfléchi sur son objet qu'un homme d'esprit se trouve embarrassé, et ne sait par où commencer à écrire. Il aperçoit à la fois un grand nombre d'idées ; et, comme il ne les a ni comparées ni subordonnées, rien ne le détermine à préférer les unes aux autres ; il demeure donc dans la perplexité. Mais lorsqu'il se sera fait un plan, lorsqu'une fois il aura rassemblé et mis en ordre toutes les pensées essentielles à son sujet, il s'apercevra aisément de l'instant auquel il doit prendre la plume, il sentira le point de maturité de la production de l'esprit, il sera pressé de la faire éclore, il n'aura même que du plaisir à écrire : les idées se succéderont aisément, et le style sera naturel et facile [...]. (Buffon, 1926: 13)
>
> It is due to the lack of a plan, it is for not having sufficiently reflected on his object that a thinking man finds himself perplexed, and does not know where to start to write. He perceives simultaneously a great number of ideas; and since he has not compared nor subordinated them, nothing compels him to prefer the one to the other, and so he remains in perplexity. But when he will have made himself a plan, when he will have assembled and ordered all his thoughts essential to his subject, he will easily distinguish the moment at which he should take up his pen, he will feel the point of maturity of the mind's production, he will be in a hurry to make it open, he will even find it a pure pleasure to write. The ideas will easily succeed one another, and the style will be natural and easy [...].

Buffon's advice on writing can very well be compared to what Nicolas Boileau had written in his classical literary manifesto *L'Art poétique* from 1673:

> Avant donc que d'écrire apprenez à penser.
> Selon que notre idée est plus ou moins obscure,
> L'expression la suit, ou moins nette ou plus pure.

> Ce que l'on conçoit bien s'énonce clairement,
> Et les mots pour le dire arrivent aisément. (Boileau 1960: 69)[4]

Buffon thus agrees with Boileau that a writer is more effective in his work if he is mentally in full control of what he wants to communicate. If such considerations can be so important in a discussion of the issue of style, it is because a scientific writer should have no other concern than to truthfully render his vision of reality. "Good style" can be understood as correct use of language, and there is consequently no better style in scientific writing, according to Buffon, than a use of words that permits us to perceive reality as it is.

Actually, if it is relevant to speak about seventeenth-century literary aesthetics when reading Buffon, it is primarily because he argues that style and beauty in literature (including scientific literature) are the result of a close connection between content and expression. "Bien écrire," he writes, "c'est tout à la fois bien penser, bien sentir et bien rendre" [To write well is at the same time to think well, to feel well and to render well] (Buffon, 1926: 15). This means that we not only have to think carefully before writing, but must also make every possible effort to translate, that is to express with exactitude, what we have thought. Buffon argues eloquently that concrete writing should always consist in letting the pen travel through ("parcourir") different spaces of premeditated ideas:

> Pour bien écrire, il faut donc posséder pleinement son sujet, il faut y réfléchir assez pour voir clairement l'ordre de ses pensées, et en former une suite, une chaîne continue, dont chaque point représente une idée; et lorsqu'on aura pris la plume, il faudra la conduire successivement sur ce premier trait, sans lui permettre de s'en écarter, sans l'appuyer trop inégalement, sans lui donner d'autre mouvement que celui qui sera déterminé par l'espace qu'elle doit parcourir. (Buffon 1926: 15)

> In order to write well, one must hence possess one's subject completely; it is necessary to reflect upon it enough to see the order of one's thoughts, and to form a sequence, a continued chain of which every point represents an idea. And when one will have taken up the pen, it will be necessary to conduct it successively on this first draught, without permitting it to deviate, without laying an unequal stress upon it, without giving it another movement but the one determined by the space through which it has to pass.

This is what real style in scientific literature is all about, Buffon says, and this closeness of fit between thought and expression will make our style precise, simple, clear and sustained.

In the same way, the so-called "tone" is according to Buffon nothing but conformity between the way we write and the subject we write about: "Le ton

---

4. The quote is taken from the first part, verses 150–154.

n'est que la convenance du style à la nature du sujet, il ne doit jamais être forcé; il naîtra naturellement du fond même de la chose" [The tone is nothing but the conformity of style to the nature of the subject, it should never be forced; it will emerge naturally from the very bottom of the thing] (Buffon 1926: 16). Expression should in other words always be adapted to content, and good style and real artistic beauty will emerge only when harmony reigns between the two. Referring to a typical French seventeenth-century idea of a close connection between truth and beauty, Buffon also writes:

> [...] il n'y a que la vérité qui soit durable, et même éternelle. Or un beau style n'est tel en effet que par le nombre infini des vérités qu'il présente. Toutes les beautés intellectuelles qui s'y trouvent, tous les rapports dont il est composé, sont autant de vérités aussi utiles, et peut-être plus précieuses pour l'esprit humain que ceux qui peuvent faire le fond du sujet. (Buffon 1926: 16–17)
>
> Only truth is stable, and even eternal. But a beautiful style is such only by the infinite number of truths it represents. All the intellectual beauties which it embodies, all the relations of which it is composed, are as many beauties just as useful as, and maybe even more precious to the human mind than those which make up the subject-matter.

Only what is true can possibly be beautiful, and the task of every real artistic creator is to describe nature without altering it. "Ce style figuré, dont on fait vanité, /Sort du bon caractère, et de la vérité; / Ce n'est que jeux de mots, qu'affectation pure / Et ce n'est pas ainsi que parle la nature" (Molière 1971, 2: 158)[5] says the misanthrope in Molière's famous play from 1666 about what he considers to be a bad poem. The classical idea of *vraisemblance* ("likelihood" rather than "probability") is essential here: a work of art must resemble to nature, it must be an expression of nature, and it must not deviate from our natural perception of nature's appearance.

It might seem obvious that the task of every scientific work is to give an accurate image of the objects it is meant to describe. The real problem here is not the general question of accuracy between scientific language and nature, but the more precise one of accuracy between scientific language and our conventional or common sense-based conception of nature. Linnaeus' mistake, according to Buffon, is to present nature in his scientific work in a way that strongly deviates from the way nature presents itself to us in direct reality.

It must be remembered here that French classical writers were strongly attached to the idea that common sense ("le bon sens" or "le sens commun") is the most effective instrument for perceiving and understanding reality. In *La*

5. Act I, scene 2, verses 385–388.

Critique de L'École des femmes from 1663, Molière has one of his characters say: "Je regarde les choses du côté qu'on me les montre, et ne les tourne point pour y chercher ce qu'il ne faut point voir" (Molière 1971, 2: 648)[6]. This phrase can be seen as a recapitulation of some very important aspects of the method advocated by Buffon. If we want to be able to distinguish the significant properties of things and to appreciate their real value, we must look at them, Buffon says, as they are shown to us. What is important is for example – as we have already seen – that a scientist should observe the objects of nature in their integrity and not focus on a limited number of aspects of their being. If we call attention to only one side of an object we will inevitably fail to give an exact and comprehensive account of reality:

> [...] l'on ne présente qu'un côté de l'objet, on met dans l'ombre toutes les autres faces ; et ordinairement ce côté qu'on choisit est une pointe, un angle sur lequel on fait jouer l'esprit avec d'autant plus de facilité qu'on s'éloigne davantage des grandes faces sous lesquelles le bon sens a coutume de voir les choses. (Buffon 1923: 14)
>
> One presents only one side of the object, one puts all the other aspects in the shade; and ordinarily the side which one chooses is a point, an angle on which one lets the mind play with all the more ease as one deviates from the large aspects in which common sense usually look at things.

The use of common sense, that is the use of a conventional way of understanding and judging reality, permits us to get the whole picture of nature rather than only a limited aspect of it.

3. Giving meaning to nature

This idea of a common sense-based description of nature also emerges in Buffon's conviction that all elements of nature should be considered in their relations to man. When he describes the kingdom of animals, for example, he finds it "natural" in the sense of "conformable to nature" to start with the animals that man has domesticated. This is because man is the centre of creation. In Buffon's vision of nature there is no kinship between man and animals. Man was created to rule the universe and the natural history of mankind is to Buffon the history of how man became the master of nature. "Ce qui est en jeu, en relation avec le projet et l'ordre de l'*Histoire naturelle*, c'est moins la place de l'Homme dans la nature que sa position et son statut par rapport aux autres espèces vivantes"

6. Scene 3.

[What is at stake, in relation to the project and the order of l'*Histoire naturelle*, is less the place of Man in nature than his position and status in relation to other living species] (Tinland 1992: 543). And since man is the centre of creation and scientific knowledge is human knowledge, it must be considered legitimate to describe nature not only as it appears to man, but also as man relates to it. Any other depiction of nature would be inaccurate and scientifically unjustified. To describe nature in accordance with the principles of common sense is thus at the same time to describe nature in the way it makes sense to us, that is, in the way we give meaning to its objects in their relationships to us.

The importance given by Buffon in his work to common sense and thereby to meaning and meaningfulness may surprise us. We are dealing with one of the great French eighteenth-century scientific writers, and we are used to thinking of the French eighteenth century as a period when new empirical scientific ideals were emerging and when ancient forms of rationalist, metaphysical and meaning-oriented thinking were being severely questioned. It is a generally accepted idea today that the eighteenth century marks a breakthrough for modern Western society, and that modernism has made us more practical and pragmatic than before. The problem is, however, that modernism has also made it more difficult for us to claim the possibility of a meaning-oriented description of reality. Modern man can use reality, he can describe it in terms of precise knowledge, but he cannot assign it an inherent meaning.

It becomes clear that, in this perspective, Buffon can be understood as a sort of opponent of modernism. As we have seen, Buffon is an eighteenth-century scientist who rejects abstract systems of classification based on a limited number of criteria. He is a scientific writer who prefers characterization to precise definition, who believes the task of language is to reproduce the entirety of the reality it designates and not to provide designations for abstract concepts. He is, moreover, a scientific writer who relates to classical aesthetics, according to which a writer should reproduce nature as man naturally perceives it and according to which common sense is the most effective tool to gain and communicate knowledge about nature. But Buffon is also anti-modern in the sense that he refuses to abandon a meaning-oriented explanation of nature, in which animals and plants are described through their signifying differences, but also through a historical evolution having resulted in the nature we can observe today. In Buffon's world, meaning is thus produced thanks to synchronic and diachronic systems of classification whose terms can always be related to the place man himself occupies in the space and time of reality.

Objects of nature are in other words described as meaningful in Buffon's system. We read earlier in a quotation from the preliminary discourse of the *Histoire naturelle* that we can perceive the existence of things only when we distinguish their relations and properties ("Les choses par rapport à nous ne sont rien en elles-mêmes, elles ne sont encore rien lorsqu'elles ont un nom, mais elles commencent à exister pour nous lorsque nous leur connoissons des rapports, des propriétés" [Buffon 1749, 1: 25]), and we said that Buffon accuses Linnaeus of providing definitions instead of knowledge, and that he wants scientists to describe the elements of nature and not to name them, since names never can take the place of true representations. Let us now add that Buffon implicitly also accuses Linnaeus of doing *meaningless* science, since the latter refuses to describe elements of nature as they appear to us in relation to each other. Nature should according to Buffon be described as an organic system where everything can be related to something else and where, in one way or another, all elements can finally be related to man and to his existence in the world.

4. Conclusions

To conclude, I would say that we are dealing here with an opposition between two opposite conceptions of scientific method and language. On the one hand we have what we may regard today as real science, involving the establishment of and reference to a precise system of explanation, use of a specialized (and hence limited) perspective, and the use of a specific (and often abstract) and partially invented terminology. On the other hand we have a common sense-based method and use of language, characterized by a rejection of limited systems, a belief in a comprehensive description of reality, and the idea that abstract systems are meaningless and even ridiculous. Perhaps we believe too easily today that we all automatically agree with Linnaeus in saying that science should of course abandon common sense and that it should of course use precise and abstract language. The fact is, I would say, that what Buffon discusses in his preliminary discourse is still a burning question, and especially within the field of the humanities.

We can for example ask ourselves to what degree it is generally accepted to speak about literature with the help of abstract systems. Is it really always well accepted to speak about human action, in real life or in literature, with the help of specialized theories such as the sociology of Pierre Bourdieu[7] or the psychoanalysis of Jacques Lacan[8]? If a doctoral student today considers it relevant to exploit these kinds of theories when reading and explaining fictional literature, he or she will inevitably at some point in his or her work be accused by someone

7. See for example Bourdieu 1979 and Bourdieu 1998.
8. See Lacan 1966.

of doing exactly what Buffon accuses Linnaeus of doing, that is, simplifying a complex reality, using terms that have no meaning, forgetting reality itself and producing a purely artificial work that will soon lose its interest, since its theories will be replaced by others. And the alternative way of working is always the one advocated by Buffon, that is, to describe reality as it simply and naturally appears to us, to use a common sense-based language in our analyses, and of course to focus on the aesthetic aspects of our own presentation.

My conclusion, then, is that although we may believe that we always defend modern science and that we consequently embrace Linnaeus' position against Buffon's, practically it is often, if anything, the other way round. I am not saying that we should do one or the other (we should probably do both), I am only saying that it is often difficult, and that it takes a lot of courage – and probably a good theory – to abandon common sense-based rationality in favour of pure scientific discourse. The scientific controversy between Buffon and Linnaeus, which took place in the middle of the eighteenth century when abstract science had just emerged, may very well be of little interest to modern natural science (although a French biologist like Hervé Le Guyarder has claimed the opposite[9]), but it is still highly relevant in the humanities. We are still hesitant in the face of modern science's demand for precision and specialization, and we still often resist giving up the idea of a synthetic, common sense-based depiction of reality as it appears to us in its entirety.

And perhaps we are right in not giving up the ideal of a "natural" method of describing reality. In his interesting article "Linnaeus and the Natural Method", James L. Larson remarks that Linnaeus was very well aware of the artificiality of his sexual system and that he wished to supplement his purely scientific system with a system based on some sort of more "natural" method: "Linnaeus' reflections on the natural method [...] form a complement to his sexual system of classification. Aware of the difference between a practically commodious system and a method consonant with affinities observable in nature, Linnaeus conceived a natural method which would represent all natural affinities fundamental to botany" (Larson 1967: 312). This natural method was outlined by

9. "Les Bases scientifiques de son [de Buffon] antagonisme avec Linné, paraissent scander l'histoire de la biologie, des balbutiements de l'anatomie comparée jusqu'à l'actuelle anatomie moléculaire. Elles correspondent donc, en biologie, à un des invariants de la pensée conceptuelle qui réapparaissent quels que soient les hommes et les techniques." [The scientific bases of his antagonism towards Linnaeus seem to accentuate the history of biology, from the early days of comparative anatomy all the way to current molecular anatomy. They thus correspond, in biology, to one of the invariants of conceptual thought that re-emerge independently of men and techniques] (Le Guyader 1992: 491).

Linnaeus in his *Fragmenta methodi naturalis*, first published in *Classes plantarum* in 1738 and later on in a revised version in *Philosophia botanica* in 1751. Larson writes: "The groups [of criteria for distinction of plants] listed in the *Fragmenta methodi naturalis* are based, not upon the reproductive function, but upon the general outward appearance of plants [...]. By means of such general resemblances plants could be grouped, as Linnaeus admitted, without injustice to actual, observable affinities" (Larson 1967: 313). According to Larson, however, Linnaeus failed to invent a natural method because of his inability to see other definitive characters in plants than their parts of fructification: "The artificial sexual system of classification, which satisfied limited professional requirements, could not attain any significant contact with knowledge of natural objects. [...] Linnaeus' arbitrary limitation of definitive characters to the parts of fructification made it impossible for him to discover any grouping of genera less artificial than the system of classification based upon stamens and pistils" (Larson 1967: 320).

This is a perspective that to a certain extent justifies the critique formulated by Buffon and that permits us to consider the scientific achievements of Linnaeus and other eighteenth-century scientists in a somewhat different light. The invention of the sexual system was one of many significant steps made in the eighteenth century towards modern science and towards the use of abstract scientific language, but it may also be regarded as one of many examples of the *alienation* which modern science implies from a sometimes more spontaneous and thus "natural" perception of nature.

References

Atran, Scott
1992 The common sense basis of Buffon's "*méthode naturelle*". In Jean Gayon (ed.), *Buffon 88 Actes du Colloque international pour le bicentenaire de la mort de Buffon*, 223–240. Paris: Vrin.

Bourdieu, Pierre
1979 *La Distinction, critique sociale du jugement*. Paris: Minuit.

Bourdieu, Pierre
1998 *Les Règles de l'art, genèse et structure du champ littéraire*. Paris: Le Seuil. First edition 1992.

Buffon, Georges Louis Leclerc de
1749–1789 *Histoire naturelle, générale et particulière avec la description du Cabinet du Roy*, Paris: Imprimerie nationale, 36 volumes.

Buffon, Georges Louis Leclerc de
1926 *Discours sur le style*, Paris: Société d'édition "Les Belles Lettres". First published Paris: Pierre Brunet [1753].

Boileau, Nicolas
1960 *Le Lutrin et L'Art poétique.* Paris: Larousse. First published in *Œuvres diverses*, Paris: Bilaine [1674].

Lacan, Jacques
1966 *Écrits*. Paris: Le Seuil.

Larson, J. L.
1967 Linnaeus and the natural method, *Isis* 58: 304–320.

Le Guyader, Hervé
1992 Linné contre Buffon: une reformulation du débat structure-fonction. In Jean Gayon (ed.), *Buffon 88 Actes du Colloque international pour le bicentenaire de la mort de Buffon*, 491–501. Paris: Vrin.

Molière (Jean-Baptiste Poquelin)
1971 *Œuvres completes*, Paris: Gallimard, 2 volumes.

Sloan, P. R.
1976 The Buffon-Linnaeus controversy, *Isis* 67: 356–375.

Tinland, Franck
1992 Les Limites de l'animalité et de l'humanité selon Buffon et leur pertinence pour l'anthropologie contemporaine. In Jean Gayon (ed.), Buffon 88 *Actes du Colloque international pour le bicentenaire de la mort de Buffon*, 543–555. Paris: Vrin.

From theory of ideas to theory of succedaneum: The Linnaean botanical nomenclature(s) as "a point of view on the world"

Philippe Selosse

1. Introduction

Botanical nomenclature did not arise by spontaneous generation: its elaboration is the continuation of the Greek and Latin folk nomenclature, which was based on processes of classification and categorization called folk processes (Berlin 1972), and of the logical nomenclature of the sixteenth century (Selosse 2004, 2007). On the whole, the elaboration of nomenclature took a great deal of time, more or less twenty centuries. The consequence is that nomenclature is not the fact of one speaker, but the product of many speakers over a very long period of time, connected with several epistemic contexts. Moreover, these contexts are characterized by non-specialization with regard to disciplines: philosophy, language and botanical research were simultaneously considered.

The aim of this paper, therefore, is not to represent Linnaeus as a pioneer or a maker of binominal nomenclature out of nothing. On the one hand, Linnaeus appears, not as person, but through his works, as the representative, the paragon, of the turning point in another epistemic context: his work constitutes a genuine synthesis of earlier theories and those that followed. On the other hand, the Linnaean nomenclature is, for the first time, one which is self-conscious, that is, which is conscious of being a system as such: in this scientific consciousness we can paradigmatically observe the articulation of plant classification in nomenclature, and vice versa. My aim is to draw from Linnaeus' works his theory of nomenclature within the epistemic context of the Enlightenment, that is, by taking into consideration its own complexity of blended scientific, philosophical and linguistic conceptualizations. My hypothesis depends on the use of the Latin word "succedaneum", which is almost synonymous with "substitute": a succedaneum is a thing taking the place of another. The word "succedaneum" was used in particular in pharmacopoeias, where it applied to a drug substituted for another because they shared common properties, even if the second was more complete, appropriate and natural than the first. "Succedaneum"[1] frequently occurs in Linnaeus' hundreds of aphorisms, not with medicinal but with conceptual applications. In my view, the concept of succedaneum allows us to understand the unity of these hundreds of aphorisms[2], that is, how the various taxonomic and linguistic concepts are conceived in a homogeneous epistemic frame.

2. The theory of ideas

The period of the Enlightenment is characterized by a semiotic theory, later called the theory of ideas (Auroux 1979), and developed by many philosophers, such as Locke, Leibniz and the Encyclopedists (Beauzée, Rousseau). According to this nominalist theory of representation, we, unlike God, have no direct perception of a thing: we merely perceive the idea of a thing, an objective but incomplete idea, which is given by the name.[3] The nomenclature also appears as a representation of the "nominal essences" (Locke 1755), conceived in the mind but founded in reality. One of the first and main Linnaean aphorisms refers to this theory: "Knowledge consists in having a *true idea* of things"[4]. In this epistemic frame, Linnaeus considers the idea as a go-between, and that is why he says that "plant and name are two ideas"[5]: the name simply refers to an idea, which is related to another idea, the idea of the plant. Hence, the name does not mean – but refers to – the essence, conceived as the idea of a character connected to its own idea (see Figure 1).

Name > Idea $_2$ > Idea $_1$ > Character > Plant

Figure 1. The theory of ideas

In this way, it is permissible for a generic name to be drawn out from a proper name: a proper name does not signify the essence of the plant, it is just the idea

1. I keep the Latin word in my paper instead of an English word such as substitute, for it was often used in English from the beginning of the seventeenth century, especially in medicine and botany, with the meaning previously defined.
2. To understand the unity of these aphorisms in another perspective, see Han-Liang Chang's article in this volume (section 5.2).
3. Cf. Linnaeus' poem *The Real*, translated into English by Pr. Pettersson: "all that we see are shadows of reality".
4. "Notitia consistit in vera idea objectorum" (Linnaeus 1735: § 10; italics mine). In this paper, all Linnaean quotations are given in my own translation.
5. "Nomen et planta sunt duae ideae" (Linnaeus 1737: § 238).

of a man, a botanist or a king linked by convention to the idea of the character of the plant. That is to say, the arbitrary human idea carried by the name is a substitute for (or succedaneum to) the natural idea suggested by the plant.

This theory of ideas explains one aim of the *Species Plantarum*, which is "to cover all the peculiar *Ideas* and to distinguish them by distinct *Names*, so that we can KNOW God's works"[6]. These ideas, created by God (Linnaeus 1735: § 14; Linnaeus 1736b: § 157) and referred to by distinct names, are two sorts of entities: species and genus. Being divine, their essences are perceived by God, but not by botanists, who only can distinguish or discriminate between them. Hence, species and genus are simultaneously natural (they are created by God) and artificial (they are represented through human perception).

For all these reasons, the plant in itself cannot be known. At worst, a botanist discriminates a plant only thanks to a "factitious", i.e. practical and artificial, character (Linnaeus 1751: § 188); at best, he might manage to discriminate the "essential" character of the plant (Linnaeus 1751: § 187). But in general, he only perceives, simultaneously and indiscriminately, the essential and accidental properties of a plant, that is, a blending of factitious and essential characters, named "natural character" (Linnaeus 1751: § 189). To sum up, "the Factitious Character is the succedaneum"[7] to the natural character, which is itself the succedaneum to the essential character.

3. The concept of "disposition"

Linnaeus does not use the term "classification". He speaks very often of "disposition", that is, the manner of ordering the several specific and generic ideas. According to his concepts of species, genus and character, he distinguishes three sorts of dispositions[8]:

– To the worst degree, there is the "synoptic disposition" (Linnaeus 1736b/1751: § 154). This is a method based on a descending practice of logical division. Logical division is an Aristotelian metaphysical method to define the essence, in a dichotomous way: something is defined by a genus and its difference, the whole (genus and difference) being a species, which becomes a genus at the following step of division of the definition, and so on, down to

6. "COGNITIONEM horum ut rite acquiramus, singula distincta *Idea* & distincto *Nomine* complecti oportet" (Linnaeus 1753: *Lectori Aequo*; italics and block capitals in the original).
7. "Character Factitius succedaneus est" (Linnaeus 1751: § 190).
8. Cf. Richard Sörman's paper in this volume (section 4).

the ultimate species, which involves essence and individuals. As an example (see Figure 2, after Bauhin 1623: 187), we can see how this division is applied to the genus *Gentiana* in the sixteenth century, so that the different species of Gentian are all given by a dichotomous or trichotomous division. At the first step of this division, three species are defined by differences of habitat: the Alpine Gentian, the Meadow Gentian and the Marsh Gentian. Inasmuch as it is itself divisible, the species Alpine Gentian is given as a genus containing two species, defined by differences of height: the Greater Alpine Gentian and the Lesser Alpine Gentian. At the ultimate step of division, the species Greater Alpine Gentian becomes a genus, containing in its turn two species, defined by differences of colour: the Yellow Greater Alpine Gentian and the Purple Greater Alpine Gentian. This kind of disposition is conceived by Linnaeus as arbitrary (to proceed with the logical division, it was necessary only to choose a good criterion, whatever it might be: habitat, height, colour, frequency…) and artificial (it is a *logical* method applied to *natural* beings), and as pertaining to *practical* knowledge (Linnaeus 1751: § 152).

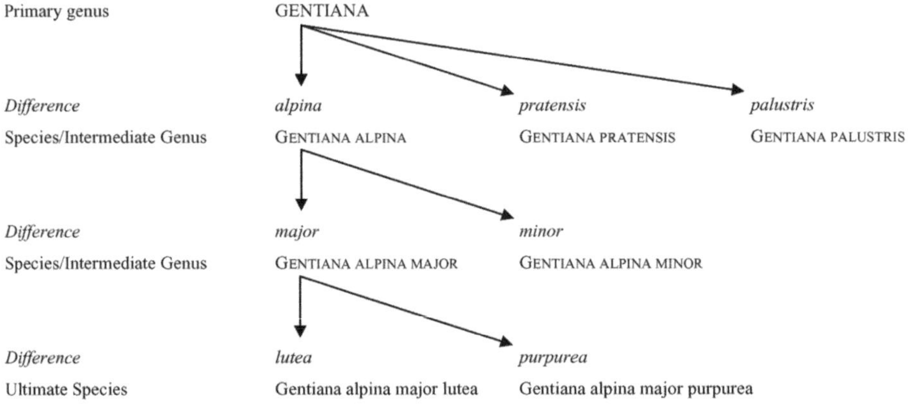

Figure 2. Logical division applied to the genus *Gentiana* (after Bauhin 1623: 187)

– To a better degree, there is the "system" or "artificial method" (Linnaeus 1736b: § 155–156). This is a method based on a descending practice (often illustrated by keys), which is characterized by using the same single principle for all the plants (Lamarck 1792: 301). The Linnaean sexual system is an example of this sort of disposition: it is based on the single principle of the physiological importance of plant sexuality, considered from various points of view (place, figure, proportion and number of the plant reproductive organs). This disposition is conceived as artificial (it is a rational human method applied to natural beings created by God) but not arbitrary (the importance of plant sexuality is founded in reality and is not the result of a practical choice), and as pertaining to *theoretical* knowledge (Linnaeus 1751: § 152) – theoretical, because it is not the world but a representation of the world.

– Finally, to the best degree, there is the "natural method" (Linnaeus 1751: § 77). This is an ascending method of reassembling plants into groups, on the basis of several principles. The choice of principles is determined by the aim of not separating the natural classes. "Prescribed by nature itself"[9], this method is natural and is the world itself.

Practical knowledge	**Theoretical knowledge**	**World**
IDEAS		OBJECTS
	Species > Genus > Order > Class	Natural Orders
Synoptic disposition	*Artificial methodical disposition*	*Natural methodical disposition*
SYNOPSIS	SYSTEM	METHOD

Figure 3. The theory of "succedaneum" applied to the concept of "disposition"

To sum up (see Figure 3), the concept of disposition is structured by what we can now call the theory of succedaneum. The synoptic disposition, being arbitrary, is a substitute for (succedaneum to) the system; and the system itself, being artificial, is a substitute for (succedaneum to) the natural method (Linnaeus 1738: *Praefatio*).

But botanists use various criteria, i.e. various parts of plants, to classify synoptically or systematically, so that there are as many systems as there are botanists (Cesalpino, Ray, Knaut, Hermann, Boerhaave, Tournefort, Dillen, Micheli, Haller, Linnaeus, etc.). So many, that the question appears to be: how could the unity of Creation be revealed in such a multiplicity of systems? In the epistemic frame of the "System of Nature", inherited from Leibniz (1994), just as the universe is composed to infinity of monads which separately show, as Leibniz says, "a point of view" on the world and whose "summation" constitutes the world (Leibniz 1991: § 57), likewise nature is represented in its totality by the different botanical systems which are so many "points of view" (Leibniz 1990: III, VI, § 13). Every system used by a botanist (Cesalpino, Tournefort, Haller, Linnaeus, etc.) reveals the same reality in terms of essence, but differently in terms of representation and structuration (one system, one principle). Hence, at this moment, the order of Nature is given only by the summation of

9. "C'est la nature qui prescrit ici" (Adanson 1763: xciv).

the several systems, that is, every system is artificial, because it is partial, but essential, because it answers for a view on the world, a view which God previously makes firm in its objectivity in the botanist's theory. For Leibniz, only God has a comprehensive view of the universe. In the same way, for Linnaeus, only a botanist, compared to Apollo, that is, as having a divine essence, could have such a whole view of Nature (Linnaeus 1738: 486, § 4). As for Leibniz, the more variety there is, the more reasons we have to revere God's splendour – hence the variety of systems. But behind this variety there is a crucial principle, which re-establishes the unity of Creation: the Leibnizian principle of continuity, inherited from Ray (1682). Every species is close to another, there is always a species which is closer to another, and so to infinity: there is no empty space, "Nature does not take a leap"[10].

Because of this principle of continuity as well, inquiry into the natural order – called the natural method in opposition to the artificial method (the system) – appears to be the asymptotic aim of research: "it is an eternal law that the artificial method is only a succedaneum to the natural method"[11]. Every system comes nearer and nearer to reality, gives a clearer perception of it, but remains a succedaneum, that is, as in Baconian inductivism, a luminous, illuminating hypothesis of another more heuristic system; there is always another system which gives a clearer perception of reality – and so on, until the natural method is reached. The succedaneum theory explains the degree of truth contained within the Linnaean sexual system: this system is the result of all the former systems of the sixteenth and seventeenth centuries (Linnaeus 1738), which are, when referred to it, a host of succedanea, just as the Linnaean system is the succedaneum to other systems and to the natural method. The system, though theoretical, is not arbitrary: a coherent system reproduces the constantly reopening, recursive structure of the monadological world. In this world, every cell of every plant or animal is itself a world, with its plants, animals, marshes, forests, and so on, to infinity (Leibniz 1991: §§ 65–67). For example, some of the twenty-four classes of the Linnaean sexual system distinguished in the vegetable kingdom are reduplicated in many classes, which contain some orders having the same names as the classes (see Figure 4: for example, orders of class 20 reflect classes 2 to 8, 10, 12 and 13; orders of class 16 reflect classes 5, 10 and 13).

10. "Natura non facit saltus" (Linnaeus 1751: § 77).

11. "Perpetuum est, quod methodus artificialis sit tantum naturalis succedanea" (Linnaeus 1738: *Praefatio*).

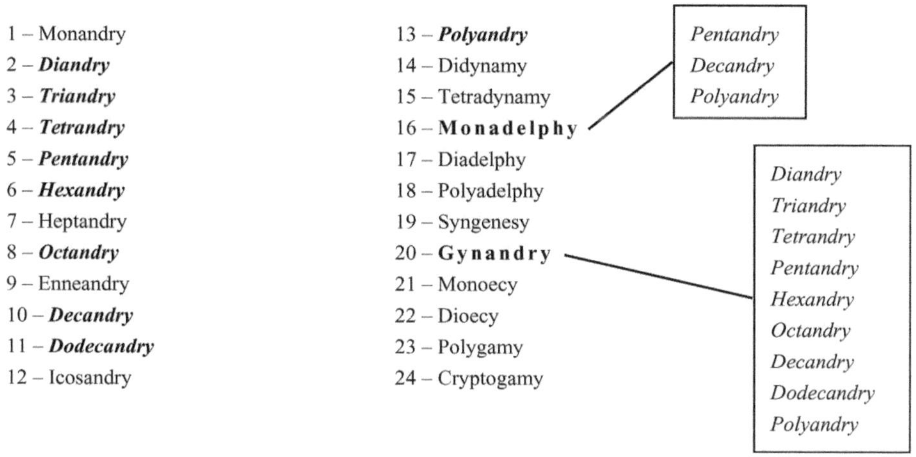

Figure 4. Recursive structure of the classes in the sexual system

4. The concept of "name"

In the sixteenth and seventeenth centuries, nomenclature was thought, thanks to logical division, to be able to give the essential being through words. Names recapitulated the process of logical division, down to the ultimate species, so that nomenclature was made up of genuine sentences, including the essential being of a species. For example (see Figure 2), in the case of an ultimate species such as *Gentiana alpina major lutea* (Yellow Greater Alpine Gentian), the essence was given by the last Latin word, that is, *lutea* (yellow). All this means that in those centuries language reflected the world, the world could be known through language, and the classification of plant beings, based on logical division, was given by language: world, classification and language were mutually consistent. Nomenclature was a set of signs, the world of plants was a world of signs (or "characters", which means the same thing etymologically), signs given or left by God: natural theology made the homogeneity firm.

But in the Enlightenment period and especially in Linnaeus' theory, logical division can no longer give the essence of a body: it merely allows the recognition, the distinction of essences, not the cognition of essences. It is restricted to a practical role. The nominalist criticism, through Hobbes (1971), Locke (1755), Berkeley (1991) and Condillac (1749), explains this limitation of its power: abstraction, general ideas, cannot provide essence and found a system. Influenced by Bacon, Linnaeus proceeds in an ascending and inductivist way, rising up in the search for causes and categories: species, genus, order, class – and the nomenclature reflects this inductivism. This conception is evident in the tripartition into three sorts of names:

– The "Specific Synoptic Name", which distinguishes "congenerous plants by semi-dichotomous notes" and which is "a succedaneum to the essential difference"[12]. The words "synoptic" and "dichotomous" recall the earlier method of logical division and the synoptic nomenclature inherits its practical, succedaneous nature.

– The "Description", which constitutes "the natural character of the species"[13]. For Linnaeus, even if a description is a text, it pertains to nomenclature, since a name is defined by the unicity of the idea it means. As it includes the natural character, it is a mixing of accidental and essential criteria: on the one hand, this means that it is a step after the specific synoptic name, which is only accidental; on the other, it means that description, including essential properties, is a succedaneum to the "Specific Essential Name", which will be easy to find, when all the plants of a genus have been discovered.

– The "Specific Essential Name" (Linnaeus 1751: § 290), the aim of nomenclature, which is polynomial and gives only but all the essential properties of a species.

To sum up, a plant is defined arbitrarily by a synoptic character, which means a synoptic idea, linked to a specific synoptic name, e.g. *Salix foliis subintegerrimis lanceolato-linearibus longissimis acutis: subtus sericeis ramis virgatis* (see right-hand part of Figure 5 for this development). From the idea of the specific synoptic name, and according to his observations in reality and about the other close species, the empiricist will draw out a natural character. The idea of this natural character will be linked to a description, which will provide another idea, from which, one day, a divine botanist may draw out the essential character carrying the essential idea of the plant – and, finally, the specific essential name *Salix foliis linearibus revolutis*. Then, thanks to the specific essential names in their entirety, the idea of species can be conceived, and hence the idea of a generic natural character, that is, mixing accidental and essential criteria. And again (see left-hand part of Figure 5), according to observation, and from one succedaneum to another, one will make out the idea of a generic essential character, thus giving the idea of a generic essential name, and finally, the essential idea of the genus. In this theory of succedanea, there is also a 'Trivial Name" (Linnaeus 1751: § 257), that is, one that is entirely factitious (see *Salix incana*,

12. "Nomen Specificum Synopticum plantis congeneribus notas semidichotomas imponit [...] Synopsis est succedanea differentiae essentialis" (Linnaeus 1751: § 289).
13. "*Character Naturalis* speciei est *Descriptio*" (Linnaeus 1751: § 258; italics and black capitals in the original).

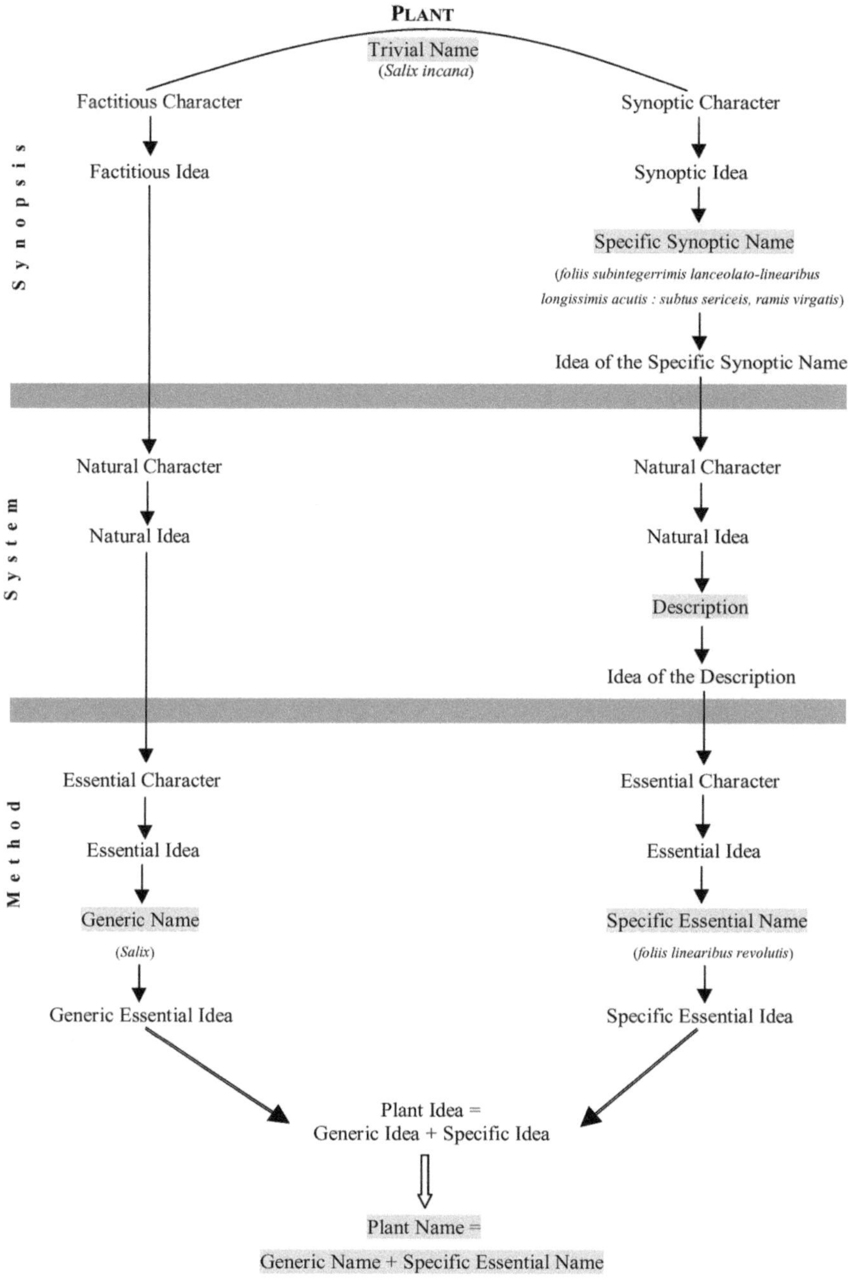

Figure 5. The theory of succedaneum applied to Genus and Species

at the top of Figure 5). This binominal name had to be a fixed landmark, standing before the several changes of specific synoptic names into specific essential names. It was thought by Linnaeus to be the most arbitrary and, as such, to have no value.

5. Conclusions

The succedanea, or operations of substitution, are repeated from the whole species to the whole genera, from the whole nomenclature to the sexual system, and, through the system, if allowed, to the hoped-for natural method, out of reach. Step by step, by refuting logical division, Linnaeus adjusted a complex cognitive method, closely connecting language and classification, nomenclature and system within the theory of succedaneum: science begins to be structured as a "well-made language", as in Condillac's theory (Condillac 1780: II, IX, 149), a structure which fascinated many scientists of the French Revolution (Duris 1993, 2006). This method is characterized by an ascending practice and by a series of steps of validations, confirmations and investigations: every empirical character appears to be the succedaneum to an idea, itself the succedaneum to another name, etc. That is a typical scientific conception, integrated in a Leibnizian epistemic context, for it recalls the increasingly perfect perceptions of monads, nearer and nearer to essence, without reaching it. It is also a perfect integration of the philosophical conceptions of language. Aristotle, Leibniz and especially Condillac (Auroux 1979: 133) conceived the successivity of two or three steps in the defining process, from a "conjectural" definition (often called "description") to a more certain one. The different Linnaean specific names are only so many definitions of the idea of the plant; like the various definitions (Leibniz 1990: III, III, § 15) or like the different botanical systems, they are "points of view" on the world, reflecting its essence more or less exactly.

Let knowledge be conceived as entirely phenomenal, let the theory of ideas disappear, let the idea of science as a well-made language vanish, and nomenclature will be a simple way to refer to the objects of nature. So the Linnaean nomenclature will develop in another epistemic context. At the end of the Enlightenment, "cause" and "sign" had been distinguished (Hume 1983), that is, a sign, contrary to a cause, could not be defined *temporally*: nature was the world of causes, heterogeneous to nomenclature, a world of signs. Nomenclature was then no more than a way of referring. That is the reason why the most arbitrary and specific Linnaean name, the "trivial" one, which was only used to label a species, became the basis for nomenclature, which became binominal and practical. That explains too why the main Linnaean polynomial specific names changed their original systematic and essential function for a descriptive and practical one, according to the empirical data they contain: on the one hand, description and on the other hand, diagnosis (this latter mixing Linnaean synoptic and essential names).

References

Adanson, Michel
1763 *Familles des Plantes*. Paris: Vincent.

Auroux, Sylvain
1979 *La sémiotique des Encyclopédistes*. Paris: Payot.

Bacon, Francis, Baron de Verulam
1986 *Novum Organum*. Edited and translated into French by Michel Malherbe and Jean-Marie Pousseur. Paris: Presses Universitaires de France. First published as: *Instauratio magna*. Londini: apud J. Billium [1620].

Bauhin, Caspar
1623 *Pinax Theatri Botanici*. Basel: König.

Berkeley, George
1991 *Principes de la connaissance humaine*. Edited and translated into French by Dominique Berlioz. Paris: Flammarion. First published as: *A Treatise concerning the Principles of Human Knowledge*. Dublin: Rhames [1710].

Berlin, Brent
1972 Speculation on the growth of botanical nomenclature. *Language and Society* 1: 51–86.

Condillac, Etienne Bonnot, abbé de
1749 *Traité des systèmes*. La Haye: Neaulme.

Condillac, Etienne Bonnot, abbé de
1780 *La logique*. Paris: L'Esprit et de Bure l'aîné.

Duris, Pascal
1993 *Linné et la France* (1780–1850). Genève: Droz.

Duris, Pascal
2006 Linné. *Pour la Science*, *Hors série* 26.

Hobbes, Thomas
1971 *Leviathan*. Edited and translated into French by François Tricaud. Paris: Sirey. First published London: Crooke [1651].

Hume, David
1983 *Enquête sur l'entendement humain*. Edited by Michelle Beyssade and translated into French by André Leroy. Paris: Flammarion. First pub-

lished as: *An Enquiry concerning human understanding*. Edinburgh: A. Millar and Donaldson [1758].

Lamarck, Jean-Baptiste
1792 Sur les Systèmes et les Méthodes de Botanique, et sur l'analyse. *Journal d'Histoire Naturelle* 1/8: 300–307.

Leibniz, Gottfried Wilhelm
1994 *Système nouveau de la nature*. Edited by Catherine Frémont. Paris: Flammarion. First published Paris: Journal des Savants [June-July 1695].

Leibniz, Gottfried Wilhelm
1990 *Nouveaux Essais sur l'Entendement Humain*. Edited by Jacques Brunschwig. Paris: Flammarion. First published (manuscript) [1703].

Leibniz, Gottfried Wilhelm
1991 *La Monadologie*. Edited by Jacques Rivelaygue. Paris: Le Livre de Poche. First published (manuscript) [1714].

Linnaeus, Carl von
1735 *Systema Naturae*. Lugduni Batavorum: Haak.

Linnaeus, Carl von
1736a *Bibliotheca Botanica*. Amstelodami: Schouten.

Linnaeus, Carl von
1736b *Fundamenta Botanica*. Amstelodami: Schouten.

Linnaeus, Carl von
1737 *Critica Botanica*. Lugduni Batavorum: C. Wishoff.

Linnaeus, Carl von
1738 *Classes Plantarum*. Lugduni Batavorum: C. Wishoff.

Linnaeus, Carl von
1751 *Philosophia Botanica*. Stockholmiae: G. Kiesewetter.

Linnaeus, Carl von
1753 *Species Plantarum*. Holmiae: impensis L. Salvii.

Locke, John
1755 *Essai philosophique concernant l'Entendement Humain*. Edited (5th ed.) and translated into French by Pierre Coste. Amsterdam / Leipzig: J. Schreuder / P. Mortier le Jeune. First published as: *An Essay concerning Human Understanding*. London: T. Ballet [1690].

Ray, John
1682 *Methodus Plantarum Nova*. Londini: impensis H. Faithorne & J. Kersey.

Selosse, Philippe
2004 Prinzipien der botanischen Nomenklatur in der Renaissance. *Linguistica Antwerpiensia, New Series* 3: 133–152.

Selosse, Philippe
2007 Die botanische Nomenklatur der Renaissance im Licht ihrer *Épistèmè*. *Zeitschrift für romanische Philologie* 123/2: 171–194.

Section 3.
The spread of scientific ideas

Linnaeus's international correspondence. The spread of a revolution

Ann-Mari Jönsson

1. Introduction

The Linnaean Correspondence is an important source both for Linnaeus's scientific work and for the history and ideas of his time.[1] Linnaeus had about 600 correspondents. The correspondence can be estimated to have originally consisted of approximately 10,000 letters. Some 5,500 letters from the period 1728 to 1778 are known today. There are more than 2,000 letters from him and more than 3,000 written to him. About 45% of the correspondence is in Latin and 45% in Swedish; the rest is in other European languages. The first letters from Linnaeus were written in 1728, to his teacher in Lund, Professor Kilian Stobaeus the Younger (1690–1742); the first known letter to him is from the German alchemist Gottfried Jacob Jänisch (1707–1781), penned in Hamburg on 9 June 1735, when Linnaeus was on his way to Holland. The last known letter from Linnaeus is to his closest friend Abraham Bäck (1713–1795), physician and praeses of the Collegium Medicum in Stockholm, dated December 1776. From the 1730s there are around 300 letters (250 from Linnaeus's years abroad). During the next decade there are about 700 letters. In the 1750s the correspondence grew by some 1,500 letters. During the 1760s the correspondence reached its peak, with some 1,650 letters. And from the last decade, the 1770s, only around 950 letters have been preserved. Finally, there are some 400 undated letters (Nyström 2006:119–120). The letters thus bear witness to a world important to scholars of both the Swedish and Latin languages and cover a material that is a veritable gold mine for dissertations and articles.

Through a collaboration between the Swedish Linnaean Society, the Royal Swedish Academy of Sciences, Uppsala University and its library, the Linnean Society of London and the Centre international d'étude du XVIIIe siècle, all the surviving letters to and from Linnaeus are now being published electronically.

1. I have previously discussed Linnaeus's scientific revolution in Jönsson 2006a and 2007.

The Linnaean Correspondence, www.linnaeus.c18.net, is financed by the Bank of Sweden Tercentenary Foundation.[2]

2. Linnaeus's international correspondence

It was in Holland, the centre of the cultural network of the time, that Linnaeus first gained international fame as a promising young botanist. In June 1735 he had defended a dissertation in which he asserted that malaria was caused by the presence of clay in the drinking water. Linnaeus knew that this was not a very good contribution to science, but he needed the doctor's diploma that would entitle him to lecture at universities all over the world.[3]

What was of immense importance, though, was the international network that he built up in Holland. He befriended the great Hermann Boerhaave (1668–1738) in Leiden, the foremost professor of his time in natural history, chemistry and physics. Linnaeus also found himself two Dutch benefactors, the amateur botanist Johan Frederik Gronovius (1690–1762), who proofread Linnaeus's works and saw to it that they were actually published,[4] and George Clifford (1685–1760), who employed Linnaeus as the curator of his large estate, Hartecamp, a magnificent botanical foundation, until spring 1738. It was at Hartecamp that Linnaeus laid the world before his feet. In two years he had acquired a worldwide reputation. When he left Sweden he had in fact brought with him manuscripts of his famous early works.[5]

It was during his time at Hartecamp that Linnaeus began his international correspondence. Linnaeus corresponded with the most learned and notable people in Europe, who informed him of new discoveries and regularly sent him new books, seeds and plants. Linnaeus was well aware of the value of his correspondence. To have a scientific correspondence was very prestigious, and Linnaeus considered himself the equal of Galileo, Newton and Leibniz. In one of his autobiographies he writes that his valuable collection of letters ought to be made

2. The project includes an inventory with digitized images of the letters and critical editions with summaries, together with biographical, bibliographical and historical comments. Further information can be obtained from the website. All the letters referred to can be found at http://linnaeus.c18.net/Letters/index.
3. Linnaeus to Ludwig, 22 January 1737.
4. See Nordström 1955 concerning Linnaeus and Gronovius.
5. *Systema naturae* (1735), *Fundamenta botanica* (1736), *Bibliotheca botanica* (1736), *Hortus Cliffortianus* (1737), *Genera plantarum* (1737), *Flora Lapponica* (1737), *Critica botanica* (1737), *Methodus sexualis* (1737) and *Classes plantarum* (1738) followed one another in rapid succession.

public as it contains many hundreds of letters, which treat everything of note that occurred from 1735 until his death. He lists seventy correspondents – from Russia and Turkey in the east to America in the west. The ones that he valued most were the physician François Boissier de la Croix de Sauvages (1706–1767) of Montpellier, his benefactor Gronovius, mentioned above, and Johann Georg Gmelin of St Petersburg (1709–1755), famous for having explored Siberia.[6] The number of correspondents grew constantly. When Linnaeus died, more than 170 Swedes and 400 foreigners had written to him. It should be emphasized that Linnaeus had a global network. Over three thousand letters had been sent from Europe, America, Asia and Africa by colleagues and students, counts and dukes, as well as by admirers such as Jean-Jacques Rousseau. In particular Linnaeus's own disciples, "the apostles" as he preferred to call them, reported back to their professor in Uppsala from expeditions all over the world – expeditions that were sometimes so dangerous that they never returned (Jönsson 2004: 352–360, Nyström 2005, 2006).[7]

Linnaeus was a man who sparked one of the greatest scientific revolutions ever. To start a revolution is the greatest achievement in science. His contemporaries were well aware of his importance. The French botanist Louis Gérard (1733–1819) calls Linnaeus a new Newton in a letter dated as early as 2 January 1757. In fact, Gérard considers Linnaeus even greater than Newton, since Linnaeus was able to confirm his revolutionary ideas ("... erisque historiae naturalis alter Newtonius cum eo nempe discrimine, quod vera certaque hipothetibus ac figmentis addideris").[8]

It may be asked what is required to have one's discoveries classified as a true scientific revolution. Someone who has thought about this is the historian

6. *Vita Caroli Linnaei* (1957:141). Linnaeus wrote five autobiographies (in the third person), one in Latin, the other four in Swedish. Regarding Linnaeus and Gmelin, see Sydow 1978.
7. The apostles' journeys have also been translated into English. See *The Linnaeus Apostles* (2007). Cf. Gotti 2006, where the important role played by communal correspondence in the spread of specialized news in the seventeenth century is examined. This correspondence was not always intended for personal purposes only, but had a more official function, since it offered opportunities to keep abreast of the times. Scholars could thus find out about work in progress, new publications and how controversially they were received.
8. 'And you will be a second Newton in natural history, with this difference of course that you have been able to confirm your hypothesis and inventions.' Also, see Malmeström (1954), who discusses the similarity between a passage in Linnaeus's introduction to the 10th edition of *Systema naturae* and the last words of Newton's *Principia*.

of science, Tomas S. Kuhn, who in 1962 published *The Structure of Scientific Revolutions*, a work that has now become a true classic. Kuhn discusses under what circumstances a revolution occurs. He investigates, for example, Copernicus's revolution in astronomy, which changed for ever our idea of the universe. Copernicus says himself in the preface to *De revolutionibus orbium caelestium* [On the revolutions of the heavenly spheres] (1543) that the astronomy that he inherited had in the end turned into a monster. By this Copernicus means that the prevailing Aristotelian-Ptolemaic system, which put Earth in the centre, was no longer able to deal with the problems that confronted astronomy. The result was chaos in science. It was thus time for a new paradigm, to use Kuhn's terminology.

Kuhn talks about different sorts of revolutions, some big, others small. Some affect many people, others few. A revolution passes through various stages, which take different lengths of time depending on the revolution. But there are a number of common characteristics. A revolution requires a person with an unusual ability to carry out his intentions. This man is either very young or completely new in the science he wants to change. The young revolutionary realizes that his new system will bring great advantages compared to the old. At the same time he knows that he cannot solve every problem, but only some obvious disadvantages of the old system. In the initial stage it is important that some other scholars at least accept the new system. They may do so for various reasons, including more or less dubious ones. They may, for instance, just be very good friends. But if they are competent, they see the potential of the new system. At this stage the followers of the old system fight desperately to defend their long-established system, although they realize its evident disadvantages. We are then at a stage where the new system balances between breakthrough and total oblivion. The strength of the new system is that, when it has been used for some time, its advantages will be apparent to a larger group of people. The system can then be developed and improved. At this stage the revolution has spread so widely that it cannot be stopped and the number of followers multiplies. The old system gradually loses its attraction, and a new generation of scholars takes over the new system, which continues to develop through new instruments and in new ways, so that in some cases it will be different from what it was intended to be in the beginning. Thus the revolution can change its course through new reformers and in the end become a completely new system. Finally, a scientific revolution is characterized by the fact that there is no counter-revolution. The new thoughts are accepted, although many try for various reasons to fight back and stop them. If these people refuse to accept the revolution they and their work will fall helplessly into the pit of oblivion and become mere history of science.

We will now try to analyse Linnaeus's science in the light of Kuhn's model of a revolution. What was Linnaeus's revolution? How did it happen? What elements did he use? What resistance did he have to fight? How did he prevail? What auxiliaries did he have at his disposal? We shall thus try to understand how Linnaeus's new ideas developed and spread.

3. The classification of plants

One of the great problems in Linnaeus's time was to get an overview of the growing numbers of plants that were being found and then to classify them. The fact that there were various systems – and not a single one – of course caused great confusion. Let me take an example: The German August Quirinus Rivinus (1652–1723) and the Frenchman Joseph Pitton de Tournefort (1656–1708) based their systems on the petals and the forms of the leaves, respectively. The Englishman John Ray (1627–1705), meanwhile, classified plants on basis of their flowers, seeds, fruits and roots. All these systems were artificial and difficult to use. Nomenclature, the naming of plants, created further problems. Every plant was given a distinguishing description, or diagnosis, which goes back to *Pinax Theatri Botanici* (1623) [A catalogue of the botanical theatre], a work that Caspar Bauhin (1560–1624), professor of botany at Basel, published in 1623. These diagnoses could also be written in the vernacular, which further complicated matters. Furthermore, two different plants could, for example, end up with the same name, because one botanist did not know what the other was doing. To sum up: when Linnaeus entered the scene there were no stringent principles for how to classify, describe and name plants and how to refer to previous, older, works. Here we have the chaos in science that Kuhn mentions.

4. Linnaeus's new paradigm

Linnaeus was the man who would introduce a new system that brought order to botany. In his revolution, Linnaeus laid the foundation of a completely new system of classification, the sexual system, which he introduced in his *Systema naturae* of 1735. This system is based primarily on the number of the stamens. It is an artificial system just like the previous ones, but very easy to use. As noted earlier, Kuhn claims that a young revolutionary in the early stages of his revolution can only remedy certain obvious disadvantages of an old system. This is true of Linnaeus, who would later bring about another necessary reform. In *Species plantarum* (1753), Linnaeus introduced his binary (or binomial) nomenclature, which meant that he gave a one-word name, for example *Primula*, to a genus and a two-word name, for example *Primula veris,* to an individual species within that genus (Stearne 1998:4 and Stafleu 1971:118–124). Later, in the tenth edition of *Systema naturae*, I (1758), Linnaeus extended the use of binary names to animals. Binary nomenclature is a kind of identifying code, which shows relationships between plants or animals.[9] It is a true milestone in botany and zoology.

Linnaeus also standardized the terminology of botanical Latin, which is a technical language for describing plants. He looked to Latin not as an instrument of style, but as an instrument of science. And what an instrument he makes of it. He demonstrates both his own eminence and that of the Latin language. Nobody has in fact shown more splendidly than Linnaeus what an excellent scientific tool Latin is. He offers proof that a Latinist and a botanist can make an excellent combination, if you have an open, creative mind.[10] Here we thus have one of those instruments which Kuhn says are developed during a revolution.

It cannot be emphasized enough that it was the simplicity of Linnaeus's system of classification, together with its carefully standardized nomenclature, that made possible a scientific understanding of the vast collections of plant and animal specimens that resulted from the expeditions of the eighteenth century.

5. Linnaeus's revolution

5.1. The first stage of Linnaeus's revolution

According to Kuhn, a revolution develops through various stages. In Linnaeus's international correspondence, we can distinguish three distinct stages and see in detail how his ideas were received by the international scientific world.

The first stage appears in 1735, when Linnaeus published his *Systema naturae*. He was then only 28 years old. We can see in the correspondence how his ideas were well received in a personal circle of Dutch botanists, for example the leading young professors of botany, Johannes Burman (1707–1779) in Amsterdam and Adriaan van Royen (1704–1779) in Leiden. That these two men were young, 28 and 31 years old respectively, has a certain relevance in this context; it meant that they were keen to try something new and had not already

9. Thus, Bauhin's *Valerianella nemorosa repens, foliis vitis idaeae minoribus, flore subrubente, semine umbilicato oblongo* became *Linnaea borealis* according to Linnaeus's binary nomenclature.

10. There is a detailed description of Linnaeus's botanical Latin in Stearn 1998. See also Jonsell 2000:323–332 and Jönsson 2001:49–76.

committed themselves to any of the prevailing systems. Burman, for example, classified rarer African plants according to Linnaeus's system in his publication *Rariorum Africanarum plantarum decades* [Rarer African plants, in ten books], which appeared in 1738–1739. Here we can remember that Kuhn says that a scientific revolution is started by a young man who is new in his science and that it is very important that other scientists take over and use the new system.

The German-speaking countries, England, France and Italy were strongholds of the classification systems of Rivinus, Tournefort and Ray that I have previously mentioned. Linnaeus was much criticized, especially by the leading German botanists, such as Christian Gottlieb Ludwig (1709–1773) in Leipzig, Albrecht von Haller (1708–1777) in Göttingen, Lorenz Heister (1683–1758) in Helmstädt, and Johann Georg Siegesbeck (1686–1755) in St Petersburg (Jönsson 2003 and Jönsson 2004:360–382).

In a revolution, the infrastructure is of great importance. In this respect, scientific publications play a major part, since it is important that the new thoughts are mentioned and discussed, so that other scholars can relate to them. Linnaeus constantly informed Johan Peter Kohl (1698–1778), who published *Hamburgische Berichte von den Gelehrten Sachen* (1732–1759), about his latest discoveries. The distribution of books is another important factor. On 1 January 1736 Linnaeus sent his *Systema naturae* to Johan Ernst Hebenstreit (1703–1757), famous for his travels in northern Africa, and asked him to review the book in German magazines. At first Linnaeus had no success and complained that Hebenstreit had never even mentioned *Systema naturae* by name. Linnaeus wrote expressively that botanists just laughed at his book.[11] He knew that his ideas needed to be quickly spread and tested. There was always the fear that somebody else might steal a march on him.

The situation is easy to understand. In the German-speaking countries there was a strongly established tradition in the history of science. Linnaeus's works arrived early. But his problem was that he was almost unknown in these countries. Linnaeus's correspondence with the German critics thus contains important documents, offering a unique glimpse into how the international world regarded the young Linnaeus and how his revolution was resisted at an early stage. We can now take a closer look at how the discussion in the letters went. First of all – there are many ways to kill a young scientist. Linnaeus had to deal with everything from serious scientific arguments to hatred, ridicule, personal attacks and complete silence. Linnaeus had in fact promised his old teacher Boerhaave in Leiden never to take part in any scientific fights, but he was forced to, and he

---

11. Linnaeus to Johann Ernst Hebenstreit, 1 January 1736.

does so in his correspondence with leading scholars of Europe (Jönsson, 1997, 2003, and 2004).

The letters between Linnaeus and Christian Gottlieb Ludwig, who was to become professor of botany in Leipzig in 1740, give us a picture of the young Linnaeus sure of his success. On 10 May 1737 Linnaeus wrote to Ludwig that we were created human beings in order to dissent, so that through this dissension truth would finally appear. Botanists lived in a free republic. Everyone could decide what he wanted to say, what he thought. Only time would judge them. The background is that Ludwig had dared to criticize Linnaeus's sexual system in a previous letter. Ludwig had predicted the difficulties Linnaeus would have, saying that very few German botanists would follow Linnaeus because of the subtlety of his method and the changes in well-established plant names. Linnaeus could not just rush in and overturn everything in botany. The argument was to continue. On 12 August 1737 Linnaeus vigorously refuted Ludwig's prophecies of failure. In ten years' time, he declared, Ludwig would have changed his opinion of him. What now caused Ludwig to vomit, he would defend in the future. And that was indeed what happened.

In Sweden, another German, Johann Georg Siegesbeck (1686–1755), is known to posterity as something of a ridiculous figure, since Linnaeus named a smelly little weed *Sigesbeckia*. But in reality Siegesbeck was a very dangerous critic. In *Epicrisis in clar. Linnaei nuperrime evulgatum systema plantarum sexuale, et huic superstructam methodum botanicam* [A critical valuation of the illustrious Linnaeus's recently published sexual system of plants and the botanical method based on it'] (1737), he attacked Linnaeus from a religious point of view and thus threatened the very basis of his science. In retrospect we can laugh at this criticism and consider it very trivial, hardly worth mentioning. But at the time it was a severe blow. To have the wrong religious opinions could be devastating to one's career. We can think of the great German philosopher Leibniz who had to leave his professorship. Linnaeus could never forget Siegesbeck's attack. On 24 April 1745 Linnaeus wrote to his friend Count Sten Carl Bielke (1709–1753) at Lövsta: "... en publique malice utan orsak ..., den kan ej utplånas i dödeligt hjerta", i.e. being publicly ridiculed without reason could never be eradicated from a human heart.

However, there was a rare exception. On 21 March 1739 Johan Gottlieb Gleditsch (1714–1786) in Berlin wrote to Linnaeus and complained greatly about the poor state of natural history in Germany. Those who taught botany were poorly paid, since the subject was not recognized as a scientific one. Some of the German botanists quarrelled about the more conspicuous exotic plants and out of absurd pride despised indigenous plants, ferns, algae, fungi and mosses, which they considered nature's waste products; they rejected excursions. Gleditsch was surprised that botanists extolled such people in their works and named new genera after them, ranking them as stars of the first magnitude in the heavens of the learned. Clearly they had no just claim to fame. In Leipzig it was commonly held that those who studied medicine should not learn about herbs, or look for plants in woods and swampy places and thereby destroy their polished shoes, clothes and powdered hair. But Gleditsch devoted himself to natural history, studying all the methods available and learning their principles and also their defects, which confused his disciples, because there was no clear definition of the genus. He learnt that botany was separate from medicine in its aims and that botany rested solely in plants. Hence the botanist was a philosopher: he could examine plants, and it was his task to create distinct concepts; he should neither accept vague terms nor dispute anyone's authority without sufficient reason. Gleditsch meticulously examined species *in situ*, recording everything in a field journal. He could then collect characters of the single genus and subordinate the species.

In England and France, negative attitudes towards Linnaeus were represented, for example, by Johan Jacob Dillenius (1684–1747), professor of botany at Oxford, and Antoine Laurent de Jussieu (1686–1758) and his two brothers Bernard de Jussieu (1699–1777) and Joseph de Jussieu (1704–1779) in Paris. Antoine Laurent de Jussieu was also Tournefort's successor.

The strong position of earlier-established systems was thus hard to overturn for a new botanist. However, Linnaeus was, as we have seen, convinced that only time would judge him. He therefore started to develop and consolidate his revolution. We now come to the second stage.

5.2. The second stage of Linnaeus's revolution

The second stage of Linnaeus's revolution can be dated to the 1740s. During this period many of his correspondents made it clear to him that he was now in the middle of an open war. In 1738 Linnaeus had returned to Sweden, and in 1741 he had been appointed professor of medicine and thus had a firm base for his scientific activities. He published a second edition of *Genera plantarum* (1742), which strengthened his position considerably. On 3 February 1742 Siegesbeck's colleague, Traugott Gerber (1710–1743), supervisor of the medical garden of Moscow, wrote to Haller in Göttingen that he feared that Linnaeus's new ideas could no longer be stopped.[12]

Linnaeus maintained his hold on Holland. Burman had, as mentioned above, published *Rariorum Africanarum plantarum decades* (1738–1739) with the

12. *Repertorium* (2002: I, 349–350).

same nomenclature as Linnaeus had used in his *Hortus Cliffortianus* [Clifford's Garden] (1737) and with the same division of genera as in *Genera plantarum*. This was the first time another scholar actually used Linnaeus's new ideas. In 1705 the English physician John Clayton (1686–1773) had travelled to North America to collect plants. These plants were later published by Gronovius in *Flora Virginica*, I–II (1739–1743). In reality this was the first work that was based on Linnaeus's sexual system. Gronovius was also one of the first to spread his revolution to American botanists, such as Cadwallader Colden (1688–1776) in New York and John Bartram (1699–1777) in Pennsylvania. In a letter dated 17 October 1743, Gronovius told Linnaeus that Colden, who was a judge in Coldingham in New York, had been interested in botany for more than thirty years. But since he had been working according to the method of his fellow-countryman Ray, he had not been able to correctly classify a single plant. However, something extraordinary had now happened. One of Linnaeus's friends in Leiden, a man called Du Bois, had taught Colden Linnaeus's sexual system and introduced him to his works. After only a few days' conversation with Du Bois and having studied Linnaeus's works for just a few weeks, Colden had learnt how to classify plants. Using Linnaeus's sexual system Colden was now able to describe correctly all the plants in Coldingham. Here we can easily understand why Linnaeus's new system came to dominate. The simplicity of the method that made it possible to arrange a large collection of plants in a very short time was evident. The sexual system was in fact so easy to use that anybody could take it and get results straightaway. Colden would later be the one who introduced Linnaeus's system in the North American colonies.

The revolution then spread through other Dutch botanists, for example Burman's colleague, Evert Jacob Wachendorff (1703–1758), and Linnaeus's friend, David de Gorter (1717–1783), in Harderwijk. Wachendorff, who in 1743 had been appointed professor of botany and chemistry in Utrecht, published *Horti Ultrajectini index* [A catalogue of the garden of Utrecht] (1747), in which the plants were described according to the Linnaean tradition and the same names of genera were used as in Linnaeus's second edition of *Genera plantarum* (1742). This work was in fact considered the only one of importance published in Utrecht during the eighteenth century. Gorter, who was the son of Linnaeus's promotor in Harderwijk, Johannes de Gorter (1689–1762), wrote to Linnaeus on 7 June 1747 that it was actually the excursions that the latter had arranged outside Leiden that had aroused his interest in botany. Gorter was important to Linnaeus, because he went to the Imperial Academy of Sciences at St Petersburg, where he influenced Russian scientists (cf. Knoespel, this volume). Gorter decided to test Linnaeus's sexual system. In a letter from Petersburg, dated 16 June 1759 (Old Style), Gorter told Linnaeus that in 1743 he had acquired two bushes of *Clutia*, one female and one male plant. That year the two plants flowered in the Academic Garden of Harderwijk. The female yielded fruit. However, the following year Gorter separated the two plants and placed them at a distance of twenty feet. Both plants flowered again. But the female plant yielded no fruit. Gorter proudly reported that this experiment in practice proved Linnaeus's theories on the sexuality of plants. Gorter later published *Flora Gelro-Zutphanica* (1745), in which the plants in the county of Zutphen in Gelderland were arranged according to Linnaeus's sexual system.

Linnaeus's revolution then spread to Austria through two other Dutch scholars, Gronovius's son, Laurens Theodor Gronovius (1730–1773), and another friend of Linnaeus's, Gerard van Swieten (1700–1772), in Vienna. Swieten had been called to Vienna as a physician-in-ordinary to Empress Maria Theresa to improve the health care system. On 12 March 1746 Swieten wrote to Linnaeus that he was very pleased that Linnaeus suffered from *botanomia*, 'a manic love for botany'. He was of the opinion that Linnaeus was now in the middle of an open war. He therefore encouraged him to fight his war well prepared, but without bitterness, though he had been attacked and wounded. Swieten meant that Linnaeus should not see himself as a victim. Swieten in his turn had a disciple by the name of Nicolaus Joseph von Jacquin (1727–1817). Later Jacquin was to be the first in Austria to use Linnaeus's new ideas in his works. From the letters it is apparent, I must say with terrible clarity, that we must never underestimate the power there is in the fact that X knows Y who knows Z. We think the world we live in is small, but we must remember that in those days the world was even smaller.

In a letter dated 13 January 1748, the Swiss botanist and physician Johannes Gessner praised Linnaeus's *Flora Zeylanica* (1747), *Hortus Upsaliensis* (1748), *Amoenitates academicae* (1749) and *Systema naturae* (1748). In 1746 Gessner had founded *Die Naturforschende Gesellschaft* in Zurich. Not surprisingly, during his time in Leiden Gessner had befriended Gronovius. Gessner was of the opinion that there was nothing more fruitful for botany than Linnaeus's *Systema naturae*. On 30 October 1748 Gessner wrote that Linnaeus's works had finally begun to arrive in Switzerland. He then immediately realized how wonderfully these reformed natural history. Gessner subsequently started to arrange his own plant material, a herbarium of more than 6,000 plants, according to Linnaeus's method. After that he continued with the plants in Caspar Bauhin's *Pinax theatri botanici* (mentioned above) and Johan Bauhin's (1541–1612) *Historia plantarum universalis* [A Universal History of Plants] (1650–1651). This was a decisive breakthrough for Linnaeus, since the works of the Bauhin brothers were the early standard works in the botanical literature. Here we also see how essential the infrastructure was for the spread of a revolution.

The men of the second stage of Linnaeus's revolution saw themselves as reformers and missionaries as they spread Linnaeus's new ideas to wider circles.

5.3. The third stage of Linnaeus's revolution

The third stage of Linnaeus's revolution comes in the 1750s. His correspondents then wrote that he had finally and decisively won his war. Linnaeus had laid the theoretical ground for his revolution in *Philosophia botanica* in 1751. The revolution actually reached its peak with the publication of his *Species plantarum* in 1753, in which, as noted, the binary nomenclature for plants was introduced. Linnaeus's revolution was now so firmly established that it was spread by a constantly increasing number of followers. Among them were Linnaeus's own disciples from abroad whom he had taught in Uppsala; the German-speaking countries, England, France, Switzerland, Italy and so on now finally surrendered to Linnaeus. Jacquin, for example, published *Enumeratio systematica plantarum quas in insulis Caribaeis vicinaque Americes continente detexit novas* [A systematic enumeration of the new plants which he discovered in the Caribbean islands and on the adjacent continent of America] (1760) and *Selectarum stirpium Americanarum historia* [A history of selected American plants] (1763). Here Jacquin used Linnaeus's binary nomenclature for the first time and made himself a name as a taxonomist. In 1762 Jacquin's *Enumeratio stirpium ... in agro Vindobonensi* appeared, applying the binary nomenclature for the first time to the Austrian flora. Jacquin praised the usefulness of Linnaeus's nomenclature. Around Jacquin there were important botanists such as Giovanni Antonio Scopoli (1723–1788), who became professor of botany and chemistry in Pavia in 1776. He published *Flora Carniolica* in 1760, in which the flora of Carniola was described according to Linnaeus's system of classification.

In the letters we see how Linnaeus attracted large numbers of students not only from Sweden, but from all over Europe, who wrote to him and asked to be allowed to come to Uppsala and become his disciples.

As an example of the third stage we can take the situation in England. Two of Linnaeus's most important English correspondents were the merchants and amateur botanists Peter Collinson (1694–1768) and John Ellis (1711–1776). In his first letter to the Swede in the mid-1750s, Ellis wrote that Linnaeus's sexual system was superior to all other systems and that he worked hard to establish it in England. Ellis also emphasized the simplicity of the system. If there had been too much to learn, he said, nobody would have had the strength to use it. Ellis also advocated a translation of Linnaeus's system into English, at least so much of it that amateurs would also become interested. Here Ellis raised an essential point. It was of great importance to publish Linnaeus's works in the vernacular to improve knowledge of them among amateurs. James Lee (1715–1795), a gardener in Hammersmith outside London, published *An Introduction to Botany, containing an explanation of the theory of that science, and an interpretation of its technical terms, extracted from the works of Linnæus ... With ... an appendix containing upwards of two thousand English names of plants* (1760). However, introducing Linnaeus's binary nomenclature in England was difficult at first. On 20 April 1754 Collinson wrote and reproached Linnaeus in the following terms: "my dear friend, we that admire you are much concerned that you should perplex the delightful science of Botany with changing names that have been well received, and adding new names quite unknown to us. Thus Botany, which was a pleasant study and attainable by most men, has now become, by alterations and new names, the study of a man's life, and none now but real professors can pretend to attain it. As I love you, I tell you our sentiments." Collinson is saying here that, through his new names, Linnaeus has made botany so complicated that it has become a matter for professors only.

The first British Linnaean was a man by the name of Patrik Browne (1720–1790), whom Linnaeus had befriended in Holland. Browne published *The Civil and natural history of Jamaica* (1756), in which he followed Linnaeus in all essentials except as regards the binary nomenclature. In a letter dated 31 May 1757 Ellis told Linnaeus that those who used to laugh at his sexual system had now changed their opinions and that they were in fact ready to change old obsolete names for those used by Linnaeus. Ellis would also try to persuade the botanists at the British Museum to use Linnaeus's new names.

Not until around 1760 did Linnaeus achieve his real breakthrough in England, mainly through his correspondents Philip Miller (1691–1771), William Hudson (1730–1793) and John Hill (1716–1775) in London. Miller was the curator of the Chelsea Physic Garden and he finally accepted Linnaeus's sexual system in his seventh edition of *The Gardener's Dictionary* (1759). Hill and Hudson were both pharmacists and began to publish works in the Linnaean genre. Hill published *Flora Britannica* in 1760 and Hudson *Flora Anglica* in 1762. Now Linnaeus also gained a few female adherents, such as Jane Colden (1724–1759), the daughter of Cadwallader Colden, who was the first female botanist in America. Linnaeus also had a female adherent in England, Lady Anne Monson (1727–1776) (Jönsson 2006b). On 21 September 1764 Adam Kuhn (1741–1817), Linnaeus's only American disciple, wrote to Linnaeus that he had visited Lady Monson, who was much devoted to botany. She had a beautiful collection of plants and insects. In England it was customary to drink the health of the monarch immediately after dinner. Lady Monson, however, first drank to Linnaeus, since he was the King of Nature. Linnaeus also owed much of his breakthrough in England to his disciple Daniel Solander (1733–1782), who arrived in London in 1760, where he was received by Ellis.

In France the real capital of botany was considered to be Montpellier. It was here that Pierre Magnol (1638–1715) had published his *Prodromus historiae generalis plantarum* [Introduction to a general history of plants]. Here we also have a disciple of the previously mentioned physician Sauvages, Antoine Gouan (1733–1821), who became an important French correspondent and the one who introduced Linnaeus's ideas in France. In his *Hortus regius Monspeliensis* [The Royal Garden of Montpellier] (1762), Gouan was the first in that country to use Linnaeus's binary nomenclature. He did so gladly. Friendship, Gouan wrote on 12 September 1763, was the true joy of the heart, love the delirium. However, it was not an easy task to take on. An ardent opponent of Linnaeus's in Paris was Bernard de Jussieu's disciple, Michel Adanson (1727–1806), who was famous for his travels to West Africa. Adanson had published *Familles des plantes*, I–II (1763–1764), in which he had launched a natural system based on Tournefort. This system, however, was openly mocked by the adherents of Linnaeus's sexual system. On 12 September 1763 Gouan wrote to Linnaeus that Jussieu's natural method had been published in the name of his follower Adanson. Furthermore, on 20 May 1764 the same correspondent described Adanson's *Familles des plantes* as a concoction of Linnaeus's *Critica botanica* and *Classes plantarum* and the work of another Frenchman, Jean François Séguier's *Biblioteca botanica*. Gouan frankly concluded that it was a work characterized by delirium and theft. What was new was stolen from Linnaeus and others and called Adanson's own work. On 27 July 1764 Linnaeus wrote ironically to Gouan that he wondered at (*admiratus sum*) Adanson's natural method, because it hardly resulted in a single natural class. Adanson was unable to show any distinguishing characteristics. He had scattered all the natural genera, and changed almost all names for the worse. There was nobody Adanson was more angry with than the one he had benefited most from, namely Linnaeus himself. Linnaeus concluded that no sane man could act in this way. In another letter, dated 22 September 1764, Gouan expressed the view that Adanson had written a dishonouring work; regarding Jussieu it could only be said that he had written the work, while another had taken the credit.

Linnaeus's impact on Holland, France and England was evident, but the German-speaking parts of Europe, apart from Austria, were still hostile towards him. On 1 February 1753 Gleditsch in Berlin reported to Linnaeus that total ignorance prevailed regarding Linnaeus's *Philosophia botanica* (1751). Those interested in botany still followed the old Rivinus and Tournefort. There was only one real systematic, Christopher Karl Strumpff (1711–1754). He was a professor in Halle who in 1752 had published the fourth edition of Linnaeus's

Genera plantarum. Nevertheless, in the letters to Linnaeus we can see that there were people who followed Linnaeus's new ideas. They included pharmacists, for example, who had collections of around a thousand plants that could now easily be arranged using Linnaeus's system. Above all they did not need to defend the established system out of prestige, having instead an obvious interest in a system that was easily learnt and easily used.

It was not until Linnaeus had published his *Species plantarum* that he finally made his breakthrough also among the Germans. On 2 March 1757 Friedrich Wilhelm von Leysser (1731–1815) in Halle wrote to Linnaeus telling him that he had studied all of his works and regarded himself as his true follower. Leysser also said that, in a marvellous way, he had succeeded in enriching the flora of Halle. This flora had been described earlier by Christopher Knaut (1638–1694) and Johann Christian Buxbaum (1693–1730), but Leysser had now written a new kind of flora based on Linnaeus's sexual system. In 1761 Leysser thus published his *Flora Halensis*. He also started to teach Linnaeus's works to German students and to arrange excursions. Linnaeus was delighted that his new system and names were finally being used.

Eventually Linnaeus's revolution was spread by his German disciples, such as Johann Christian Daniel von Schreber (1739–1810) of Halle and Paul Dietrich Giseke (1741–1796) of Hamburg. On 5 January 1758 Schreber wrote to Linnaeus that he had in reality been his only teacher, since there was nobody in Halle who could teach according to a scientific system. On 15 June 1759 Schreber was in urgent need of Linnaeus's *Systema naturae* of 1758, containing the binary nomenclature for animals, in order to be able to complete his work on German insects. He wanted to know how to constitute genera and species and boldly sent Linnaeus some work to be corrected. Schreber later published *Nova species insectorum* (1759). He also confided in Linnaeus: it was not without problems being his follower in Germany. Thus Michael Alberti, professor of medicine at Halle and a malicious critic of Linnaeus, had eradicated a moss, *Lycopodium clavatum*, out of hatred because it was one that he knew Linnaeus wanted. Furthermore, Strumpff, mentioned above, was also an object of Alberti's hatred, since he had edited Linnaeus's *Genera plantarum* (1752), as was Strumpff's student Leysser, who taught according to Linnaeus's new thinking. Schreber accordingly wished to drink from the living fountain of botany. After many complications he arrived in Uppsala in June 1760 and studied under Linnaeus until November 1761. He successfully completed his studies with the dissertation *Theses medicae* [Medical theses] (1760). When he returned home he was almost immediately appointed professor of natural history in Erlangen. He translated Linnaeus's journey to Öland and Gotland in 1764 and his Västergötland journey in 1765. He updated these works with regard to nomenclature and they achieved an enormous circulation in Europe. He wrote to Linnaeus that being a professor was nothing. What really mattered was that he could call himself Linnaeus's disciple. At least that is what he wrote. Schreber is also one of the very few in the Linnaean correspondence who are concerned about opportunities for women to study botany. In 1771 he wrote to Linnaeus (in an undated letter) that his wife had started to learn Linnaeus's system. However, she would have been able to learn more, if she had had a book in German or French containing Linnaeus's *Fundamenta botanica* and *Genera plantarum*.[13] Schreber's correspondence with Linnaeus is one of the largest; it began in 1758 and continued until 1772. And throughout this time Schreber also maintained a correspondence with Linnaeus's son, Carl von Linné the Younger (1741–1783).

Linnaeus's *Species plantarum* was also successfully introduced in Russia. Most of the scholars at the Imperial Academy of Sciences at St Petersburg, inaugurated in 1725, were Germans. Linnaeus had tried to launch some of his disciples in Russia. He attempted energetically to make Solander professor of natural history in St Petersburg. When Solander refused to go Linnaeus tried to send some of his other disciples, including Jonas Theodor Fagraeus (1729–1797) and Peter Hernquist (1726–1808). Finally, Johan Petter Falck (1732–1774) was appointed professor of botany and medicine at St Petersburg (Nyström 2004, 16–21).[14]

In a letter dated 28 September 1759, Constantin Scepin (b. 1727), who was one of Linnaeus's very few Russian disciples, described the situation of natural history in Russia. He told Linnaeus that Panajota Condoidi, the praeses of the Imperial Academy of Sciences in St Petersburg, was an excellent man, but that it was just as difficult to create a people of botanists out of the Russians as it had once been to found Rome. Aeneas was the founder of Rome, and the Russians needed Linnaeus or a man of similar diligence and outstanding reputation to reform botany in their country. Some of the men of the third and final stage, such as Gouan, went as far as to turn Linnaeus's sexual system into a new, hybrid system; that is to say, Gouan took Linnaeus's system and combined it with some thoughts of his own.

---

13. Goerke (1978:227–237) has dealt with Schreber's time in Uppsala. More information can be obtained from Schreber's letters to Linnaeus, *The Linnaean Correspondence* (see n. 1 above).

14. See also *The Linnaeus Apostles. Global Science and Adventure*. II:1. Johan Peter Falck's Journal. Russia Kazakhstan. Translated from the German by Tom Geddes. Transcribed by Viveka Hansen.

Real proof that Linnaeus had really achieved his breakthrough in Uppsala as well was a dissertation, *De variolis curandis* [On the curing of smallpox] (1754), under the presidency of Nils Rosén von Rosenstein (1706–1773), professor of medicine, in which his student, Jonas Peter Bergius (1730–1790), tried to apply Linnaeus's binary nomenclature to diseases (Örneholm 2003:152 §2.1). Here, we can recall what Kuhn says: namely, that during a revolution adherents multiply and the system can develop, begin to live a life of its own, end up as something it was not intended to be, and finally also spread to other sciences.

In Italy, too, Linnaeus had no success to begin with. Giulio Pontedera (1688–1757), professor of botany in Padua and a good friend of Haller, had totally rejected Linnaeus's sexual system and based his classifications on Tournefort. Joseph von Rathgeb (d. 1753), an Austrian diplomat in Venice, reported to Linnaeus that there was great chaos in Italian botany. On 24 September 1751, Rathgeb wrote that Saverio Manetti (1723–1785) of Florence was considered the King of Botany in Italy and concluded that he who was blind in just one eye was considered king among the blind. This criticism was prompted by the Italian professors' refusal to accept Linnaeus's sexual system and their continued use of the obsolete phrases instead of the binary nomenclature, which had severe consequences for the identification of plants and the creation of new genera.

In fact it was not until the beginning of the 1770s, in Rome, that Linnaeus's revolution could celebrate its final triumph. It was Linnaeus's young Portuguese correspondent, José Francisco Correia da Serra (1751–1823), later an important explorer and politician, who paved the way for Linnaeus in the Papal States. Correia da Serra had met Linnaeus's Swedish disciple Johan Jakob Ferber (1743–1790) in Rome and complained about the low standard of botany in Italy. Ferber then simply recommended his countryman Linnaeus. Accordingly Correia da Serra wrote to Linnaeus and a correspondence began. Correia da Serra in his turn contacted Cardinal Zelada (1717–1801) and showed him a letter in which Linnaeus was horrified by the low standard of natural history in Rome. In his letter Linnaeus had also wisely, and grossly, flattered the Pope. Zelada was so strongly influenced by Linnaeus's criticism that he decided to reform the study of botany. He contacted the Pope, who was very touched by Linnaeus's kind words, and the Pope then saw to it that the professor of botany at the time was granted a pension and replaced by a young lion, Antonio Minasi (1736–1806), who was a leading representative of the Italian Enlightenment. Minasi then taught according to Linnaeus's sexual system the following year (Jönsson 2005).

The men of the third stage finally confirmed Linnaeus's revolution. Linnaeus's critics had now definitely lost the battle, since they had refused to accept the binary nomenclature. The old systems of classification were relegated to the history of science. Haller had tried to classify the whole Swiss flora with the help of a natural system. In 1742 he published *Enumeratio methodica stirpium Helvetiae indigenarum* [A methodological enumeration of the indigenous plants of Switzerland], but in that work he only succeeded in creating a system for the cryptogams. On 29 November 1769 Schreber complained in a letter to Linnaeus that Haller had written a disparaging review of Schreber's *Beschreibung der Gräser* in *Göttingische Zeitungen von Gelehrten Sachen*, saying that Schreber should have used Haller's method and not Linnaeus's terminology. But Schreber considered this strange for a man who wrote "Wer frei darf denken, denket wohl!"

Another tragic example of how the development of science can result in an important work falling rapidly into oblivion is the French giant in zoology Mathurin-Jacques Brisson (1723–1806), who in 1756 published his magnificent *Regnum animale* (1756–1762). But this work was to have no lasting impact on zoology, since the binary nomenclature for animals was, as we already have seen, only established in *Systema naturae*, I, in 1758.

6. Conclusions

Linnaeus was unique in many respects. He sparked one of the greatest scientific revolutions ever – a revolution that spread throughout Europe. It began in Holland in 1735 and celebrated its final triumph in Italy in 1773, paving the way for Darwin and his successors. The whole story of Linnaeus's revolution splendidly demonstrates the position Linnaeus came to occupy in the European consciousness. Ultimately, we reached that stage in the development of a scientific revolution when the penetration of a new idea was so all-pervading that there was no room for those who did not accept it.

References

Goerke, Heinz
1978 Linnaeus' German Pupils and their Significance. *Svenska Linnésällskapets årsskrift* 1978: 223–239.

Gotti, Maurizio
2006 Communal correspondence in Early Modern English: *The Philosophical Transactions Network*. In: Marina Dossena and Susan M. Fitzmaurice, *Business and Official Correspondence: Historical Investigations*, 17–46.

Jonsell, Bengt
2000 Botanical Latin – Still Alive! In: Arne Jönsson and Anders Piltz (eds.), *Språkets speglingar, Festskrift till Birger Bergh*, 323–332. Lund: Skåneförlaget.

Jönsson, Ann-Mari
1997 The Early Correspondence between Carolus Linnaeus and Christian Gottlieb Ludwig. An Example of an Early German Criticism. *Svenska Linnésällskapets årsskrift* 1996–1997: 131–178.

Jönsson, Ann-Mari
2001 Linnæus's *Svartbäckslatin* as an International Language of Science. *Svenska Linnésällskapets årsskrift*, 2000–2001: 49–76.

Jönsson, Ann-Mari
2003 The Reception of Linnaeus's works in Germany. In: E. Kessler and H. C. Kuhn (eds.) *Germania Latina*, II, 721–739. München: Wilhelm Fink Verlag.

Jönsson, Ann-Mari
2004 Carl Linnaeus and his Early International Correspondence. The Making of a Botanist. In: Marianne Wifstrand Schiebe and Pär Sandin (eds.), *Studies in Honour of Professor Staffan Fogelmark. Presented on the Occasion of his 65th Birthday 12 April 2004*, 352–382. Uppsala: Dahlia Books.

Jönsson, Ann-Mari
2005 Linné och påven. Om Correia da Serra, en bortglömd länk i Linnés vetenskapliga revolution [Linnaeus and the Pope. On Correia da Serra, a Forgotten Link in Linnaeus's Scientific Revolution]. *Svenska Linnésällskapets årsskrift* 2004–2005: 57–78.

Jönsson, Ann-Mari
2006a Om konsten att göra revolution. Om Linné och hans vetenskap såsom den speglas i hans brevväxling och i samtida litteratur [On the Art of Revolution. Linnaeus and his Science as Mirrored in his International Correspondence and Contemporary Literature]. In: Gunilla Ransbo (ed.), *Humanister forskar*: Humanistdagen vid Uppsala universitet 2006, 89–104. Uppsala: Studentbokhandeln.

Jönsson, Ann-Mari
2006b Botanik – inte erotik. Linné och Lady Monson [Botany – not Eroticism. Linnaeus and Lady Anne Monson]. *Svenska Linnésällskapets årsskrift* 2006, 79–112.

Jönsson, Ann-Mari
2007 Mitt i ett öppet krig [In the Middle of an Open War]. *Forskning och Framsteg*, 2007:1. Stockholm.

Kuhn, Thomas S.
1962 *The Structure of Scientific Revolutions*. Chicago: University of Chicago Press.

The Linnaeus Apostles
2007 *The Linnaeus Apostles. Global Science and Adventure*, I–VIII. London & Whitby: Works issued by the IK Foundation & Company.

Malmeström, Elis
1954 Carl von Linné och Isaac Newton [Carl Linnaeus and Isaac Newton]. *Svenska Linnésällskapets årsskrift* 37–38: 80–82.

Nordström, Johan
1955 Linné och Gronovius [Linnaeus and Gronovius]. *Svenska Linnésällskapets årsskrift* 37–38: 7–22.

Nyström, Eva
2004 Om Natural Historiens tilstånd i Ryssland. Johan Anders Lexells brev till Linné 1772–1776 [On Natural History in Russia. Johan Anders Lexell's Letters to Linnaeus 1772–1776]. *Svenska Linnésällskapets årsskrift* 2004–2005: 7–56.

Nyström, Eva
2005 Species Plantarum and Linnaeus's Correspondence Network. *Symbolae Botanicae Upsalienses* 33:3: 61–71.

Nyström, Eva
2006 Carl von Linnés korrespondens – ett vetenskapligt nätverk från 1700-talet [The Correspondence of Carl Linnaeus – A Scientific Network of the Eighteenth Century]. In: Gunilla Ransbo (ed.), *Humanister forskar*: Humanistdagen vid Uppsala universitet 2006, 119–127. Uppsala: Studentbokhandeln.

Örneholm, Urban
2003 *Four Eighteenth-century Medical Dissertations under the Presidency of Nils Rosén* – Edited and translated, with an introduction and commentary. PhD dissertation, Department of Linguistics and Philology, University of Uppsala.

Repertorium
2002 *Repertorium zu Albrecht von Hallers Korrespondenz* 1724–1777, I. Hrsg. von Urs Boschung, Barbara Braun-Bucher, Stefan Hächler, Anne Kathrin Ott, Hubert Steinke, Martin Stuber. Basel 2002: Schwabe & Co. Ag. Verlag.

Stafleu, Franz A.
1971 *Linnaeus and the Linnaeans. The Spreading of their Ideas in Systematic Botany, 1735–1789*. Utrecht.

Stearn, William T.
1978 *Carl Linnaeus. A Bicentenary Guide to the Career and Achievements of Linnaeus and the Collections of the Linnean Society*. London.

Stearn, William T.
1998 *Botanical Latin. History, Grammar, Syntax, Terminology and Vocabulary*. London.

Sydow von, Carl-Otto
1978 Linnaeus and Gmelin. *Svenska Linnésällskapets årsskrift* 1978: 212–222.

Vita Caroli Linnaei.
1957 *Vita Caroli Linnaei. Carl von Linnés självbiografier.* På uppdrag av Uppsala universitet utgivna av E. Malmeström och A. Hj. Uggla. Uppsala.

The influence of Carl Linnaeus on the *Encyclopaedia Britannica* of 1771

Rosemarie Gläser

1. Introduction

The distinguished Swedish biologist, physician and mineralogist, Carl von Linné (1707–1778), known to the Latin-speaking scientific community of his time as Carolus Linnaeus – later admirers even hailed him as "the Pliny of the North" and "a new Newton" – had a considerable influence on the development of the sciences in the British Isles throughout the eighteenth century. Besides his lively correspondence with English and Scottish scholars and his short visits to botanical gardens in Chelsea and Oxford in 1736, he was acknowledged as an authority in the first *Encyclopaedia Britannica* of 1771. The breakthrough of the Linnaean classification system for plants and animals and the nomenclature for their denomination came with the decision of the editors and authors of this nationwide reference work to prefer Linnaeus's system to similar attempts by competing contemporaries.

It is to be assumed that Linnaeus was unaware of the impact his work had on the *Encyclopaedia Britannica.* English reference books had a rather limited circulation in the Latin-speaking international community of experts. Moreover, their distribution over great distances across the continent took a long time. Linnaeus did not know English, and in the 1770s he was suffering from ill health (Broberg 2007).

But Linnaeus most likely received reports of the preparations for the voyage of James Cook (1728–1779), astronomer and captain of the exploration ship the *Endeavour,* bound for Tahiti to observe the transit of Venus in 1769 and continuing from there into the South Sea. On this expedition, which resulted in the discovery of the east coast of Australia (in April 1770), Cook was accompanied by the English botanist Joseph Banks (1743–1820) and the Swedish botanist Daniel Solander (1735–1782), a well-known disciple of Linnaeus. Banks paid for all the botanical research equipment out of his own resources. Later on, Banks and Solander became members of the Royal Society in London.

On his second expedition to the southern hemisphere (1772–1775), James Cook as captain of His Majesty's (George III) sailing ship the *Resolution* was once again accompanied by the botanists Banks and Solander, as well as by two younger disciples. Of great benefit to the crew and to their research work was the participation of the two German scientists Johann Reinhold Forster (1729–1778) and his 18-year-old son Johann Georg Forster (1754–1794), who later published the travelogue *Reise um die Welt* in English and German (1778–1780). The second ship that accompanied the *Resolution* from the very start was the *Adventure,* commanded by Captain Tobias Furneaux. (Cf. Forster 2007.)

Joseph Banks, on his first expedition, discovered between 1,200 and 1,500 new species in Botany Bay and described them according to the Linnaean system. Johann Georg Forster's travel report was a further confirmation of this established practice of botanical denomination. It is worth mentioning that Forster's travel reports were the first publication in this genre to be based on strictly scientific observation and description. Moreover, they were enriched by realistic illustrations depicting ethnic groups, local scenery, and indigenous flora and fauna named according to the Linnaean system.

Against this background, the present article sets out to elaborate the repercussions of Linnaeus's work as a biologist and physician in the *Encyclopaedia Britannica* of 1771. This will be exemplified by instructive passages drawn from articles and treatises included in that work of reference. After a brief comment on the Swede's connections with English and Scottish scholars, and a general presentation of the first *Encyclopaedia Britannica*, the discussion will focus on prominent articles in the fields of botany, zoology, and medicine which reveal the influence of Linnaeus's work.

2. Linnaeus's connections with British scholars

When Linnaeus left Sweden for the European continent ("Ad Exteros") in February 1735, in the company of his friend Clas Sohlberg, his main concern was to acquire the academic degree of a doctor of medicine at the University of Harderwijk in Holland. This accomplishment was achieved with a disputation on "cold fever" on June 12, 1735. (Cf. Schwede 1980.)

During the three years of his stay in Holland, Linnaeus expanded his contacts with leading botanists. He visited H. Boerhaave in Leiden. Of great importance to Linnaeus was George Clifford in Hartekamp, a versatile personality who was not only a doctor of law, but also a successful banker and the director of the East India Company. He made Linnaeus the supervisor of his private zoological and botanical gardens. The outcome of this work was Linnaeus's survey entitled *Musa Cliffortiana florens Hardecampi* (published in Leiden in 1736).

In 1735 Linnaeus succeeded in getting his major work *Systema naturae* published in Holland, which enabled him to send copies to scholars abroad, including in England. In the same year Clifford commissioned Linnaeus to travel to England to acquire some rare plants for his own botanical collection and to extend his scholarly contacts. In great detail Linnaeus recorded his stay in England in his later *Vita*. In Chelsea (London) he met the well-known botanist Dr Hans Sloane (1660–1753), a scientist who had travelled widely and even regarded himself as one of Linnaeus's disciples, as he had applied observations from the *Systema naturae* in his own studies of plants. Moreover, he was president of the Royal Society in London between 1727 and 1740.

Linnaeus's contact with the distinguished botanist Johann Jakob Dillenius (1687–1747), whom he visited in Oxford, was very productive. The first meeting of the two scholars was to some extent overshadowed by misunderstandings, because Dillenius, who had only read part of the *Systema naturae* at that time, was of the opinion that Linnaeus had directed his book against him. After this initial disharmony was overcome, their cooperation became fruitful. Dillenius even invited Linnaeus to stay in Oxford and to become a university professor. The Swedish botanist was offered the demanding and meritorious project of revising and enlarging the *Pinax theatri botanici*, written by William Sherard in 1623. The Greek word *Pinax* in the title ('writing tablet', 'picture') was a metaphorical description for a scholarly work. The original botanical survey included 6,000 plants. Dillenius, however, intended a new edition that was to include systematic botanical knowledge of 16,000 plants. (Cf. comment of Sieglinde Mierau in Carl von Linné 1980: 352.) For a number of personal reasons, Linnaeus felt obliged to decline this generous offer and returned to Sweden via Holland, where Clifford had covered the expenses for his journey to England (cf. Goerke 1989).

Throughout his lifetime, Linnaeus sustained a lively correspondence in Latin with scholars in many countries; 5,500 of his letters have been preserved (cf. Jönsson, this volume). In his *Vita,* Linnaeus enumerated his correspondents according to their academic degree and their place of work. The column "Anglia" includes the following names: Petrus Collinsonus (Mercator curiosus, Londini); Joh. Mitchell (Medicus ex Virginia, Londini); Isacus Schotus Lawsonus (Med. exercitus); Johan Andrew (Medicinae doctor Oxoniis); Joh. Jacob Dillenius (Prof. botan. Oxonii); Humphredus Sipthorn (Prof. botan. Oxonii); and Jacobus Donell (Medicus Irlandiae Cashel). (Cf. Carl von Linné 1980: 230.)

Linnaeus was acknowledged in academic circles of England and Scotland and honoured in many ways. In 1762 he was appointed an honorary member of the Royal Society of London, and in 1773 of the Medical Society in Edinburgh ("Edinburgensis Coll. Med.").

3. The first *Encyclopaedia Britannica* of 1771

The British society in the second half of the eighteenth century witnessed the appearance of two most influential works of reference, intended for a broader readership beyond the limited circles of the educated middle classes and the aristocracy. These books were the *Dictionary of the English Language*, designed and compiled, as an individual achievement, by Dr Samuel Johnson and appearing in 1755, and the *Encyclopaedia Britannica*, elaborated by "a Society of Gentlemen in Scotland" and published in three volumes in 1771. While Dr Johnson's work *A Dictionary of the English Language* codified the contemporary literary and colloquial vocabulary as to its meaning, etymology, usage and regulated spelling, the *Encyclopaedia Britannica*, also called *A Dictionary of Arts and Sciences* […], was a huge thesaurus of theoretical and practical knowledge, and a reflection of dominant paradigms in particular areas of research. The extended title on the frontispiece of the *Encyclopaedia Britannica* presents the editors' goal of achieving a compilation of knowledge *upon a new plan, in which the different Sciences and Arts are digested into distinct Treatises or Systems; and the various Technical Terms, etc. are explained as they occur in the order of the Alphabet.*

The idea of the wording of the title page – conveying both information and an implicit advertisement – is resumed in the preface. The authors have a definite pragmatic aim in mind which is put forward as quite a rigorous principle: "Utility ought to be the principal intention of every publication. Wherever this intention does not plainly appear, neither the books nor their authors have the smallest claim to the approbation of mankind. To diffuse knowledge on Science, is the professed design of the following work" (Vol. I, Preface, page V).

The procedure of the authors, in pursuing this aim, is exactly that of compiling knowledge drawn from distinguished publications of European authors past and present, many of them writing in Latin, and of presenting the most recent, innovative ideas by describing the state of the art in such fields as botany, zoology, and medicine, where Linnaeus was an undisputed authority.

In its essence, the *Encyclopaedia Britannica* mirrors the spirit of enlightenment and confidence in the progress of mankind. This conviction was also expressed in the famous contemporary French work of reference, the *Encyclopédie française ou Dictionnaire raisonné des sciences, des arts et des métiers* (1751–1777). The editors of the *Encyclopaedia Britannica*, in their preface, made a direct reference to this source. The French encyclopedia was composed of 28 volumes. It was designed, prepared and supervised by Denis Diderot (1713–1784) and Jean-Baptiste le Rond d'Alembert (1717–1783). In their depth of knowledge and presentation of the progress made in science and technology, both the French and the English encyclopedias demonstrate intellectual vigour, pride in the achievements of the unlimited human mind, and confidence in the advancement of the human race.

In the field of biology, the French encyclopedia propagated the ideas of Buffon (1707–1788) and disregarded those of Linnaeus altogether (cf. Sörman, this volume). On the other hand, its English counterpart mentioned Buffon's main work, *Histoire naturelle,* only briefly in the "List of Authors", whereas Linnaeus is represented by no less than six titles (*Systema naturae*; *Amœnitates academicae*; *Philosophia botanica*; *Genera plantarum*; *Species plantarum*; and *Fundamenta botanica*). This bibliography is highly incomplete; there is no reference to time and place of publication. Moreover, the French botanist appears under the spelling *Bouffon*.

4. Repercussions of Linnaeus's work in prominent articles of the *Encyclopaedia Britannica*

In the eighteenth century, the specific fields of the modern natural sciences were not yet clearly defined. The three kingdoms of nature at that time were those of plants, animals, and minerals. Their corresponding scientific disciplines, botany, zoology and mineralogy, however, were still included in the all-embracing study of natural history – an umbrella term with far-reaching theoretical and practical implications. The spelling of the headwords in capital letters follows the original usage of the first *Encyclopaedia Britannica.*

4.1. Botany

The *Encyclopaedia Britannica* of 1771 contains a long, independent treatise of 24 double- column pages on BOTANY (Vol. I, 635 ff.), whereas the field of ZOOLOGY, "the science of animals", has only a short reference to the article on NATURAL HISTORY (Vol. III, 953), and the term MINERALOGY is not even referred to the entry on MINERAL, "in natural history [...] used, in general, for all fossil bodies [...] dug out of a subterraneous mine from which it takes its denomination" (Vol. III, 248).

Strong evidence of the impact which Linnaeus had on the development of botany in the eighteenth century is the fact that the authors of the treatise on BOTANY in the first *Encyclopaedia Britannica* deliberately decided in favour of his system for the classification of plants and animals. This decision was an overt preference for Linnaeus over his predecessors and contemporaries, and further disseminated his work in western Europe. The writers of the article on BOTANY enter into an implicit dialogue with their readers:

> We observed in the former section, that in the progress of this part of botany many different methods had been followed by different authors. Caesalpinus, Ray, Bauhinus, Van Royan, Ricinus, Tournefort, Linnaeus, Sauvages have each adopted a peculiar method of characterizing and classing plants. *It would be foolish to distract the attention of the reader by an explanation of all these methods. We shall therefore proceed to explain that of Linnaeus, which is perhaps the only one now taught in Europe.* (Vol. I, 635, my emphasis)

It is interesting to note that the name of the French natural scientist, Buffon, the great rival of Linnaeus, is not even mentioned in the list of classification systems quoted above.

The following passage of the text elaborates the survey of the sexual system presented by Linnaeus. It is illustrated by a plate of 24 types of plants. The authors of the article on BOTANY provide a condensed survey:

> This method of reducing plants to classes, genera, and species, is founded upon the supposition that vegetables propagate their species in a manner similar to that of animals. Linnaeus endeavours to support this hypothesis by the many analogies that subsist between plants and animals. (…) It is from this circumstance that Linnaeus's system of botany got the name of the *sexual system*. The names of his classes, orders & are all derived from this theory. He calls the stamina of flowers the *males*, or the male parts of generation; and the pistils *females*, or the female parts of generation. Plants whose flowers contain both male and female parts, are said to be *hermaphrodites*, etc. His classes, orders, and genera, are all derived from the number, situation, proportion, and the other circumstances attending these parts, as will appear from following scheme. (Vol. I, 635. Italics in the original)

Linnaeus's botanical system is elaborated in greater detail on 19 printed pages (in double columns) and in six plates containing drawings of the essential features of plants.

4.2. Zoology

As mentioned above, the subject of zoology is not treated in an independent article in the encyclopedia. It constitutes a separate part of the treatise on NAT-

URAL HISTORY. In this context, the authors emphasize the analogy between the classifications of plants and animals in Linnaeus's system, with reference to their use of the same terminology and logical principles: "NATURAL HISTORY, is that science which not only gives compleat [*sic*] descriptions of natural productions in general, but also teaches the method of arranging them in Classes, Orders, Genera, and Species. This definition includes Zoology, Botany, Mineralogy, etc." (Vol. III, 362).

The headword for individual animals is "their proper name" (in English), with a reference to the Latin term for the genus or species.

In the following passage, the authors give an abridged survey of Linnaeus's system of animals:

> Linnaeus divides the whole animal kingdom into 6 classes. The characters of these 6 classes are taken from the internal structure of animals, in the following manner.
> Class I. MAMMALIA (mammals)
> Class II. AVES, or Birds
> Class III. AMPHIBIA, or Amphibious Animals
> Class IV. PISCES, or Fishes
> Class V. INSECTA, or Insects
> Class VI. VERMES, or Worms
> (Vol. III, 362)

The whole article on the subject of NATURAL HISTORY is limited to three and a half pages. The authors, however, present basic definitions of the key concepts which underlie the categories of Linnaeus's classification system and are derived from propositions of logic: "GENUS in natural history, a subdivision of any class or order of natural beings, whether of the animal, vegetable, or mineral kingdoms, all agreeing in certain common characters" (Vol. II, 673). SPECIES as defined in different entries (in Vol. III, 620) is an ambiguous term, as it is related to several fields of knowledge, ranging from logic to optics and commerce. In the system of biological entities, the concept SPECIES is understood in a logical sense: "a relative term, expressing an idea which is comprised under some general one called a genus" (Vol. III, 620). The concept of the next higher category, class, is clearly defined: "CLASS, an appellation given to the most general subdivisions of any thing; thus, animal is subdivided into the classes quadrupeds, birds, fishes, & which are again subdivided into series or orders, and these last into genera. See NATURAL HISTORY, and BOTANY" (Vol. II, 207).

Throughout the *Encyclopaedia Britannica,* these defined categories are consistently applied by the authors of the entries on items of biology.

4.3. Medicine

To his contemporaries Linnaeus was not only the famous botanist, zoologist and mineralogist, but also a knowledgeable doctor of medicine who had graduated in Holland. His reputation in this field was soon acknowledged in the British Isles. Consequently, the authors of the *Encyclopaedia Britannica* paid particular attention to Linnaeus in the comprehensive treatise on MEDICINE (in Vol. III, 58 ff.). The pertinent passage reads:

> Of late several attempts have been made to reduce medicine into the form of a regular science, by distributing diseases into classes, orders, genera, and species. Sauvage was the first, and indeed the only person who ever attempted to complete this great work. Others, as Linnaeus, Vogel, Dr Cullen, etc., have since endeavoured to improve Sauvage's method of classing; but they have contented themselves with an enumeration of the characters and arrangement of the different genera, without entering into their history or cure. Sauvage enumerates 315 genera, Linnaeus 325, Vogel 560, and Dr Cullen has reduced them to 132. The bare inspection of these numbers shews [*sic*], that physicians are far from being agreed with regard to what constitutes the generic or specific characters of a disease. (Vol. III, 58)

This passage gives evidence of the state of the art in applying a biological nomenclature of genera to diseases. It is noteworthy that Linnaeus had made a significant contribution to this debate.

5. Articles on agricultural animals and domestic pets

The subject areas AGRICULTURE and HUSBANDRY in the *Encyclopaedia Britannica* of 1771 include a number of most informative articles on farm animals used for tillage, and of others raised for meat, milk, wool or leather production. All these animals are described with regard to their utility to the farmer. This idea is also evident in the articles on domestic animals kept for man's company, and naturally in the description of detrimental rodents and other vermin occurring in the farmyard.

The headword of each article consists of the Latin genus name of the particular animal according to the Linnaean system. As a rule, the English common (trivial) name is mentioned in the following explanatory text. Additional articles relating to the male, female and young animal are included under the English animal name as the headword. This procedure may be illustrated by the following examples: EQUUS (*horse, mare, stallion, foal, colt*); BOS (*ox, cow, calf*); OVIS (*sheep, ram, ewe, lamb*); CAPRA (*goat*); SUS (*pig, sow, swine*);

PHASIANUS (*poultry, cock, hen*); ANAS (an umbrella term for both *goose* and *duck*). The domestic animals with a more social function are CANIS (*dog, bitch*) and FELIS (*cat*). However, the designations *puppy* and *kitten* for the young of these animals are missing.

In general, the articles are introduced by a description of the anatomical and physiological features of the animal. This text segment offers essential zoological facts in an unemotional, neutral style and is strikingly different from the *character images* that follow. In the latter, the author usually describes the properties and behaviour of the particular animal in relation to human attitudes and moral standards, using emotive means of expression. This overt style shift suggests the conclusion that another person was in charge of these character sketches. From an anthropomorphic viewpoint, the author expresses his personal value judgement of a farm animal by a wide variety of figurative stylistic means. (Cf. Gläser 2006.)

It goes without saying that animal character images like these are not the result of theoretical principles or psychological case studies (not yet applied in the eighteenth century), but rather reiterate popular clichés and stereotypes about domestic animals. Moreover, the authors of the articles suggest a certain *hierarchy* of character images. The highest position of appreciation is occupied by the horse, the lowest by the pig. In these descriptions, the animal is often personified, and grammatically the male gender (*he, his*) prevails.

The horse, EQUUS (genus name), belongs to the order of BELLUAE and shares genetic properties with the ass (genus ASINUS) and the zebra. The domesticated horse receives an excellent character image in the following description:

> The horse, in a domestic state, is a bold and fiery animal; equally intrepid as his master, he faces danger and death with ardour and magnanimity. He delights in the noise and tumult of arms, and seems to feel the glory of victory; he exults in the chase; his eyes sparkle with emulation in the course. But though bold and intrepid, he is docile and tractable; he knows how to govern and check his natural vivacity and fire of his temper. (Vol. II, 506)

This heroic type of horse, reminding the reader of a charger, differs greatly from the horse as a pack, riding or draught animal:

> He not only yields to the hand, but seems to consult the inclination of his rider. Constantly obedient to the impressions he receives, his motions are entirely regulated by the will of his master. He in some measure resigns his very existence to the pleasure of man. He delivers up his whole powers; he reserves nothing; he will rather die than disobey. *Who could endure to see a character so noble abused! Who could be guilty of such gross barbarity!* (Vol. III, 506, my emphasis)

By the standards of a modern encyclopedic reference book, this exalted and pathetic presentation of a working horse and the anthropomorphic interpretation of its behaviour would be totally inappropriate. But animal protection campaigners might draw supporting arguments from the author's defence of the horse.

A much less favourable character is attributed to the ass (genus ASINUS):

> The ass is as humble, patient, and tranquil, as the horse is bold, ardent and impetuous. He submits with firmness, perhaps with magnanimity, to strokes and chastisement; he is temperate both as to the quantity as to the quality of his food; he contents himself with the rigid and disagreeable herbage which the horse and other animals leave to him, and disdain to eat; he is more delicate with regard to his drink, never using water, unless it be perfectly pure. (Vol. II, 510)

The ox (genus BOS), too, is measured against the quality scale of the male horse. However, the bull fighting in the arena differs greatly in appearance and performance from the ox drawing the plough. The author of the article presents an ideal picture of this farm animal:

> A bull, like a stallion, should be the most handsome of his species. He should be large, well made, and in good heart; he should have a black eye, a fierce aspect, but an open front; a short head; thick, short and blackish horns, and long shaggy ears; a short straight nose, large and full breast and shoulders; thick and fleshy neck, firm reins; a straight back; thick, fleshy legs, and a long tail well covered with hair. (Vol. I, 624)

This description of a desirable type (or specimen) of an ox could be understood as a practical guide for a farmer intending to purchase oxen at a cattle market or auction. The main purpose of the ox in the British Isles in the author's opinion is "to propagate the species; although he might be trained to labour, his obedience cannot be depended on".

The goat (genus CAPRA) seems to be held in higher esteem by the article writer than the sheep, being regarded as more intelligent, more independent, less timid, more sociable and more tameable. Moreover, the goat seems to like being petted (like indoor cats). The author presents the following character image:

> The goat is an animal of more sagacity than the sheep. Instead of having an antipathy at mankind, they voluntarily mingle with them, and are easily tamed. Even in uninhabited countries, they betray no savage dispositions (...) Goats are sensible of caresses, and capable of a considerable degree of friendship. They are stronger, more agile, and less timid than sheep. They have a lively, capricious, and wandering disposition; are fond of high and solitary places, and frequently sleep upon the very points of rocks... (Vol. II, 30)

The author's description of the sheep follows the utility principle entirely. This farm animal is a provider of milk, meat, wool, leather and tallow. It is susceptible to heat and cold. Its blind herd instinct may expose it to danger in mountainous areas. The author introduces his observations on the ram (or ARIES) as follows: "This animal is perhaps the most gentle and inoffensive, and at the same time the most timid and stupid of all quadrupeds" (Vol. III, 447).

Placed at the bottom end of the likeability scale of farm animals, the pig (genus SUS) is given a very unfavourable character image. Its ugly appearance, dirty way of feeding and unsavoury housing conditions are disgusting to the writer of the article, and the quality of pork is not recommended:

> The sus scrosa, or common sow, is a native of the southern parts of Europe. It feeds coarsely, digs up roots, etc., from the ground with its snout. It is a fat, sleepy, stupid, dirty animal, wallowing constantly in the mire. The sow brings forth a great number of young at a litter, and is therefore considered as a profitable animal. The flesh, however, is not reckoned to be of the most wholesome kind. (Vol. III, 879)

From a quite different perspective, the dog and the cat as familiar housemates of man do not enjoy equal recognition by the authors of the respective encyclopedia articles. The dog (genus CANIS, order FERAE), originally stemming from the domesticated wolf, is characterized by his fidelity and reliability to his master:

> In a savage state, he is fierce, cruel, and voracious, but, when civilized and accustomed to live with men, he is possessed of every amiable quality. He seems to have no other desire than to please and protect his master. He is gentle, obedient, submissive, and faithful. Those dispositions, joined to his almost unbounded sagacity, justly claim the esteem of mankind. Accordingly no animal is so much caressed or respected: He is so ductile, and so much formed to please, that he assumes the very air and temper of the family in which he resides. (Vol. II, 23)

In this respect, the dog and the horse may be regarded as man's most reliable friends. On the other hand, the author's value judgement of the domestic cat (genus FELIS) is far from favourable. The author concedes that the cat is useful for catching mice and rats in the house, garden, barn and storehouses. But it is ascribed human properties such as unreliability, unpredictability and cunning:

> Of all domestic animals, the character of the cat is the most equivocal and suspicious. He is kept, not for any amiable qualities, but purely with a view to banish rats, mice, etc., and other noxious animals from our houses, granaries, etc. Although cats, when young, are playful and gay, they possess at the same time an innate malice and perverse disposition, which increases as they grow up, and

> which education learns them to conceal, but never subdue. Constantly bent upon theft and rapine, though in a domestic state, they are full of cunning and dissimulation; they conceal all their designs, they seize every opportunity of doing mischief, and then fly from punishment. They easily take on the habit of society, but never its manners, for they have only the appearance of friendship and attachment (…) In a word, the cat is totally destitute of friendship; he thinks and acts for himself alone. (Vol. II, 586)

In a different frame of reference, parts of this extremely anthropomorphized description of an animal's character could evoke associations with a human criminal who constantly varies his tactic and strategy to do harm to society. This one-sided and biased description of the "prototypical" domestic cat is an exception in eighteenth-century English literature. There are numerous opposite examples of a happy community of single persons or families and cats, and Samuel Johnson was known as a great cat-lover.

On the whole, from a scientific perspective, the passages on the character of domestic animals are untenable and incompatible with the anatomical descriptions provided by Linnaeus and his contemporaries. Most likely, they result from the author's intention to include popular rural knowledge for the reader's entertainment.

6. Summary

After the appearance of the first *Encyclopaedia Britannica* in 1771, profound changes occurred in the political, economic and cultural life of British society, giving rise to diversified demand on the book market and eventually preparing the ground for a second, considerably enlarged edition of this comprehensive work of reference. Its editors thought it essential that the articles and larger treatises kept pace with current improvements in British husbandry, with technological progress in industrial production, and with the advancement of the sciences, arts, and trades promoted by the Royal Society in London, whose president was Joseph Banks, an ardent adherent of Linnaeus.

Another strong impetus came from the records of the voyages (in 1769–1770 and 1772–1775) of Captain James Cook and the scientists and naval officers accompanying him, reporting on the discovery of Australia and describing its remote flora and fauna according to the Linnaean system. A wealth of topical information was thus available to the authors and editors, lending further confirmation to the utility principle, the pragmatic guideline of their procedure.

The second *Encyclopaedia Britannica* appeared between 1777 and 1784, a huge reference work in 10 volumes, covering 8,595 printed pages and including many illustrations. The editor was James Tytler (1747–1804). Among its considerable improvements were the biographical and geographical articles reporting on the latest transatlantic and trans-Pacific voyages of explorers and scientists.

The third *Encyclopaedia Britannica* followed as early as 1788, comprising 18 volumes and 14,573 pages. The editor was Colin Macfarqhar, who remained in charge of this project until his death in 1793. It was completed by the Scottish clergyman George Gleig (1753–1840) in 1797. With these three editions, the tradition of the *Encyclopaedia Britannica* lasting to the present day was inaugurated. To the English-speaking readership of the eighteenth century, it became a universal thesaurus of knowledge, and one in which Carl Linnaeus, too, has sunk deep roots.

References

Broberg, Gunnar
2006 *Carl von Linné.* Übersetzung: Margaretha Tidén. Stockholm: Schwedisches Institut.

Broberg, Gunnar
2007 Linnaeus, Scientist or Man of Letters? Plenary Paper at the International Symposium on the Languages of Science in the Time of Linnaeus, June 7–9, 2007, Uppsala, Sweden. Collected Abstracts, pp. 14/15 and oral discussion contribution.

Emersleben, Otto
1998 *James Cook.* Reinbek bei Hamburg: Rowohlt.

Encyclopaedia Britannica; or, *A Dictionary of Arts and Sciences, Compiled upon a new plan, in which the different Sciences and Arts are digested into distinct Treatises or Systems, and the various Technical Terms are explained as they occur in the order of the Alphabet.* By a Society of Gentlemen of Scotland. In Three Volumes, Edinburgh MDCCLXXI. Reprint 1989. (Original title.)

Encyclopaedia Britannica
2003 *Macropedia. Ready Reference.* Fifteenth edition. Article "Encyclopædia Britannica", p. 487. Chicago etc.

Forster, Georg and Georg Christoph Lichtenberg
1976 *Cook der Entdecker. Schriften über James Cook.* Leipzig: Verlag Philipp Reclam jun.

Forster, Georg
2007 *Reise um die Welt. Audiobook*, read by Frank Arnold. Eichborn Lido.

Gläser, Rosemarie
2006 Die Themenkomplexe Gartenbau und Landwirtschaft in der Encyclopaedia Britannica von 1771 – eine fachsprachengeschichtliche Untersuchung. In: Petr Rösel (ed.), *English in Space and Time. Englisch in Raum und Zeit. Forschungsbericht zu Ehren von Klaus Faiß*. Fokus. Linguistisch-philosophische Studien, 31, Trier: Wissenschaftlicher Verlag, 28–50.

Goerke, Heinz
1989 *Carl von Linné. Arzt, Naturforscher, Systematiker*. 2. Auflage. Stuttgart: Wissenschaftliche Verlagsgesellschaft.

Linné, Carl von
1980 *Lappländische Reise und andere Schriften*. Leipzig: Verlag Philipp Reclam jun.

Schwede, Alfred Otto
1980 *Carl von Linné. Der Blumenkönig des Nordens*. Berlin: Union Verlag.

Linnaeus and the Siberian expeditions: Translating political empire into a kingdom of knowledge

Kenneth J. Knoespel

Knowledge in the European eighteenth century cannot be separated from geography. Inherent in the efforts to control territory through warfare and political negotiation, we find an interest in developing practices that permit measurement of land and the determination of resources. The science of cartography hardly develops in abstraction but as an effort to gather and retain information necessary for military defense and economic development. Exploration of the East Indies, Africa, and North and South America in the eighteenth century is balanced by an expanding interest in taking stock of the natural resources within the borders of a particular country. Spheres of control that at first might be associated with military firepower reach outward and inward in ways that often intersect or become superimposed on each other like interconnecting canopies. The Russian dissertations associated with Linnaeus and published in the eighteenth century by the University of Uppsala provide a remarkable example of the ways that local and global exploration are combined to shape networks of information that come to shape what we think of as universal systems of knowledge.

The transformation of Sweden from a great military power to a nation that seeks to redefine her ambitions after the Battle of Poltava (1709) has become a historical trope within Linnaean studies. From the heroic image of Carl XII and his military conquest of territory, we move to the peaceful image of Carl Linnaeus (1707–1778), the Swedish Saint of the plant kingdom. Russia, so it would seem, sinks below the horizon and becomes a thorn in the side of Sweden or a rude nation that Sweden might interpret to the rest of the world. Although attractive, this simple narrative, ideological in its own right, unravels when we consider the continuous interaction between Sweden and Russian at the beginning of the eighteenth century. The redefinition of Sweden's role in natural philosophy during the time of Linnaeus involves a continuous interaction with Russia on multiple levels. Of course, in part this interaction is defensive and involves military technology. But on other levels it involves the mutual development of Russian and Swedish scientific academies devoted to economic development.

For both Sweden and Russia alike development involves learning how to use natural resources (Koerner 1994, 1999).

Linnaeus's multifaceted response to information supplied by the Siberian expeditions that begin in 1724 offers an opportunity to follow his interaction with the newly founded Imperial Academy of Sciences at St Petersburg (1725) through his botanical research at the University of Uppsala. Even more substantively, Linnaeus's interaction with Russian correspondents in Latin and German as well as through drawings, diagrams and physical specimens shows a growing awareness of the ways a universally shared language of natural history may replace a political empire with an empire of knowledge. After noticing some of the ideological structures that underlie the approach to the northern landscape, I turn to some of the rhetorical strategies employed by Linnaeus especially in his relation with his Russian correspondents and students. Finally, I consider ways that Linnaeus's work participates both in the stabilization of codes and the development of strategies that could be used to share information and gather reconnaissance for the use of a particular nation.

1. Reading the Northern Landscape

The northern landscape is an essential part of Linnaeus's work both in regard to his own expeditions in Sweden and also within the setting of Swedish efforts to understand the unique features of the north. The study of the landscape in Scandinavia was accompanied by the discovery of northern antiquities that reinforced an idea of Scandinavia's past in ways that could be ideologically productive. Antiquarians such as the seventeenth-century Dane Olle Worm provoked archaeological and early linguistic research into the significance of the Viking sites (Scheffer 1698; Björner 1676, 1742; Bureus 1731, 1744; Loccenius 1676; Peringskiöld 1710; Malm 1996). The Swedish polymath Olof Rudbeck sought to integrate a northern vision of European history with the dominant Mediterranean narratives of Greece and Roman. By comparing the Baltic basin and the Mediterranean through language, geography, and artifacts Rudbeck sought to establish a northern origin for all known cultures (Eriksson 2002; Knoespel 1994). Although Rudbeck's historiographic research was quickly dismissed, it retained a force that extended beyond ideological pretention. By grounding his work in the study of ancient languages he envisioned a system of historical linguistics in which the Indo-European languages stemmed from Icelandic. While his effort was doomed to failure, it documented an array of northern archaeological material that could be compared with the textual record of Mediterranean cultures. The comparative methodology used by Rudbeck gave precedence to northern data because it was nearby in contrast to the more remote textual records of the south. I emphasize the "literal" quality of the investigation because it has an important bearing on the attention Rudbeck the Younger gives to his exploration of Lappland and the direction it gives to the ongoing documentation of the landscape.

Sweden's defensive military posture in regard to Russia as well as her reassessment of Russia after Poltava describes a significant part of Sweden's northern foreign policy when Linnaeus arrives in Uppsala. Rather than demonstrating her strength through military conquest, Sweden has begun to look at her strength as part of a European network in which she may both represent and present the north. Linnaeus contributes to this network through his early travels in Holland and Germany and through his detailed description of Sweden. Just as Olaus Magnus had represented the north for Europe in *Carta Marina* (1539) and *Historia de gentibus septentrionalibus* (1555), Linnaeus came to be recognized by his own countrymen for describing the land in which they lived in Swedish (Frängsmyr 2000; Lindroth 1967; Magnus 2001, 2005). While Gustavus Vasa could be celebrated for creating a unified kingdom and while Gustav II Adolph and Carl XII could create a Swedish military mythos for Europe, through his travels in Sweden, Linnaeus creates an account of the land in which people live.

While the Battle of Poltava (1709) dramatically halted Swedish territorial expansion in the east, it also challenged Russia to undertake her own exploration of Siberia. The eighteenth-century Russian expeditions, referred to as the "Siberian Expeditions", surveyed the massive Russian continent and provided vast amounts of information essential for governing the other "new world". The expeditions also expanded European interpretation of natural history and the development of organizations that could sustain research projects over an extended time period. Although many smaller Russian expeditions were undertaken during the century, it is three in particular that are especially relevant here. The First Kamchatka Expedition (1724–1730), led by the Danish navigator Vitus J. Bering, discovered the coast of Alaska. The Second Expedition (1733–1743) gathered an immense amount of data on Russian anthropology, archaeology, flora and fauna. The Third Expedition (1768–1774) that sought to gather astronomical data also refined previously gathered information. Initially encouraged by Leibniz and authorized by Peter the Great, the expeditions resulted in the publication of documents that promoted further research much as the exploration of North and South America. Philipp Johann Tabbert von Strahlenberg's *Der nord- und östliche Teil von Europa und Asien* appears at the end of the first expedition; Johann Georg Gmelin's *Flora Siberica* (4 vols. 1747–1749; 1770) emerges from the second; and Peter Simon Pallas's *Flora*

Rossica (1784–1788) from the third (Donnert 1986:95–114). From the vantage point of Sweden's orientation to Russia, the work of Strahlenberg is all the more important because it is based on his detailed reconnaissance of Siberia after his capture at Poltava. Even in defeat, educated military engineers such as Strahlenberg carried on detailed reconnaissance that interpreted Russia to the rest of Europe. While Russia could push Sweden away from the Neva Delta and the Finnish Gulf, it also affirmed the ways that Sweden fit into network of European knowledge production. There is a sense in which Russia becomes even more important for Sweden after Poltava. Instead of being a set of Russian provinces that needed to be controlled militarily and politically, after 1709 Russia becomes a territory that Sweden could translate to the rest of Europe.

Linnaeus would have had his first indirect encounters with Russia through his family and regional background. Southern Sweden until the mid-seventeenth century remained under Danish rule. When Carl XI established control over the Danish provinces of Skåne and Blekinge (now the southern most provinces in Sweden), the Swedish territory of Ingermanland (now a Russian province south of the Gulf of Finland) became a place of forced exile for population that had difficulty in shifting their loyalty toward the Swedish crown. Movement to the east would have been regarded as a move toward unsettled and conflicted territory. When Linnaeus moves to Uppsala in 1730, he moved to a major northern university where Rudbeck the Younger and others, especially after the disastrous Swedish defeat at Poltava, had conveyed a strong dose of Swedish prejudice against Russia. Linnaeus's view of Russia would have been shaped by Poltava and the kind of geographic reconnaissance displayed by Strahlenberg. Even though the Russians controlled territory, it could be said that, because of prior reconnaissance and ongoing research, Sweden knew more about the Russian territory than the Russians themselves (Lindroth 1967).

2. The Russian dissertations of Linnaeus

The Russian dissertations referred to below belong to the 186 academic dissertations which are associated with Linnaeus during the period 1743–1776. Unlike the common practice today in which doctoral students write their own theses and defend them, in eighteenth-century Sweden the candidates defended the ideas of their professor. "Thus what are usually referred to as the Linnaean dissertations are attributed to Linnaeus, even though they are also associated with the respondents who defended them and whose names appear on their title pages." (Hunt Library) Overall, the dissertations address an extensive range of subjects and suggest the topics that especially interested Linnaeus. Access to all the dissertations is provided by the superlative digital archive of the Hunt Library at Carnegie Mellon University (Graham 1966; Sörlin and Fagerstedt 2004; Svanberg 2006).

2.1. *Betula nana* (1743)[1]

The dissertation entitled *Betula nana* (1743) and listed under the name Laurentius Klase, demonstrates Linnaeus's early effort to assimilate information regarding a northern birch tree (Figure 1). The fundamental objective is to affirm that the *Betula nana*, or the dwarf-birch tree, may be found in Sweden, Norway, Greenland, Moscovy and other parts of Russia. Johann Amman, the Russian botanist working in the St. Petersburg Academy, is cited twice. First it is confirmed that Amman has seen the northern birch tree in the marshland of Ingermanland and Karelia: "Amannus in paludibus Ingermanniae & Careliae eam vidit" [Amman has seen it in the marshy areas of Ingermanland and Karlia]. The second reference to Amman confirms that he found the plant in Russia in 1739 after reading Linnaeus's *Flora lapponica* (1737). Although the dissertation gives Amman credit for identifying Russian synonyms for the *Be-*

Q. F. F. Q. S.
DISSERTATIO BOTANICA
De
BETULA
NANA,
Quam,
Consensu Amplissimæ Fac. Medicæ
in Reg. Acad. Upsaliensi,
PRÆSIDE
VIRO Experientissimo & Celeberrimo
D. D. CAROLO LINNÆO,
Med. & Botan. Prof. Reg. & Ord.
Publico Bonorum Examini modeste subjicit
Sæ. Ræ. Mtis.
ALUMNUS
LAURENTIUS MAG. KLASE,
SMOLANDUS.
In Auditorio Gustav. Maj. ad d. 30 Junii
An. MDCCXLIII.
Horis, ante meridiem, solitis.
STOCKHOLMIÆ
Typis Historiographi Regni
Hr. Candidat Broling

Figure 1. Dissertatio Botanica de betula nana (30 June 1743). Lidén 001.

1. *Dissertatio botanica de Betula Nana...Laurentius Klase* (30 June, 1743) Dissertations consulted are from the on-line Linnaean Dissertation Database at the Hunt Institute, Carnegie Mellon University [Lidén 001] Page numbers in the text refer to page numbers in the dissertations.

tula nana, it also reports that the tree was not included in his contribution to the *Proceedings of the Imperial Academy of Sciences at St Petersburg.*

The dissertation displays Linnaeus's practice of documenting synonyms and stories associated with plants as a means for showing the problems associated with botanical taxonomy. While languages vary widely in the names given to natural phenomena, it is the task of natural philosophy to identity fundamental, shared structure. In the midst of Latin we are given multiple European names for betula: *björk, birk, birken, birchtree, brozoza, Säke* (Lapp) but then multiple names in Swedish: *Hängbjörk, kartbjörk, fjällbjörk, Klingriss (Västbothnia), fjälllrapa (Dalarna); fredagsbjörk,(Småland); Ryprijs (Lapp) dwargbjörk, Ryelträ.* One name in particular draws attention, namely *fredagsbjörk.* This name was used to commemorate the passion of Jesus in Småland but, as Linnaeus observes, the association should be taken as a simile because the tree does not grow in the Holy Land. Even more, it seeks to affirm the existence of the tree as a unique natural object.

In a letter to Linnaeus (15 November 1737), Johann Amman poses a question encountered frequently in the collection of plant names. How does one build a taxonomic nomenclature that takes into account the multiple local vernacular names for plants (Blunt 2001:120)?

> You promise to account, in your *Critica Botanica*, for your numerous alteration of names. I presume that you have followed the rules you laid down in your *Fundamenta Botanica*. But many of these rules may not be universally approved, any more than your change of names. I beg you to consider what would happen if everyone were to lay down such laws and regulations whenever he felt so inclined, overturning names already known and approved by the best authors just for the sake of making new ones. Would it not lead to worse than the confusion of Babel?

Amman's question reminds us that the Siberian expeditions comprised elementary fieldwork in linguistics. From the outset, the expedition demanded that the multiple languages of the empire be recognized. The documents and samples of writing that were returned to St. Petersburg revealed that the Russian territories contained an array of languages that had never been the object of European study. The information collected by G.F. Müller during the second expedition was enormous. "His completed and catalogued collections included forty-two books of chronicles, and books of description of Siberia (ten prepared by Müller himself and three compiled by his students and overseen by him) and a large quantity of maps, drawings, and city plans. Further, he delivered fourteen thick files of reports, documents, letters, orders, and other communications between his group and St. Petersburg between 1733–1743." (Black 1986:76) The assimilation of languages and dialects was also taking place among an international group of explorers who themselves were continually engaged in translation. For Germans such as Müller, the expedition also served as an intensive course in Russian.

The reports sent back to Moscow and Petersburg in German and Russian contributed to the urgency for establishing a Russian language that could be relied upon for the purpose of collecting and maintaining information. Not surprisingly, the practice of using Greek and Latin in an increasingly chauvinistic Slavic setting became an issue in itself. Such nomenclature worked but it also carried a non-Russian stamp. Almost immediately conflict emerged as Russians worried that the collection of data was largely in the hands of non-Russians. An indication of how sensitive non-Russian involvement had become appears in a 1746 letter to Linneaus in which Gmelin reports that it has become a crime to export plants beyond the borders of Russia:[2] "Magnum enim crimen habetur plantam Russicam alibi vulgandam concedere." [It has even been decreed that it is a major crime to send Russian plants to other places.] The prohibition on sending plant specimens abroad suggests more than national chauvinism. It indicates that Russian academicians were increasingly aware that a foreign taxonomy of Russian resources amounted to surveillance. It reminds us as well of the ways military conflict could be replaced by competition between scientific academies.

The Klase dissertation raises questions that will appear in the late eighteenth century as an effort to the connection between the creation of plant taxonomies and taxonomies for folktales. It is interesting to consider how twentieth-century Russian linguistic research devoted to the multiple levels of language carries the traces of the impact of the Siberian Expeditions on the study of linguistics. (Jakobson 1990 (1956): 69–79) In concurrent research, I am exploring the ways taxonomic process may be traced in the catalogue of constellations and stars. (See for example Ryan 1999: 133–135.)

2.2. *Plantae rariores Camschatcenses* (1750)

The dissertation, *Plantae rariores Camschatcenses*, published in 1750 under the name Jonas P. Halenius, advertises the Russian flora collected by Linnaeus since 1743 (Figure 2). The first three sections of the dissertation offer an encomium to recent European botany. The third part celebrates the work of Linnaeus: "Haec omnia commilitones meos ad imitationem Magistri sui clarissimi ita excitarunt, ut plurimi, amoenitate scientiae capti, in aperto hoc Florae the-

2. Letter from Johann Georg Gmelin to Carl Linnaeus, dated 21 October 1746 (All references are to the on-line *Linnaean Correspondence* LO741).

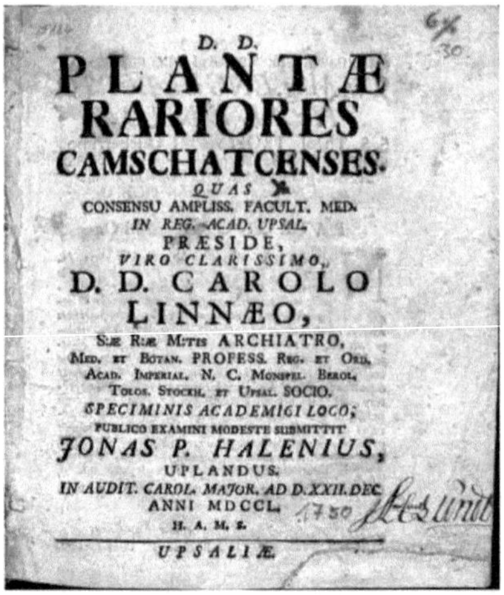

Figure 2. Plantae rariores camschatcenses (22 December 1750) Lidén 030.

atro vires suas experiri, non sine fructu, tentaverint." [All my remarks are a mere imitation of my highly celebrated teacher who has opened for those drawn to the love of knowledge a floral theatre that contains more than a little fruit.] The fourth section is devoted to the new discoveries made in Siberia. Samuel Georg Gmelin is singled out above all others: "qui novem annorum indefesso itinere, totam fere penetravit Sibiriam, quin Publico satisfaciat Botanices non dubito, cum Floram suam Sibiricam, omnibus numeris absolutam, in lucem edere valeat, in quae nescio, utrum magis admiranda sit Autoris plantarum rariorum ingens atque stupenda collectio" [For nine years with indefatigable energy he traveled in the heart of Siberia with little doubt that his botanical work *Flora Siberia* (1748–49) would satisfy the public and bring to the light and admiration many rare plants that had been unknown].

Credit is given next to the collection of plants made by Lercheus and used by Pavel Grigorij Demidoff.[3] His collection is of interest because it shows similarities with plants collected in Lappland: "Not only a few but many of (the plants he collects) are identical to plants found in our Lappland – some completely ignored or given little attention –. Indeed, it may be argued that some of the plants found in Kamchatka are hardly distant from those found in

3. The reference to 'Lercheus' is probably a reference to J.J. Lerche. See Alan Graham, *Plantae rariores camschatcenses: A Translation of the Dissertation of Jonas P. Halenius, 1750* (*Brittonia* 18:2 New York: Springer Verlag 1966 [2008], 131–139.

Canada"[4]. The dissertation shows Linnaeus's interest in identifying the connection between northern flora in Lappland, Siberia, and Canada. While *Betula nana* (1743) focuses on a single plant, this dissertation concentrates on the assimilation of multiple herbaria as well as examination of individual plants that have been grown in Uppsala. For example, references to *Kleinia* (Hort. Ups. 254), *Spiraea* (Hort. Ups. 131) suggest an *iter plantarum* that includes plants that have been grown in Uppsala. Overall, the dissertation reveals the development of an expanding network of exploration and comparative research that is being centralized in Uppsala through documentation, herbaria, and actual garden plots.

While the dissertation at first appears as a list of plants, it also includes detailed information of the network used to bring plants to Uppsala. Gmelin often receives praise for his discoveries. For example, his description of *GENTIANA* (item 5) closes with "Inventionis gloria debetur Cl. Gmelino" [Credit for this splendid discovery is due Gmelin]. Item 6 listed as "*SWERTIA*" (item 6) is described as "Inventa primum a Cl. Gmelino, qui eandem ad D. Raesidem misit; deinde a Stellero, quam D. Demidoff misit." [having first been discovered by Cl. Gemlin who sent it to D. Raesidus; Stellero sent it next to D. Demidoff]. *SPIRAEA soliis pinnatis*, item 15 in the dissertation "Frutex missus fuit ad Hortum Upsaliensem a Nobi. D. Demidoff, sed periit ex itinere" [was sent to the garden in Uppsala by his excellency D. Demidoff but died as a consequence of the trip]. *HELLEBORUS* (item 18) is described as "the smallest of its kind but was still able to inspect. *The Flores Siberiae* includes Fumaria, which has an affinity with a bulb-like plants, has cone-like flowers and is the largest within this genus"[5]. *BARTSIA* (item 19) is described as "Hanc plantam legit etjam Cl. GMELINUS" [read by Cl. Gmelin]. The network of personal connections, individual herbaria, and books generated from the herbaria serves as an invitation to add to the growing body of knowledge. Through Linnaeus and others the collected plants are being given a language.

2.3. *Necessitatem promovendae historiae naturalis in Rossia* (1766)

The 1766 dissertation published under the name of a 17-year old A. M. Karamyschev (1744–1791) recites the Siberian plants in the Uppsala garden (Figure 3).

4. "non solum multas, cum rarissimis nostris plantis Lapponicis communes sed etjam alias, partim ignotas omnino, partim minime tritas; & denique quasdam etjam cum Canadensibus easdem, argumento Canadam a Kamsk chatca non longe distare..." (5)
5. "Minima est hae planta in su genere, attamen spectabilis; inter Flores Sibiriae speciosos & maxime singulars est etiam quaedam Fumaria bulbosis affinis, floribus condecorata, in so genere maximis." (23)

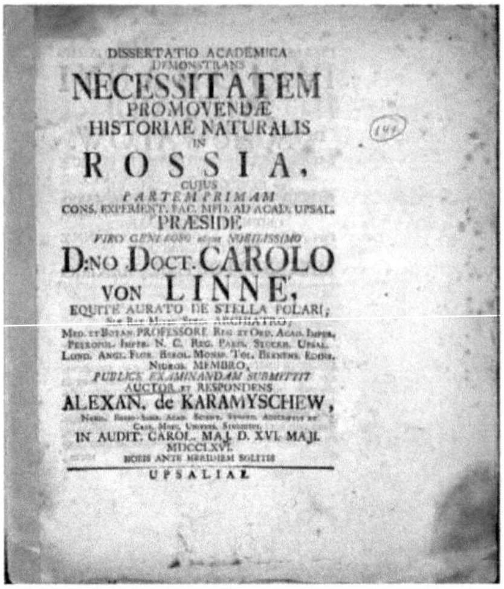
DISSERTATIO ACADEMICA
DEMONSTRANS
NECESSITATEM
PROMOVENDÆ
HISTORIAE NATURALIS
IN
ROSSIA,
CUJUS
PARTEM PRIMAM
PRÆSIDE
D:NO DOCT. CAROLO
VON LINNE,
EQUITE AURATO DE STELLA POLARI;
PUBLICE EXAMINANDAM SUBMITTIT
AUCTOR ET RESPONDENS
ALEXAN. de KARAMYSCHEW,
IN AUDIT. CAROL. MAJ. D. XVI. MAJI.
MDCCLXVI.
UPSALIAE.

Figure 3. Dissertatio academica demonstrans necessitatem promovendae historiae naturalis in rossia (16 May 1766) Lidén144 and 144.1.

Well beyond the enumeration of 351 plants, the document furthers Linnaeus's fame. The dissertation provides an opportunity for Linnaeus to throw his voice into Karamyschev and speak about Russia as if he is describing his own country. The prefatory letters are directed strategically. Johannus Shuwalov [Shúvalov] was an advisor to Catherine the Great and had the support of Lomonosov; Adodurow [Adodórov] was a member of the Russian Assembly of the Academy and had been Catherine's Russian teacher. In 1762 he became the curator of the University of Moscow (Black 1986). He was especially interested in the development of manufacturing in Russia.

The preface celebrates the Linnaean system and affirms the ways that the new science transcends the imperfect memory of the ancients.

> Uti vero illi, qui scientias addiscere, operae pretium ducent, omnium primo debent litteras & syllabas sibi redere familiars, & tandem ipsa componere verba; sic etiam & qui naturae cognitione imbuti esse volunt, id in primis agree debent, ut cognitos sibit reddant terminus technicos scientiae, dehinc partes quascunque animalium, vegetabilium & mineralium considerent, & tandem expednant ipsas systematis leges, quae continuo indicant objecti cognoscendi classem, ordinem, genus, speciem, caet. (A2 verso)

> 'Those whose work contributes to knowledge and whose work becomes distinguished before anything else learn to shape familiar letters and syllables and even to shape them into words. Thus it is that they who seek to think about nature must first know how to deal with technical knowledge that involves the parts of ani-

> mals, vegetables and minerals and that includes on systematic laws which offers objective was of classifying order, genus and species.' (A2 verso)

For those who know the system, the entire world will be made know for they will be able to comprehend the synonyms that exist for every true name.

> Those who are instructed in this way of knowing are able to intuit the true name of things and make intelligible the entire world. By knowing the names of things and by seeing the way that synonyms work, those who are knowledgeable reveal the way that brings distinction to moral action, economic behavior, the nation and life itself. (A2 recto)[6]

Distinguished national collection of flora and fauna may now be joined to create a universal collection. The key to such a global vision depends on a Linnaean system that may even serve to share Russian resources with the world.

The dissertation celebrates Linnaeus for discovering an all encompassing system. While natural history may have originated in southern Europe during antiquity, such work was corrupted by the Catholic Church. Today, Russians are able to celebrate an empire that extends from the White Sea to Turkey, Persia, Mongolia, China and even to America: "Extenditur etenim hinc ab ipso mari glaciali ad limites usque Regnorum Turcici, Persici, Mogolici, & Chinici; & inde a mari Balthico ad mare usque Americanum." Subsequent paragraphs complement The Imperial Academy of Sciences at St. Petersburg and include a bibliography of those who contributed to the growth of botany in Russia; Messerschmidt, Buxbaum, Gmelin, Kranscheninikov, Martin, Steller, Amman, Heinzellmann, Gerber, Lerche, Schober, Gorter. How remarkable, notes the dissertation, that so much could be accomplished in a period of only 40 years.

Russian evidence even shows that plants have migrated through the movement of people. Plants such as *Humulus lupulus*, *Spinacia oleracea*, *Atriple hortensis*, *Artemisia Dracunculus* that are eaten in Europe were introduced when in the barbaric ages our Goths[my emphasis] occupied Italy: "ut ignotae fuere veteribus & introductae, seculis barbaris, dum Gothi nostrates occupabant Italiam." The reference to 'our Goths' [Gothi nostrates] living in Italy alerts the reader that Linnaeus has also situated his work in a setting of seventeenth-century historiography. However, instead of using linguistic evidence as Olof Rudbeck had done, Linnaeus suggests that question human migration may be answered by the migration of plants. The dissertation poses a

6. "His qui instructus fuerit mediis, primo fere intuitu agnoscit individui verum nomen, & per totum terrarum orbem intelligibile; nominee noto, synonyma autorum perspecua fiunt, quae viam monstrant omnia eas cognoscendi, quae seculi beneficio innotuere; unde mores, oeconomia, patria, usus, & caet. (A2 recto)

remarkably Baroque historical question in surprising modern terms: What do we learn about the migration of people from what they eat? While we know from the Biblical account of the Garden of Eden that our earliest ancestors lived in the tropics, we now may ask what we can learn from a Biblical report that Noah's ark settled in the mountainous region of Ararat. Indeed Ararat was part of Siberia: "Ab hoc monte catena montium facile ducit versus regions nostras inter Sibiriam Tatariamquo sitas, quae omnium altissimae sunt, unde etiam ex hisce fluvii originme ducunt, qui decurrunt in mare glacialie, in Oceanum Orientalem, Mare Caspium, immo ad omnes circumjacentes terras." Because of our geographical and botanical knowledge, we may ask if Siberia doesn't contain vestiges of the original plants collected by Noah.

> Videtur mihi itaque posse concludi, Patriam meam fuisse eam, ex qua sorte omnes post diluvium exivere mortals, & late dispersi sunt, quoniam his in regionibus, extra tropicos, primaria inveniuntur hominum alimenta....Unde & non mirum, stupendam & reliquis uberiorem copiam frumenti apud Sibiricos provenire quotanne copia tume prodiret, si natura arte insuper juvaretur? Tum certe, totus terraqueus noster globus, sola Sibiriarum naturali fertilitate impleri & ali potuisset. Quaene fuisset Gloria regni nostri, quaene fama nationis? Si solum studium nistoriae naturalis promovere in Patria carissima vellemus.
>
> 'It appears to me that one might conclude that it was my own country from which humanity emerged after the flood and from which they were dispersed since in these regions even in the outer regions one finds sustenance... Why should it surprise us if through her art nature should provide abundance in Siberia? Indeed, all of our globe could be nourished and sustained with the natural fertility of Siberia. What is the glory and fame of our kingdom and nation? We seek only to foster the study of natural history in beloved country.'

Linnaeus reinforces his hypothesis by suggesting that he is surveying his garden in Uppsala while writing: "Incidit mihi, dum haec scribo, observatio quaedam circa plantas nostras, notatu haud indigna, quae in florentissimo Musarum Upsaliensium horto, primum facta-est, quamque hoc apponere loco omnino operae pretium duco, plantae scil. ... Observatio haec summi, mea quidem sententia, momenti, indicat etenim, quod plantae seminibus vel radicibus e Sibiria in Europam accersitae, hic crescant laetius vegetiusque quam si aliunde." [It strikes me, even as I am writing, that my plants, growing in this most florid Garden of Muses in Uppsala, were placed in this space as gifts. Indeed this observation urges me to the maxim that the seeds or roots of plants obtained from Siberia or Europe, grow here as abundantly and lusciously as anywhere.] Looking into his garden he would see a microcosm of dissemination. The Siberian garden provokes a sense of wonder not only because its plants came from seeds and plantings from all over Russia but because it seems to display an almost universal capacity for plants to adapt to different climates. Siberia is the mother of the plants that growing in Uppsala: "Ita videmus in horto Upsaliensi illas plantas, quarum Sibiria mater est…". He notices that during a twenty-year period one is able to notice the emergence of new plants: "Haec si quis perpendat, fatebirtur mecum, hortos Europaeos spatio 20 annorum novam quasi indussie faciem, atque quasi Metamorphosin quondam subisse: Si causam hujus novae naturae formae quaerat quis, aliam non inveniet, quam quod nostrae opes illorum hortos ditaverint." [If one considers these things, as I have, one sees that in the period of twenty years new things appear in European gardens as if a metamorphosis had taken place. If one asks where these new things come from, one discovers nothing other than that our gardens have brought forth riches.]

The dissertation concludes with letters and celebratory poems that glorify the accomplishments of the 17-year old Karamychev. German, Swedish, French, Latin, Russian. While the poems celebrate Karamyschev, they are certainly orchestrated to celebrate Linnaeus himself. Samuel Ol. Tilas celebrates the "immanquables du Parnasse d'Upsal". Leon von Schokurow celebrates Russia. J. G. Akrel, friendship; Chretien Lado, "Le plus grand Heros dan l'Histoire de la Nature Vous y instruit.". Linnaeus as is prophesied will fill the vacancy left by the premature death of Johann Georg Gmelin.

3. Language, Taxonomy and the "Space-Between"

The dissertations considered above exhibit a process of continuous translation between languages, objects, and the languages that come to represent objects. As rhetorical structures they reveal Linnaeus 'throwing' his voice into his students. Placed side by side, the dissertations provide a glimpse into Linnaeus's evolving response to the Siberian Expeditions. The movement of students from Russian to Swedish centers of research also shows us something important about the evolving status of knowledge in the eighteenth century. The students that work with Linnaeus are part of a network of growing exchange. A. M. Karamyschev is following the example of Peter I himself as he travels to Europe.While books represent the portability of information from an earlier period, the growing movement of students in the eighteenth century shows that information also becomes increasingly situated and even anchored through eye-witness experience. The students who come to study with Linnaeus convey in their travels caution and perhaps even skepticism about the capacity of linguistic phenomena to accurately convey information. Instead of relying on Latin, German, Russian or Swedish, their presence suggests an interest in first-hand experience. While language and the practice of multiple translation obviously serves as an agent for their study, a single language is in itself insufficient.

At the same time the dissertations function as abbreviated linguistic *herbariae*, they also remind us that the plants are not words but natural objects. The question of how such literal phenomena become addressed not simply through words or other symbols but as objects engages the discourse of natural philosophy throughout the eighteenth century. It is such a question that appears again and again in the work of Isaac Newton. Whether in his mathematical work or diverse work with alchemy or the transmission of historical data, Newton works in steps to give expression to the phenomena before him. An important strategy in his methodology involves finding a common denominator that would allow him to factor out extraneous information. Newton's work does not stop with factoring, however. It continues until he is able to present the phenomena in question with what I would describe as a loyalty to representation. The hermeneutical or interpretive stance does not involve building allegories but emphasizing the interpretation of literal phenomena. I draw attention to this because it reminds us that the eighteenth century was not simply interested in the promiscuous gathering of data but was drawn to a comparative formulation of methods that would allow natural scientists to speak for the phenomena in question or even give phenomena the capacity to speak for themselves.

Formulated in the language considered a European standard for scholarship and scholarly exchange, the dissertations comprise a moveable stage for academic commentary and an experimental space for various modes of representation. It is significant that the type-font itself changes as the dissertations develop. From the Gothic type of dissertations in the 1740s one moves to the emergence of Latin and Italic type in the 1760s. Russian appears but is transliterated into Roman characters. Of course, printing technologies become a translation mechanism in themselves (Cracraft 2004). The language used by Linnaeus is not part of a fixed Latin lexicon but one which provides nouns, prefixes, and suffixes that may be fit together like *legos* to create new Latin words. Within an international context the creation of new academic vocabulary also bears a presumption that the 'new' vocabulary will take precedence over vernacular terminology.

What I have referred to as the "space-between" asks us not simply to attend to a defining structure but to see how the development of a Latin nomenclature provides a means for seeing what other wise would not be seen. The system when viewed from such a vantage point is dynamic and one that invites others to participate. Such "space-between" may be thought of as an imaginary space but we can say much more than this. The use of Latin nomenclature also enables distributed cognition because the names are deployed to invoke and test interaction with natural objects. Stamens and pistols may be identified and compared, colors registered and leaves counted. The process involves counting not in the abstract but in reference to leaves, flowers, seeds, and root systems. Words are a departure point for what we might be inclined to see as empirical. But even here the empirical that has been encoded into the formulaic words becomes unbundled as one reconstitutes the object. It is the comparative work between languages that also allows us to posit the work of distributed cognition. This is not a metaphoric structure associated with poetry but a linguistic framing device that provokes curiosity. From such a vantage point, the dissertations may be viewed as performative vehicles. As author, and perhaps even more accurately as conductor of a masque or opera, Linnaeus stages and scripts both his students and himself. There is reason to think of these dissertations as a Linnaean genre.

These Uppsala dissertations are a microcosm of what happens in the sequence of Russian northern expeditions. The Russian academy required that reports be sent to Petersburg be in Russian and Latin. (They were also sent in German and French.) What is important is that the languages function like recording instruments that must be calibrated with each other. Information from Siberia is sent like data from a Martian Rover. Building a meta-code that draws upon vernacular languages may establish a hierarchy that makes natural language less important and peripheral. In the case of Linnaeus's work on Latin taxonomic system provokes his own richly prodigious use of his own Swedish. In the case of his Russian counterparts, however, the use of Latin system raises questions about the suppression of the Russian vernacular and the rich folk nomenclature with which it is associated.

It is precisely here that the Linnaean system of taxonomy particularly in relation to Russian flora may be said to comprise a scientific kingdom of knowledge that could be viewed as superseding Swedish political designs. By supplying the taxonomic structure and language used to provide universal code, Linnaeus accomplishes what Carl XII was never able to do in his efforts to maintain geographic military control of the Russian empire. The act of naming, of course, is an act of control. Why indeed would Sweden need to maintain geographic control of Russian territory, if she were developing a system of reconnaissance that allowed her to know about the natural resources of the large neighbor across the Baltic? It is a very tiny step from botanical research to espionage, especially in regard to natural resources. But rather than thinking that a Linnaean metalanguage is being superimposed on Russian, it makes more sense to think of the ways that Linnaean taxonomy provoke an expansion of Russian language studies. Here we might compare the Linnaean influence on Swedish itself. Swedish as Russian was in the process of development. Linnaeus' own research hardly suppresses local names for plants but serves to celebrate them in all their variety. As we have seen, the long list of Swedish names for *Betula nana* is one example. It is this aspect of Linnaeus' legacy that can so easily be missed. At the same time that he enables a universal taxonomy, he provokes the collection of his own native language: a *florelegia sveciana*.

Linnaeus's interaction with Russian students and colleagues offer multiple points of entry into evolving eighteenth-century scientific networks. While we may recognize the ways in which Linnaeus reveals a Baroque sensibility that looks to the past, we also become aware of the ways he constitutes new networks. Within the setting of seventeenth and eighteenth century northern history, Linnaeus contributes to a redefinition of Sweden's interest in Russia as a source for natural resources and as a potential domain for knowledge. Given global research undertaken by his students and the adaptation of his system of classification, Linnaeus manifests strategies which mark Sweden's capacities to contribute to the creation of a new "Kingdom of Knowledge". While these strategies involve the collection and integration of data from the far corners of the earth, the Russian territories provide an expansive setting for shaping methods that are extended to all continents. The dissertations serve as the microcircuitry involved in the changing forms of empire and show how the collection of specimens participates in the development of more complex applications of language to nature. Linnaeus shapes not only translation of the plant kingdom into a universal language but also participates in the translation of Sweden from a nation engaged in an expanding political empire to a nation that comprehends the value of an expanding empire of knowledge.

References

Bergman, Karl
2002 *Makt, möten, gränser: Skånska kommissionen i Blekinge 1669–70*. [*Power Treaties Boundaries: Skåne commission in the Province of Blekinge 1669–70*] Lund: Studia Historica Lundensia.

Björner, Eric
1742 *Cogitationes Critico Philologicae de Orthographia Linguae Svio Gothicae Tam Runica Quam Vulgari*. [*Critical reflections on the philological origins of the Swedish-Gothic language*] Stockholm.

Björner, Eric
1676 *Schediasma historico-geographicum de Varegis, heroibus Scandeianis et primus Russiae dynastis, quo licuit studio exhibitum*. [*An overview of the geographical history of the heroric Varangians and also the founders of the first Russian dynasty*]. Stockholm.

Black, Joseph L.
1986 *G.-F. Müller and the Imperial Russian Academy*. Montreal: McGill-Queen's University Press.

Blunt, Wilfrid
2001 *Linnaeus: The Compleat Naturalist*. Princeton: Princeton Univ. Press.

Bromé, Janrik
1923 *Karlskrona Stads Historia Del 1 1680–1790*. [*A City History of Karlskrona, Part 1: 1680–1790*]. Stockholm.

Bureus, Johannes
1731 *Svecia sive de Suecorum regis dominiis et opibus* [*Svecia or concerning the dominions and deeds of the Swedish monarchy*]. Lyon.

Bureus, Johannes
1744 *Hyperboreorum Atlantiorum seu Sviogotorum & Normannorum Edda, hoc est, Atavia seu fons gentilis illorum & Theologiae & Philosophiae etc.* [Concerning the hyperborean Atlanteans, their northern Sveagothic Edda and the origins of their theology and philosophy]. Uppsala.

Cracraft, James
2004 *The Petrine Revolution in Russian Culture*. Cambridge: Harvard Univ. Press.

Dahl, Per
1995 *Svensk ingenjörskonst under stormaktstiden: Olof Rudbecks tekniska undervisning och praktiska verksamhet*. [*Swedish engineering during 'time time of Greatness': Olof Rudbeck's course offerings in technology and the practical arts*] Doktorsavhandling. Uppsala universitet, Institutionen för idé-och lärdomshistoria. Skrifter Nr 14.

Daniels, Rudolph L.
1973 *V. N. Tatishchev: Guardian of the Petrine Revolution*. Philadelphia: Franklin Publishing Company.

Donnert, Erich
1986 *Russia in the Age of Enlightenment*. Leipzig: Interdruck Graphischer Grossbetrieb.

Englund, Peter
1988 *Poltava: Berättelsen om en armés undergång*. [*Poltava: An account of an army's destruction*]. Stockholm: Atlantis.

Eriksson, Gunnar
1994 *The Atlantic Vision: Olaus Rudbeck and Baroque Science*. Canton, MA: Science History Publications.

Eriksson, Gunnar
2002 *Rudbeck 1630–1702: Liv, lärdom, dröm i barockens sverige* [*Rudbeck 1630–1702: Life, learning, and vision in baroque Sweden*], Stockholm: Atlantis.

Frost, Robert I.
2000 *The Northern Wars 1558–1721.* (Essex: Pearson Education Ltd [Longman].

Frängsmyr, Tore
2000 *Svensk idéhistoria: bildning och vetenskap under tusen år: del I 1000–1809. [An intellectual history of Sweden: education and science over one thousand years: part 1 1000–1809].* Stockholm: Natur och Kultur: 236–258.

Graham, Alan
1966 *Plantae rariores camschatcenses: A Translation of the Dissertation of Jonas P. Halenius, 1750. Brittonia* 18:2 New York: Springer Verlag 1966 [2008], 131–139.

Halenius, P. P.
1750 *Plantae rariores Camschatcenses.* University of Uppsala. [On-line Linnaean Dissertation Data Base, Hunt Institute, Carnegie Mellon University (Lidén 030)].

Jakobson, Roman
1990 'The Speech Event and the Function of Language,' *On Language.* Cambridge: Harvard Univ. Press: 69–79.

Jangfeldt, Bengt
1998 *Svenska vägar till S:t Petersburg [The Swedish path to St. Petersburg]* (Stockholm: Wahlström & Widstrand, 1998).

Karamyschew, Alexander
1766 *Dissertatio academica demonstrans necessitatem promovendae historiae naturalis in rossia.* University of Uppsala. [On-line Linnaean Dissertation Data Base, Hunt Institute, Carnegie Mellon University (Lidén 144 and 144.1)].

Klase, Laurentius
1743 *Dissertatio botanica de Betula Nana.* University of Uppsala [On-line Linnaean Dissertation Database, Hunt Institute, Carnegie Mellon University (Lidén 001)].

Knoespel, Kenneth J.
1994 "Reshaping the Earth: Olof Rudbeck and the Transformation of Sweden," *Cultural Exchange between European Nations during the Renaisssance* ed. Gunnar Sorelius and Michael Srigley. Uppsala: Studia Anglistica Upsaliensia 86: 193–212.

Knoespel, Kenneth J.
2004 "The Edge of Empire: Rudbeck and Lomonosov and the Historiography of the North," *In Search of an Order: Mutual Representations in Sweden and Russia during the Early Age of Reason* ed. Ulla Birgegård and Irina Sandomirskaja. Stockholm: Almqvist & Wiksell (Södertörn Academic Studies 19): 129–154.

Koerner, Lisbet
1999 *Linnaeus: Nature and Nation*. Cambridge: Harvard Univ. Press.
Koener, Lisbet
1994 "Linnaeus' Floral Transplants," *Representations* 47: 144–169.
Larson, James L.
1971 *Reason and Experience*. Berkeley: Univ. of California Press.
Larson, James L.
1994 *Interpreting Nature: The science of Living Form from Linnaeus to Kant*. Baltimore: The Johns Hopkins Univ. Press.
Lindroth, Sten
1967 Linné – Legend och verklighet" *Löjtnant Åhls äventyr.* ['*Linné: Legend and reality' Lieutenant Åhls' adventure*] Stockholm: Wahlström & Widstrand: 9–87.

Linnaean Dissertation Database, Hunt Institute, Carnegie Mellon University. http://huntbot.andrew.cmu.edu/HIBD/Departments/Library/LinnaeanDiss.shtml
Loccenius, Johan
1676 Antiquitatum seveo-gothicarum. [*The antiquities of the sveogoths*]. Uppsala.
Magnus, Olaus
[2005] *Carta Marina* ed. Elena Balzamo. Paris: Librairie José Corti.
Magnus, Olaus
[2001] *Historia om de nordiska folken*. [*A History of the People of the North*]. Hedemora: Gidlunds.
Malm, Matts
1996 *Minervas apple: om diktsyn, tolkning och bildspråk inom nordisk göticism*. [*Minervas' apple: concerning poetics, interpretation and imagery in northern Gothicism*]. Stockholm: Brutus Östlings Bokförlag Symposion.
Oredsson, Sverker ed.
1998 *Tsar Peter och Kung Karl: två härskare & deras folk*. [*Tsar Peter and King Carl: two rulers and their people*]. Stockholm: Atlantis.
Peringskiöld, J.
1710 *Monumentorum sveo-gothicorum* [*Concerning the monuments of the sveogoths*]. Uppsala.
Rudbeck, Olof, den yngre
1702 *Lapponica illustrata*. (*Lappland Revealed*) Stockholm.
Ryan, W. F.
1999 *The Bathhouse at Midnight: Magic in Russia*. University Park: Penn State Univ. Press.
Scheffer, Johannes
1698 *Svecia literata* [*Sweden Writes*]. Hamburg.

Strahlenberg, Philipp Johan
1730 *Das Nord- und Ostliche Theil von Europa und Asia, in so weit solches das ganze RuBische Reich mit Siberien under der grossen Tataren in sich begreisset, In einer Historische-Geographischen Beschreibung der alten und neuern Zeiten, und vielen andern unbekannten Nachrichten vorgestellet.* Stockholm.

Strahlenberg, Philipp Johan
1736 *An Historical-Geographical Description of the North and Eastern Part of Europe and Asia; But m,ore particularly of Russia, Siberia, and Great Tartary Both in their Ancient and Modern State: Together with An entire New Polyglot-Table of the Dialects of 32 Tartarian nations: And a Vocabulary of the Kaluckmungalian Tonges*. London.

Swedenborg, Emmanuel
1734 *Opera philosophica et mineralia*, 3 vols. Vol. 1: *Principia rerum naturalium sive novorum tentaminum phaenomnea mundi elementaris philosophice explicandi*. Dresden and Leipzig.

Tarkiainen, Kari
1974 *"Vår Gamble Arffiende Ryssen:" Synen på Rytssland I Sverige 1595–1621 och andra studier kring den svenska Rysslandsbilden från tidigare stormaktstid* [*'Our Old arch-enemy': Views of Russia in Sweden 1595–1621 and other studies concerning Sweden's image of Russian from early years of the 'Time of Greatness.'*]. (Studia Historica Upsaliensia 54).

The introduction of the Linnaean classification of nature in Portugal

Palmira Fontes da Costa

1. Introduction

In *Systema Naturae*, published in 1735, Linnaeus first presented his system of classification of the natural world to the international community. By 1770, the work was in its thirteenth edition. Some idea of its impact on a Portuguese reader from the period can be inferred from an annotated copy now held at the Portuguese National Library (Linnaeus 1793). On the one hand, the copious remarks of the anonymous reader testify to his/her eagerness to access the Linnaean language of nature. On the other hand, they reveal the reader's persistent difficulties in understanding Latin.

What kind of efforts were made by Portuguese naturalists to make the Linnaean system of classification more accessible to students and to the general enthusiast for natural history? What was the impact of the Linnaean system in the development of Portuguese botanical science? And what was the influence of presentations of the Linnaean classification in works that crossed the boundaries between science and literature?

In this paper I examine the contribution of the Linnaean method of classification to the reshaping of botanical education in Portugal. Next, I show how the publication of a Portuguese flora using the Linnaean system of classification was associated with important national aims. I also examine the importance of including illustrations in Portuguese works that first presented the Portuguese flora according to the new principles of classification. I then discuss the introduction of the Linnaean system of classification in works dealing with the flora of the Portuguese Empire and look at some of the difficulties faced by their authors. Finally, I focus on another genre of literature which contributed to the diffusion of the Linnaean ideas of classification, the poetic work of the Marquesa de Alorna's *Botanical Recreations*. In this way, I demonstrate the variety of works and places in which the Linnaean language of nature was influential not only in Portugal, but also in the Portuguese Empire.

2. Classification and botanical instruction

The Linnaean programme for the classification of nature entered the Portuguese teaching system as a result of the reform of the University of Coimbra launched in 1772 by the Marquis of Pombal (1699–1782), then the Prime Minister of King José I (1714–1777) (Carvalho 2001: 423–484). The reform contemplated the creation of a chair in natural history which, for the first time, enabled Portuguese naturalists to be educated at university level. In the study of botany, it became obligatory to follow Linnaeus's *Philosophia Botanica* (1751), to which was added the thirteenth edition of Linnaeus's *Systema Naturae* published by the University of Coimbra in 1793. The professor invited to occupy the new chair was Domenico Vandelli (1732–1815), a naturalist born and taught in Padua, where he completed his university education in medicine and natural history (Melli 1966). Vandelli had insight into the ideas of Linnaeus (1707–1778) not only through his works, but also from correspondence maintained with him for several years (1759–1773).[1] The correspondence indicates that, besides the exchange of news relevant to natural history, he was influential in sending Linnaeus various specimens, in particular from Brazil and other parts of the Portuguese Empire.

In accordance with the aims promoted by the new university's statutes, Vandelli founded the Natural History Cabinet in 1772 and collaborated in the creation of the University Botanical Garden the same year (Carvalho 1987: 44–62). Both institutions were founded with the aim of having a prominent role in the practical instruction of students enrolled on the course in natural history. Significantly, the Botanical Garden was organized according to the Linnaean principles of classification.

During this period, there was a huge lack of botanical works. The first attempts to give a Portuguese shape to this area of knowledge are to be found in two books, both published in 1788: Vandelli's *Dictionary of the technical terms of Natural History extracted from the works of Linnaeus* (Vandelli 1788) and Félix Avelar Brotero's (1744–1828) *Compendium of Botany* (Brotero 1788). If Vandelli's work is confined to a translation of some of the most frequently used botanical terms, Brotero's aimed to provide a complete introduction to botany, also including an extensive dictionary of names. In fact, this can be considered the first Portuguese treatise on the subject. Its relevance resides not only in its pioneering role but also in its intrinsic value, which was acknowledged by some

1. A summary of the letters exchanged between Linnaeus and Vandelli can be found at the Linnaean Correspondence Project (http://linnaeus.c18.net/Letters/letter_list.php).

of Brotero's colleagues, such as the German botanist Heinrich Friedrich Link (1767–1851) (Link [1801] 2005: 185).[2]

Brotero was not a university-trained naturalist. During his youth, he acquired a profound knowledge of classical languages, something that would later be of paramount importance in his career as a naturalist.[3] In 1778, at the age of 34, he fled to Paris to escape political persecution. It was here that he acquired his education in and enthusiasm for natural history. He attended public courses on the subject by Antoine Laurent de Jussieu (1748–1836), Jean-Baptiste Lamarck (1744–1829) and Jacques-Christophe Valmont de Bomare (1731–1807) (Fernandes 1988:3). Brotero wrote his *Compendium* in this city. Being aware that the statutes of the University of Coimbra required the preparation of a textbook for the new courses, he probably hoped that the publication of his *Compendium of Botany* would help him obtain a chair at the university. Indeed, the year following its publication, Brotero returned to Portugal to teach natural history and agriculture at the University of Coimbra. A few years later, in 1791, he was elected Director of the Botanical Garden of the University and, in 1811, of the Ajuda Palace Gardens and Museum in Lisbon.

In his *Compendium of Botany*, Brotero recognizes the educational value of the Linnaean system, as well as its widespread acceptance in Europe. This, and the fact that it was the system that had been adopted by the University of Coimbra, are for him sufficient reasons to dedicate the second volume of his treatise to its presentation. To help explain it, he uses illustrations taken from Linnaeus and other botanists.

However, unlike Vandelli, Brotero is not a strict follower of Linnaeus and expresses his awareness of the limitations and disadvantages of the Linnaean system. Accordingly, he also presents the views of other famous botanists on the subject in order that "the reader is made aware not only of its advantages, but also its limitations, and therefore can attribute to it its real value" (Brotero 1788: vii).

---

2. In his Notes on a Voyage through Portugal, Link stated that Brotero's Compendium of Botany "shows as much knowledge as all our German introductions and more skill than these in understanding and embarking on new perspectives" (Link [1801] 2005: 185).
3. On the life and work of Brotero, see Castel-Branco 2004.

3. Classification and national identity

The Portuguese reception of Linnaeus's system would not have reached its full significance if it had not had an impact on the very practice of botany. In fact, one of the most frequent topics in the correspondence between Linnaeus and Vandelli is precisely the urgent need for a compilation and description of the Portuguese flora.[4] Vandelli's *Florae Lusitanicae et Brasiliensis specimen* (1788) did not satisfy this desideratum, since it included only a few species and the descriptions were minimal. Vandelli was also the author of *Viridarium Grisley Lusitanicum* (1789). However, this work was considered by several of Vandelli's contemporaries to have limited value.[5] In fact, it is no more than a list of the species previously described by Grisley with the corresponding Linnaean terminology.

The task of publishing a Portuguese flora was to be undertaken by Brotero in 1804, and it was not the result of his efforts alone. Indeed, the publication of such a work was taken to be a matter of national importance. The government itself, and particularly the minister D. Rodrigo de Sousa Coutinho, encouraged Brotero and provided him with funds and a special licence needed for the herborizations that were to be conducted in various parts of Portugal. In addition, the Portuguese government bore the costs of publication. The fact that two other foreign naturalists, the Germans Comte d'Hoffmannsegg (1766–1819) and Heinrich-Friedrich Link (1766–1819), were pursuing a similar goal put nationalist pride at stake, and Brotero felt that he was under pressure to finish the project as soon as possible.

Flora Lusitanica (1804) presents a clear and meticulous description of 1,885 species from Portugal, of which more than one hundred were described for the first time. The author uses a classificatory method based on the Linnaean sexual system, but reduces the original 23 classes to 11, making it simpler and adapting it to the classification of the Portuguese flora (Palhinha 1944: 75–93). In comparison with natural systems of classification, the Linnaean method had the advantage of taking into account a relatively small number of identificatory traits, therefore enabling quick and easy identification of plants already known or not yet described. From this point of view, it had practical advantages as compared with natural methods of classification and it can be said that it was its use that enabled Brotero to finish his *Flora* in a relatively short period of time (10 years).

4. See letters from Linnaeus to Vandelli dated 12 February 1765, 15 July 1767, 13 May 1769, and 7 January 1770 (http://linnaeus.c18.net/Letters/letter_list.php).
5. Brotero describes Vandelli's Viridarium Grisley Lusitanicum as a poor work and of no use (Brotero 1804: iv–v).

It should also be noted that Brotero was aware of the slow progress in the development of natural systems of classification and that, like artificial systems, they were not without imperfections. Another element to take into account is that the establishment of natural systems would only have been possible by means of the most complete study of all the applicable classificatory groups and, since Portugal was relatively small, several of them would have been impossible to find and others would have had few representatives (Fernandes 1986: 890).

A work on natural history published in such haste was bound to have its shortcomings. The most obvious one was the absence of illustrations, something that was needed particularly for the new plants described.[6] In addition, the work left out a number of native plants (Luisier 1944: 149). Unlike Linnaeus, for whom illustrations had an insignificant role in the development of natural historical knowledge, Brotero credited visual representations with a vital role in the practice and diffusion of natural history (Brotero 1788: lxvi).

In fact, as Cristina Castel-Branco has remarked, "Brotero was forced eternally to atone for the sin of a flora written in such haste, which obliged him not to include illustrations and to leave out so many plants, although he knew they should have been included in his first work" (Castel-Branco 2004: 232). Brotero's later work, *Phytographia Lusitania Selectior*, in two volumes (1826 and 1827), was intended precisely to fill this gap. The work presents 181 copper-engraved illustrations of species native to Portugal, some of them not previously described by Brotero.[7]

In spite of all the patriotic effort put into the publication of Brotero's *Flora Lusitanica*, the work seems to have enjoyed greater success outside Portugal. Letters to the author from the Portuguese naturalist Correia da Serra, who was at the time living in Paris, provide evidence of the general acclaim given to the publication by foreign colleagues.[8] Brotero did not enjoy an easy relationship with most of his colleagues from the University of Coimbra, including Domingos Vandelli, and this certainly had a negative impact on the reception of his work in Portugal.[9] Indeed, we might say that Brotero's contribution to

6. The absence of illustrations had nothing to do with the significant costs involved in their production, since such expenses had been previously assured to Brotero by the minister D. Rodrigo de Sousa Coutinho (Fernandes 1988:6).
7. Some of the illustrations were engraved by the artist F. De Queiroz and others by artists from the publishing house Arco do Cego (Fernandes 1988: 9).
8. Correia da Serra was probably the main figure responsible for the diffusion of the work abroad. He was responsible for advertising *Flora Lusitanica* in the *Archives Littéraires*, the *Journal de Physique* and the *Magazin Encyclopedique* (Simões, Diogo, and Carneiro 2006: 102).
9. By this time Vandelli was already using the Portuguese version of his first name.

the development of Portuguese botany was only fully acknowledged in the late nineteenth century.[10]

4. Classification and empire

In his correspondence with Vandelli, Linnaeus remarked on the advantageous opportunities offered by the Portuguese Empire for the pursuit of natural historical research. He specifically wished the Portuguese to become aware of the natural richness of their realms, something that would arouse the envy of others who did not have exotic provinces.[11] To what extent were these opportunities seized by the Portuguese naturalists of the period?

Vandelli had an important role in the creation in 1768 of the Ajuda Palace Museum and Gardens in Lisbon, institutions that were focal points for Portuguese overseas collecting activity (Brigola 2003). Later on, during the reign of Queen Maria I (1777–1792), the Royal Academy of Sciences in Lisbon was founded in 1779. This institution was of paramount importance in guiding scientific research conducted overseas, as well as in promoting discussion and the publication of numerous memoirs on the subject (Cardoso 1990–1991: xvii–xxxiii; Oswaldo 1997). The anonymous instruction manual for correspondents of the Academy of Sciences from overseas, in which Vandelli most probably collaborated, does not make any specific reference to the Linnaean system of classification. It simply notes that "the indigenous name should be given, as well as the foreign name of the species, together with the name most often used by naturalists" (Real Academia das Ciências de Lisboa 1781: 39).

In spite of Linnaeus's advice to Vandelli, his *Florae Lusitanicae et Brasiliensis specimen* (1788) is a very incomplete work that presents only a limited number of species from the Brazilian flora, accompanied by detailed descriptions. It is true that Vandelli wrote many memoirs concerning the natural history of Brazil, but only in some of them are the names of living things presented according to the Linnaean system of classification.[12] Indeed, for him, "the study of natural history was not mere classification. It included observations, experi-

10. It was mainly the naturalist Júlio Augusto Henriques who contributed to this recognition, after his election to the chair of botany at the University of Coimbra (Henriques 1889).
11. Letter from Linnaeus to Vandelli dated 12 February 1765 (http://linnaeus.c18.net/Letters/letter_list.php).
12. Several of Vandelli's memoirs were published by the Royal Academy of Sciences in Lisbon (Cardoso ed. 1994).

ments to discover connections and the order of nature, its economy and policy, and the formation of the earth and the transformations it went through as well as the advantages to be gained from natural productions" (quoted in Cardoso (ed.) 2003: 13). As José Luís Cardoso has shown, Vandelli was especially Linnaean in his strong views on the close relationship between natural history and political economy (Cardoso 2003). Like Linnaeus, he considered a mastery of the products of nature the true source of economic progress.

Educational and economic reforms were closely linked, and some of the first naturalists educated at the University of Coimbra were employed by the Portuguese state on scientific voyages to the overseas territories to define a strategy for the optimal allocation of available resources (Simon 1983; Lopes, Silva, Figueirôa, and Pinheiro 2005). Vandelli believed that the international success of these trips could only be guaranteed through the use of the Linnaean system of classification. His former students Alexandre Rodrigues Ferreira (1756–1815), João da Silva Feijó (1760–1824), Joaquim José da Silva (1783–1808), Manuel Galvão da Silva (1750–?), and Manuel Arruda da Câmara (1752–1810) were involved in scientific journeys to Africa and Brazil. With minor modifications, they used the Linnaean system in their descriptions and memoirs.

Of all the scientific expeditions conducted within the Portuguese Empire, Alexandre Rodrigues Ferreira's nine-year journey to the Amazon region from 1782 to 1792 is the most wide-ranging and best-known (França 1922; Mittermeier 1992). Ferreira described and classified several animal and plant specimens that he shipped to the Ajuda Palace Gardens and Museum in Lisbon. However, the funding and the research effort invested in the expedition did not result in any publication. One of the reasons for this is that, during the Napoleonic wars, in 1808, a significant part of the collection, together with many of Ferreira's reports and memoirs, was expropriated by Étienne Geoffroy Saint-Hilaire (1772–1844) to the Muséum d'Histoire Naturelle in Paris (Simon 1983).

Manuel Galvão da Silva's expedition to Portuguese Goa was more successful in terms of publications. The species described in his *Observations on the Natural History of Goa made in the year 1784* were all classified according to the Linnaean method (Silva 1862). It should, however, be noted that this work was much more limited in scope than the observations made in Brazil by Ferreira. The same can be said of João da Silva Feijó's *Relation of the Cape Verde Islands* (1783), in which the author uses the Linnaean system. However, this manuscript remained unpublished.[13] Both Silva and Feijó sent herbaria to Lisbon, from Goa and the Cape Verde Islands, respectively (du Bocage 1862: 67). As a result of his expedition to Angola, Joaquim José da Silva also prepared a large herbarium of native plants that he sent to Lisbon (Simon 1983: 106).

The two major publications on the natural history of the Portuguese Empire can be attributed, not to trained naturalists, but to a Franciscan, José Mariano da Conceição Vellozo (1742–1811), and a Jesuit, João de Loureiro (1710–1791). Following his incursions into the region of Rio de Janeiro which lasted eight years, Vellozo finished his *Flora Fluminese* in 1790. The work comprises a description and classification of 1,640 species from the Brazilian flora, and is amply illustrated by Frei Francisco Solano. It was not published until thirty-five years later, when other floras of Brazil by foreign naturalists had already been printed (Vellozo [1825–1827] 1961). In any case, Brotero had a very critical view of Vellozo's work and considered that "in the present state of botany it would discredit the nation [...] that the title Fluminense was ambiguous and that its descriptions and drawings were very incomplete. Besides, the nomenclature of genera and species was erroneous and there were some families of plants lacking" (Fernandes 1947).

João de Loureiro was sent on a special mission to Cochinchina in 1742. Finding that missionary activities as such were not favourably regarded there, he entered the service of the King of Cochinchina as a mathematician and naturalist. Loureiro remained in Cochinchina for nearly thirty-six years, with the exception of one short interval. While in Cochinchina his chief place of residence was its capital city Hue. It was in the immediate vicinity of this city that he obtained many of his botanical specimens. In December 1777 Loureiro proceeded to Bengal, Pondicherry, Macao and Cantao. He stayed in this latter city for the next three years and it was during this period that Captain Thomas Riddel offered him the works *Systema Naturae*, *Genera Plantarum* and *Philosophia Botanica* by Linnaeus. As a result, Loureiro became a strict follower of Linnaeus. In exchange, he offered Riddel, who also put him in contact with Joseph Banks, various specimens that are now held at the British Museum.

Loureiro's *Flora Cochinchinensis* was completed in 1788 and published in 1790 with the sponsorship of the Royal Portuguese Academy of Sciences in Lisbon (Loureiro 1790). Its title is misleading in the sense that the work is not only dedicated to the flora of Cochinchina. It also includes a description and classification of several species from China, the Philippines, India and tropical

13. João da Silva Feijó, Itenerario filosofieo que contem a rellação das Ilhas de Cabo Verde pelo methodo epistolar, 1783, remains unpublished; Portuguese National Library, Codice 12984.

East Africa. The work contains original descriptions of 185 new genera and nearly 1,300 species, of which about 630 were described as new. However, as E. D. Merrill has shown, Loureiro did not always fully understood the Linnaean system and the work contains various errors, such as descriptions of genera that had already been established by other authors under different names, attributions of species to genera to which they did not belong, descriptions of species as belonging to two, three or even four different genera, the lack of detailed descriptions of new genera and new species, and the lack and imperfection of herbaria (Merrill 1935). These kinds of shortcomings limited Loureiro's impact on the history of botany, and practically all the new genera and species described by Loureiro have been redescribed under other names. However, the innovative contribution of his work should not be forgotten, nor should the fact that it created enough of a sensation in European botanical circles for Carl Ludwig Willdenow (1765–1812) to issue, in 1793, a second annotated edition of Flora Cochinchinensis in Berlin (Loureiro 1793).

5. Classification and the Muses

The introduction of Linnaeus's ideas in Portugal was also achieved through poetry. In 1811, the distinguished poet Manuel Barbosa du Bocage dedicated a poem to the "Immortal Linnaeus", in which he described him not only as a "noble soul" and a "most gentle spirit", but also as a demigod (du Bocage 1811: Preface). However, the most complete homage to Linnaeus and to his method of classification was authored by a woman, the 4th Marquesa de Alorna (Leonor de Almeida de Portugal, 1750–1839).[14] Owing to the persecution of her family by the Marquês of Pombal (1699–1782), she was imprisoned between 1758 and 1777 in the Convent of Chelas. It was here that she discovered her poetical inclinations, acquired a solid education in classical languages and developed a keen interest in scientific knowledge. After her release from prison, Leonor travelled to Madrid and then to Paris, where she had the opportunity to experience the atmosphere of the French *salons*. She lived for a few years in Vienna, where she was in touch with various members of the aristocracy and the cultural elite. Leonor was supportive of the *Ancien Régime* and strongly anti-Napoleonic. Fearing that the French would invade Portugal, she left for Britain in 1802. The French invasions took place between 1807 and 1811, but she remained in Britain for two more years. Her stay in Britain gave

14. On the Marquise of Alorna and her poetic work dealing with botany, see da Costa 2009.

the future Marquesa de Alorna the opportunity to widen her intellectual horizons by coming into contact with contemporary authors, some of whose works she translated. She wrote the poem *Botanical Recreations* when she was living in Britain, a country that was well-known for the popularization of botany (Shteir 1996).

The Marquesa de Alorna was no mere amateur of botany. The poem *Botanical Recreations* and her translation of an English botanical work, Thomas Martyn's *Letters on the Elements of Botany: Addressed to a Lady By the Celebrated J. J. Rousseau, Translated into English, with Notes, and Twenty-Four Additional Letters, Fully Explaining the System of Linnaeus* (1785), testify to the contrary. The translation included additional notes which show her interest in and understanding of Linnaeus's system of classification. In one of the notes, she comments on the popularization of his system in Britain and Portugal:

> The English or Portuguese student can only encounter advantages in the possession of many elementary books which explain terms to them in their own language. Linnaeus is translated into English. There is also the flora of Hudson, and the arrangement of Dr Wecthering which combine the English terms with those of Linnaeus, Dr Wardelly, Professor Brotero, Gresley and the Count Hoffmannsegg it is they who have introduced botany into Portugal. The Abbé Correia has written nothing on this subject as yet.[15]

The *Botanical Recreations* were written within the tradition of the didactic poem, a genre widely used during the late eighteenth and early nineteenth centuries in the popularization of botany. The poem is dedicated to Portuguese women and especially to the younger female generations. The Dedicatory Epistle invited them to study nature and to develop their passion for botany. In this invocation, the Marquesa of Alorna made explicit reference to the incomparable Linnaeus and also to the aforementioned Portuguese naturalists Brotero and Correia da Serra:

> What luxury! What wealth! If Flora bids
> You open the Linnaean Temple!
> If tireless Zephyrs whirling,
> With perfumed wings unfurling,
> Welcome you with scents of flowers!
> If Brotero and Correa invite you
> To explore Nature in glade and valley
> As well as explore gardens! If they name
> You priestesses of these floral Temples! (Alorna 1844: 8–9)

15. Arquivos Nacionais Torre do Tombo (Lisbon), Collection "Casa de Fronteira e Alorna", Codice 146, note v. This is an undated manuscript.

Linnaeus and his system of classification based on sexual characteristics had a central place in the Marquesa de Alorna's Botanical Recreations:

> Astonished the Wise, who long believe
> Linnaeus converses with the Deity:
> Or that sentient plants confide to him
> Their loves in florid eloquence (Alorna 1844: 9)
> [...]
> Linnaeus is the Polymarch of their army
> And knows the laws of mating
> And like the General whose soldiers
> Places on the battlefield; the orders and species of plants
> Linnaeus glimpsed without confusion
> That ray of light that all things illuminates
> And arcanas perceived in benign revelation (Alorna1844: 46).

In the *Botanical Recreations*, the Marqusa makes direct reference to the sexual connotations implicit in the Linnaean system of classification and even declares that "polygamy is necessary" in the plant kingdom (Alorna1844, vol. 4: 93). However, allusions of this kind do not have a central place in the text. Instead, emphasis is given to personified narratives on each plant, which lead to moral teachings. Thus, the description of the *Mimosa pudica* (the sensitive plant), whose leaves close in upon themselves and droop when touched, an example of the Linnaean class *Polygamia*, is used as a model for the behaviour of young women. In fact, the retraction of the plant when touched is used as a metaphor for the necessary retraction of women from a gender-biased society:

> This plant is worthy of the Muses,
> As it speaks to the heart and mind.
> I believe a nymph lies transformed
> Within her, whose pain, modesty and brio
> Opened the dark entrance to Orcus
> What a modest maid, the Sensitiva
> She recoils from the profane touch
> Like a virgin, frightened by a simple touch,
> She retracts from any touch, discolours and fades.
> Crown yourselves with her, choose her as your guide:
> To escape is to triumph for our sex (Alorna1844: 102).

The Marquise described Linnaeus as "the botanical genius", "with a seat in heaven" (Alorna 1844: 22). Nevertheless, she stated her awareness of the limitations of his system of classification, especially its degree of artificiality:

The Species is certain and inalterable
[…]
It is within Nature;
But further divisions are mind-designed
Methods diverse which memory holds
In orders, genera and systems
[…]
Of these species Botanists composed
New groups, designating them with the name
Of genus and genera they sought
(even Linnaeus) to discover them in Nature
But it is time and observation which all discloses
And it is before truth that conjecture ceases (Alorna 1844: 54).

Moreover, she asserted in the poem that there is more to botanical knowledge than classification:

To know the name of plants does not suffice
[…]
Analogies, habit, character
Lead Science to the greatest heights (Alorna 1844: 21).

Neither is Linnaeus the only foreign naturalist mentioned in the *Botanical Recreations*. The poet also stressed the contributions of John Ray (1627–1705), Joseph Pitton Tournefort (1628–1705) and Antoine Laurent de Jussieu (1748–1836) to the progress of this subject. Besides, Rousseau's *Elementary Letters on Botany* are presented as one of the most useful works for the improvement of botanical knowledge. Brotero and Correia da Serra, too, are praised in the poem. Despite his contribution to the diffusion of botany in Portugal, however, Vandelli's name is absent from the *Botanical Recreations*. This absence is probably related to the fact that Vandelli collaborated with the French during the French invasions and that the Marquise of Alorna was strongly anti-Napoleonic.

The *Botanical Recreations* were written between 1810 and 1813, but were only published in the first volume of the Marquesa de Alorna's *Poetical Works* in 1844. The printed version differs from the original in the notes added by the naturalist Caetano Bordão, together with those of the editor, Carlos Manuel Soye.[16] These notes expand the botanical knowledge provided by the work and include two tables on the Linnaean system of classification, taken directly from Brotero's *Compendium of Botany* (Pina 1953: 22).

It should also be noted that the fact that the *Botanical Recreations* were only published in 1844 does not imply that the work, or at least part of it, did not have

16. Only the notes to the first Canto were written by the Marquise of Alorna.

an audience before this date. It is known that various letters circulated publicly among the upper classes. Excerpts from the *Botanical Recreations* might have been diffused in this way. The circulation of copies of the *Botanical Recreations* among a restricted group of friends was probably another way in which the work was disseminated. We should also take into account that the salons organized by the Marquesa de Alorna and other members of her social network are likely to have aided the diffusion of the work. It is not strange, therefore, that although the Marquesa de Alorna published very few works during her lifetime, her literary skills were known even abroad. The fact that women writers from this period published very few works should not conceal their very important role as intellectual producers and mediators of new knowledge, including in the field of botany.

6. Conclusion

We have seen in this essay that the introduction of the Linnaean system of classification in Portugal was intimately connected with the political context of the time. It was the reform of the University of Coimbra promulgated by the Marquis of Pombal that enabled the inclusion of the Linnaean programme of classification within the university curriculum. The history of the first Portuguese flora by Brotero also reveals how the scientific value of the enterprise cannot be divorced from nationalist and political questions.

Educational, political and economic reforms were especially interrelated when the description and exploration of overseas territories were at stake. On the one hand, it is no coincidence that the majority of the naturalists educated at the new reformed University of Coimbra were employed by the state on scientific voyages to the overseas territories, with a view to defining a strategy for the optimal allocation of available resources. On the other hand, it appears that the history of botany and the Portuguese Empire seems to be a history of missed opportunities. The two major floras that were produced had various shortcomings, and one of them does not relate exclusively to the Portuguese Empire. It should also be noted that the task of producing a major natural history of Brazil would have a required a much larger number of trained naturalists, as well as a greater political and financial commitment from the state.

Finally, the diffusion of the Linnaean system of classification was celebrated in various ways, including poetry. Moreover, unlike other botanical works, the poem *Botanical Recreations* specifically invited women to join in this celebration by participating in the pleasures of contemplating and understanding the order of nature.

Acknowledgements

I would like to thank Fernando Mascarenhas for allowing me to consult some of the original manuscripts at the Fundação das Casas de Fronteira e Alorna; Francesca Rayner for her linguistic revision of this essay and translations of original quotations; and Christopher Damien Auretta for his kind assistance with translations of excerpts from the poem *Botanical Recreations*. My thanks also go to the librarians of the Arquivos Nacionais Torre do Tombo, Lisbon, and the Biblioteca Nacional, Lisbon, for their kind assistance. In addition, I would like to thank Professor Britt-Louise Gunnarsson for her very helpful suggestions in the revision of the manuscript.

References

Alorna, Marquesa de
1844 *Recreações Botânicas* [Botanical recreations]. In *Obras Poéticas* [Poetical works], 4th vol. Lisboa: Imprensa Nacional.

Brigola, João Carlos
2003 *Colecções, Gabinetes e Museus em Portugal no século XVIII* [Collections, cabinets and museums in eighteenth-century Portugal]. Lisboa: Fundação Calouste Gulbenkian, Fundação para a Ciência e a Tecnologia.

Brotero, Félix Avelar
1788 *Compêndio de Botânica* [Compendium of Botany]. Lisboa: Casa de Paulo Martins.

Cardoso, José Luís
1990–91 Introdução [Introduction]. In J. L. Cardoso (ed.), *Memórias económicas da Academia Real das Ciências de Lisboa, para o adiantamento das artes, e da indústria em Portugal, e suas conquistas (1789–1815)* [Economic memoirs of the Royal Academy of Sciences of Lisbon for the improvement of the arts and industry of Portugal and its conquests], xvii–xxxiii. Lisboa: Banco de Portugal.

Cardoso, José Luís (ed.)
1994 *Memórias económicas da Academia Real das Ciências de Lisboa* [Economic memoirs of the Royal Academy of Sciences of Lisbon] (1789–1815). Lisboa: Banco de Portugal.

Cardoso, José Luís (ed.)
2003 *Memórias de História Natural, Domingos Vandelli* [Memoirs on natural history by Domingos Vandelli]. Porto: Porto Editora.

Cardoso, José Luís
2003 From natural history to political economy: the enlightened mission of Domenico Vandelli in late eighteenth-century Portugal. *Studies in the History and Philosophy of Science* 34: 781–802.

Carvalho, Rómulo de
1987 *A história Natural em Portugal no século XVIII* [Natural history in eighteenth-century Portugal]. Lisboa: Instituto de Cultura e Língua Portuguesa.

Carvalho, Rómulo de
2001 de *História do Ensino em Portugal* [The History of Teaching in Portugal]. Lisboa: Fundação Calouste Gulbenkian.

Castel-Branco, Cristina
2004 *Félix de Avelar Brotero: botaniste portugais (1744–1828)*. Paris: Centre Culturel Calouste Gulbenkian.

da Costa, Palmira Fontes
2009 Women and the Popularisation of Botany in Early Nineteenth-Century Portugal: The Marquesa de Alorna's Botanical Recreations. In: Faidra Papanelopoulou, Agustí Nieto-Galan and Enrique Perdiguero (eds.), *Popularisation of Science and Technology in the European Periphery*. Aldershot: Ashgate, 43–63.

du Bocage, J. V. Barbosa
1862 *Instrucções Prácticas sobre o Modo de Colligir, Preparar e Remeter Produtos Zoologicos para o Museu de Lisboa* [Practical instructions on the best way to collect, prepare and send zoological specimens to the Lisbon Museum]. Lisboa: Imprensa Nacional, 1862.

du Bocage, Manuel Maria Barbosa
1811 Aos Manes do Immortal Linné [To the immortal Linnaeus], Preface. In Manuel Maria Barbosa du Bocage (ed.), *O Consorcio das Flores de Lacroix* [*The Marriage of Flowers* by Lacroix]. Rio de Janeiro: Impressão Régia.

Faria, Miguel
2001 *A Imagem Útil* [The useful image]. Lisboa: Universidade Autónoma de Lisboa.

Fernandes, Abílio
1947 Quatro cartas inéditas de Brotero para o Conde da Barca [Four unknown letters from Brotero to the Count of Barca]. *Revista da Faculdade de Ciências da universidade de Coimbra* 16: 90–120.

Fernandes, Abílio
1986 História da Botânica em Portugal até finais do século XIX [History of Portuguese botany up to the end of the nineteenth century]. In *História e Desenvolvimento da Ciência em Portugal* [History of the development of science in Portugal], 2nd vol., 851–916. Lisboa: Academia das Ciências de Lisboa.

Fernandes, Abílio
1988 Relance sobre a vida e a obra de Félix de Avelar Brotero [A brief glance at the life and work of Félix de Avelar Brotero]. *Anuário da Sociedade Broteriana* 54: 1–13.

França, Carlos
1992 *Doutor Alexandre Rodrigues Ferreira (1756–1815): História de uma missão científica ao Brasil no século XVIII* (Doctor Alexandre Rodrigues Ferreira (1756–1815): History of a scientific mission to Brazil in the eighteenth century]. Coimbra: Imprensa da Universidade.

Henriques, Júlio
1882 Felix de Avellar Brotero. *Plutarcho Portuguez* 2: 41–48.

Henriques, Júlio
1889 Felix de Avellar Brotero, *O Instituto* 37: 364–389.

Link, Heinrich Friedrich
[1801] 2005 *Notas de uma viagem a Portugal e através de França e Espanha* [Notes on a voyage through Portugal, France and Spain]. Lisboa: Biblioteca Nacional.

Linnaeus, Carolia
1793 *Systema naturae, Decima Tercima Reformata Jo: Frid. Gmelin, Editio in Lusitania Prima. Conimbricae*: Typis Academicis.

Lopes, Maria Margaret, Clarete Paranhos da Silva, Silvia Fernanda de M. Figueirôa, and Rachel Pinheiro
2005 Scientific Culture and Mineralogical Sciences in the Luso-Brazilian Empire: The Work of João da Silva Feijó (1760–1824) in Ceará. *Science in Context* 18: 201–224.

Loureiro, João de
1790 *Flora cochinchinensis: sistens plantas in regno Cochinchina nascentes: quibus accedunt aliae observatae in Sinensi Imperio, Africa Orientali, Indiae que locis variis omnes dispositae secundum systema sexuale Linnaeanum.* Lisboa: Typis et expensis Academicis.

Loureiro, João de
1793 *Flora Cochinchinensis, sistens Plantas in regno Cochinchina nascentes; quibus accedunt aliae observatae in Sinensi Imperio, Africa Orientali, Indiaeque locis variis…* denuo in Germania edita cum notis Caroli Ludovici Willodenow. Berolini: Haude et Spener.

Luisier, A.
1944 Félix Avelar Brotero. *Brotéria* 13:145–158.

Melli, G. F.
1966 Un italiano in Portugallo – Domenico Vandelli. *Estudos Italianos em Portugal*, 26: 2–18.

Merrill, E. D.
1935 A commentary on Loureiro's Flora Cochinchinense. *Transactions of the American Philosophical Society* 24:1–45.

Mittermeier, Russell
1992 *Philosophical journey: a rediscovery of the Amazon*: 1792–1992. Rio de Janeiro: Index.
Oswaldo, Munteal Filho
1997 Todo um mundo a reformar: intelectuais, cultura ilustrada e estabelecimentos científicos ilustrados em Portugal e no Brasil, 1779–1880 [An entire world to reform: Intellectuals, enlightened culture and scientific establishments in Portugal and Brazil, 1779–1880]. *Anais do Museu Histórico Nacional*, 29: 87–108.
Palhinha, Ruy Telles
1944 O sistema sexual broteriano [The Broterian sexual system]. *Boletim da Sociedade Broteriana* 19: 75–93.
Real Academia das Ciências de Lisboa
1781 *Breves instrucções aos correspondentes da Academia das Sciencias de Lisboa sobre as remessas dos produtos e noticias pertencentes à historia da natureza para reformar hum Museo Nacional* [Brief instructions to the correspondents of the Lisbon Academy of Sciences concerning the ways of sending specimens and news on natural history for the reform of the National Museum]. Lisboa: Regia Officina Typographica.
Shteir, Ann B.
1996 *Cultivating Women, Cultivating Science: Flora's Daughters and Botany in England, 1760 to 1860*. Baltimore: Johns Hopkins University Press.
Silva, Manoel Galvão da
1862 *Observações sobre a história natural de Goa feitas no anno de 1784* [Observations on the natural history of Goa made in the year 1784]. Nova-Goa: Imprensa Nacional.
Simões, Ana, Maria Paula Diogo, and Ana Carneiro
2006 *Cidadão do Mundo: Uma Biografia Científica do Abade Correia da Serra* [Citizen of the world: The biography of Correia da Serra]. Porto: Porto Editora.
Simon, W. J.
1983 *Scientific Expeditions in the Portuguese Overseas Territories (1783–1808)*. Lisboa: Instituto de Investigação Científica Tropical.
Vandelli, Domenico
1788 *Diccionario dos termos technicos de Historia Natural, extrahidos das obras de Linneo* [Dictionary of technical terms of natural history extracted from the works of Linnaeus]. Coimbra: Real Officina da Universidade.
Vellozo, José Mariano da Conceição
[1825–27] 1961 *Flora fluminensis*. Rio de Janeiro: Arquivo Nacional.

Section 4.
The development of scientific writing

Linnaeus as a connecting link in Swedish language history

Bo Ralph

1. Linnaeus and his linguistic environment

In the spring of 1732, Carl Linnaeus set out on his first expedition – to northern Sweden. In his diary he has documented everything that he observed: the nature of the soil, conceivable natural resources, plants, animals and minerals; he makes observations on the climate; he is amazed by the midnight sun; he describes the local population, whether ordinary Swedes or Sami; their tools, their living conditions, their ideas and beliefs. Everything is recorded with the utmost care; nothing escapes the young scholar's keen-sighted observation. Nature itself is his basic concern, but Linnaeus is sensitive to far more than the things traditionally considered the object of the natural sciences. His notebook begins with an exclamation in Latin: "O, ens entium miserere mei!" [Oh, being of beings, have mercy on me!]; his appeal for support is justified – the journey will contain its fair share of danger. When he returns to Uppsala in October of the same year, he humbly concludes in a similar manner: "Enti entium sit laus, honor et gloria in saeculis" [The being of beings be praised, honoured and glorified in eternity!].

The year 1732 is traditionally acknowledged as a milestone in the history of the Swedish language: it is taken to be the dividing line between Early and Late Modern Swedish. A specific text has indeed been identified as the symbolic representative of the new era, but this text is not the diary written by Linnaeus during his journey to northern Sweden. A little more than two months after his return, in the middle of December, a new weekly magazine, *Then Swänska Argus*, appeared in Stockholm – and was an immediate success. The subjects dealt with were topical, the language refreshing and intriguing. The Swedish idiom had been blessed with a new kind of artistic, but nevertheless simple, prose. The anonymous author of the magazine turned out to be a fairly young man born in 1708, a year after Linnaeus himself. He was a vicar's son, just like Linnaeus. His name was Olof Dalin.

Returning to Linnaeus's notebook, it is true that in his very first documentation of a journey Linnaeus displays most of the characteristic ingredients of his particular kind of prose, which was later to be famous. But at this time Swedish is not his first choice on every occasion. The phrases in Latin just mentioned are accompanied by almost innumerable passages in the same language throughout his diary – and not only quotations or terminology for scholarly purposes, but often ordinary descriptive prose. The reader is repeatedly reminded of the fact that the international mother tongue of scholars was still Latin, not their respective vernaculars.

Today, it may be difficult to evaluate eighteenth-century style, since, to us, everything from those days seems out of date. The author of the literally epoch-making magazine, Olof Dalin, addresses his first readers in a straightforward manner (Dalin [1732], p. 1; italics in the original maintained):

> Ingen lärer kunna neka, at ju sådane Skriffter hafwa stor nytta med sig, som, på ett angenämt och lustigt sätt, föreställa Lärdomar och Wettenskaper; Derföre hafwa och de gamla, under roliga Dikter, liufliga Samtahl eller nöysamma *Historier*, underwisat Folket om Dygden, och likasom Skiämtewijs förehållit dem alfwarsamma Sede-Läror. I nyare tider, och än i dag, se wi äfwen, hos kloka *Nationer*, sådane Skriffter med mycken nytta utgifwas och älskas.
> [...]
>
> [Probably nobody can deny that such publications that present scholarship and sciences in a pleasant and attractive way are of great use; this is why writers of old have educated the people in virtue by means of funny poems, delightful conversations or amusing stories, and remonstrated with them on ethics in something of a humorous way. More recently, and still today, we find such writings very usefully being published and loved by wise nations.]

There are occasional words or phrases in this passage that are now obsolete, and some of the words are inflected according to rules that differ slightly from present-day usage, but it is obvious even to modern readers that Dalin's language deviates from that of his contemporaries. This becomes especially clear if Dalin's text is compared to any other text from the period. In the same year, 1732, Olof Kolmodin published a book with the aim of teaching and entertaining, but his endeavour did not have the same effect as Dalin's. His *Biblisk Qwinno-Spegel* begins with a reflection on mirrors, alluding to the title of his work (italics as above):

> Speglar hafwa i alla tider warit thet Wackra Könets tyste och förtrolige Rådgifware, hos hwilka Thet sig, angående Kroppens behagligheter och vtprydande, befråga plägar. Naturen, som här, på et icke mindre angenämt än vndersamt sätt, medelst synstrålarnas tilbaka slående ifrån mörcka och glatta kroppar, leker, och

i them the förestälda synliga ting på thet artigaste och lifligaste, ei allenast til storlek och färg, vtan ock til the minste rörelser, afbildar, har i thetta mål gifwit Konsten anledning, at på flera sätt, och af åtskilliga ämnen sig härutinnan betiena och hielpa, och at sådant tingens föreställande til större fullkomlighet föra. Konsten har sedermera giordt thessa *instrument* så allmänna, och omsorgen at pryda sig skatta them så oumgängeliga, at näppeligen något hus, ther et Fruentimmer vti bor, lärer finnas, therest et sådant husgeråd icke wore at antreffa.

[At all times, mirrors have been the silent and confidential advisers of the Fair Sex, to whom It [i.e. "the fair sex"] usually poses questions about the charms and decoration of the body. Nature – which in this case, in a no less pleasant and marvellous way, plays by means of the repulsion of the beams of light [literally: sight] against dark and smooth bodies, and depicts, with respect not only to size and colour, but also to the smallest gestures, in them the visual things represented in the nicest and most lively way – has in this case given Art a cause to make manifold use of them, in order to reproduce the things with greater perfection. Art has subsequently made these instruments so common, and the care to adorn oneself so indispensible, that there is hardly any house, in which a Lady lives, where such a utensil is lacking.]

It is not necessary to be an expert on eighteenth-century Swedish to realize that here we are dealing with quite a different stylistic approach. In particular, the syntax is rather complex, albeit constructed with considerable skill. To be sure, the last two quotations both contain three sentences, but the second text is almost twice as long as the first one.

2. Linnaeus's language system

How, then, is Linnaeus linguistically related to contemporary writers? First of all, he is not an inspired activist fighting on the barricades for the use of Swedish in all possible situations. The linguistic arena of eighteenth-century Sweden is dominated by the process of standardization (see, for example, Hannesdóttir in this volume, Teleman 2002). The language debate of the time was basically preoccupied with questions concerning orthography and inflectional morphology. The development of whole new types of texts and the integration of new domains into the repertoire of the Swedish language are, of course, other important aspects. The Royal Swedish Academy of Sciences programmatically held their meetings and published their transactions in Swedish (see Teleman in this volume). As a member of the Academy, Linnaeus was only barely loyal to its language policy. He published a number of reports in the Academy's scientific series in Swedish, and his first speech to the Academy, given when he retired from his position as its first president in October 1739, is famous. It should be borne in mind, however, that it is one of a very limited number of orations held by Linnaeus in Swedish. In the light of this, it is perhaps even more remarkable how well he coped with the task. The very beginning is an oft-quoted masterpiece (italics as above):

> *Mine herrar!*
>
> 1. Alt hvad den Alsmägtige Skaparen inrättat på vårt jord-klot är gjordt i en så undersam ordning, at ej et enda fins, som ej behöfver et annats bistånd til sit underholl: *Jord-klotet* sjelft med Stenar, Malm och Grus, näres ju och födes af Elementerne: *Växter*, Trän, Örter, Gräs och Mossar, växa af jord-klotet och *Djuren* ändteligen af växterna. Alla dessa på slutet förvandlas åter til sina första ämnen, *Jorden* blifver Plantans föda, *Plantan* Matkens, *Matken* Foglens och *Foglen* ofta Rofdjurets; åter förtäres på slutet *Rofdjuret* af Roffoglen, *Roffoglen* af Matken, *Matken* af Örten, *Örten* af Jorden: Ja *Människan*, som alt vänder til sin nödtorft, blifver ofta Rofdjurets, Roffoglens, Roffiskens, Matkens eller jordens föda. Så går alt i kring.
>
> [Gentlemen!
>
> 1. Everything that the Almighty Creator has arranged on earth is made in such a marvellous order that there is not a single thing which does not need the support of another for its maintenance: The *globe* itself with stones, ore and gravel is nourished and fed by the elements: *plants*, trees, herbs, grasses and mosses, grow from the earth, and *animals* finally from the plants. All these, in the end, are transformed back to their original substances, *the earth* becomes the food of the plant, *the plant* [the food] of the worm, *the worm* [the food] of the bird and *the bird* often [the food] of the predator; further, in the end, *the predator* will be consumed by the bird of prey, *the bird of prey* by the worm, *the worm* by the herb, *the herb* by the earth: Yes, *Man*, who turns everything to his benefit, will often be the food of the predator, the bird of prey, the fish of prey, the worm or the earth. So everything goes round in a circle.]

When Linnaeus emphasizes that there is not a single thing that does not need the support of another thing for its maintenance, he is simply varying the meaning of the Latin expression *ens entium*, topicalized in his diary. As in Dalin's magazine, there are some striking forms and meanings of words, which need not bother us here. On the syntactic side, word order looks deviant here and there, according to the rules of the time, but in other cases it simply serves rhetorical ends: when the roles in the food chain in nature are reversed by passivization ("[blifver …] *Foglen* ofta Rofdjurets [föda]" into "[förtäres] *Rofdjuret* af Roffoglen" etc.), this transformation is also reflected in reversed word order: [*Alla dessa*] *på slutet förvandlas åter* is changed into *åter förtäras på slutet* [Rofdjuret]. Linnaeus soon turns to his main topic, the life of the insects, a veritable orgy of detailed observation.

Actually, there is not much in the language system reflected in this text which, on the formal side, indicates the start of a new language era. Many features would, rather, have seemed archaic to Linnaeus's contemporaries. This is true whether we consider inflectional endings, derivational endings, choice of words, word order or other syntactic phenomena.

3. Linnaeus's style

When Linnaeus produces Swedish text of great literary value, it is primarily owing to his acquaintance with the rhetorical tradition of Latin, on the one hand, and his own unbiased creativity, on the other. In a scholarly context, he prefers Latin. To be sure, he has developed a great mass of terminology in Latin, a fundamentally pioneering effort; but it is part of the Latin, not the Swedish, tradition. Here we naturally find the method of naming by two elements, such as *Primula veris* and *Homo sapiens*. His brilliance as a writer in Swedish was clearly not recognized until very late. Linnaeus is missing from the surveys of the history of literature throughout the nineteenth century, in which Dalin, among others, has a given place. It seems that the outstanding critic and writer of cultural essays, Oscar Levertin, was the first to open the eyes of the Swedes to the literary qualities of Linnaeus, in a study published in 1906.

It should be kept in mind that the language programme launched by the Royal Swedish Academy of Sciences was the basis for Linnaeus's becoming an important and influential writer in the Swedish language (see Teleman in this volume). He repeatedly plays down his own literary ambitions. Whether he is honest on that point or not, is impossible to tell. It is rather inconceivable that a passage like the one describing his departure from Gotland was written only for the sake of documentation (Linnaeus 1745, p. 302; italics as above):

> Jul. 25.
>
> Kl. half 6 om morgonen stego wi om Bord, med Lifsfara kommo wi utur Hamnen för en brännande Siö, Wännerne och Wisby förswunno, Carlsöarne stego fram, Nordan wädret begynte pipa, Wågorna blefwo rasande, Fartyget kastades emellan de brusande Böljor, Gothland förswan, *Cameraderne* blefwo siö-siuke, Taklen begynte springa, förtwiflan intog våra hiertan, och wi befalte wår sak i Guds händer.
>
> [At half past 5 in the morning we embarked, with our lives in danger we left the harbour for the breaking sea, our friends and Visby disappeared, the Karlsö

> islands stepped forward, the northerly wind started whistling, the seaway became furious, the vessel was thrown between the roaring waves, Gotland disappeared, my companions were seasick, the rigging started to burst, despair filled our hearts, and we committed our souls into God's keeping.]

Then the report moves on to a new chapter, the drama at sea suddenly put aside with no further comment! It is obvious that Linnaeus has very consciously tried to pack his story with juicy effects. But he is no doubt totally honest when he describes his favourite style as simple, using short words with a clear and unambiguous meaning. And his motivation is interesting: the soul of all scholarship is to make everything as simple as possible. His declaration can be found in the preface of a work called *Vulcanus Docimasticus*, published in 1734 (Linnaeus [1734], p. 17):

> En simpel styl, korta ord med ren mening, och undvikande af Tautologie är dett som giör ens skrifter tydelige. Mången tänker sig wara mycket klar, då han widlöftigt beskrifwer dett som kort älliest sägas kunne; hwilken doch är altid obscur att förstå, och ledsen att läsa. Men en ren och rätt method, som är siälen af all wettenskap, är just den som giör all ting så makalöst lätt. Ty har mitt förnämsta ändamål warit att skrifwa scientifice, äller en ren method.
>
> [A simple style, short words with a pure meaning, and avoidance of tautology is what makes one's publications clear. Many people believe that they are very distinct, when they give lengthy instead of concise descriptions; the one who does so is however always obscure to understand and boring to read. But a pure and correct method, which is the soul of all scholarship, is exactly the one which makes everything so incomparably easy. That is why my primary goal has been to write in a "scientific" way, or [according to] a pure method.]

This is primarily meant to apply to factual prose dealing with scientific matters.

When we ask about his literary ambitions, the question is in fact anachronistic. Not surprisingly, Linnaeus says nothing about fiction or belles-lettres. These notions were not yet part of the cultural debate in Sweden. On the other hand, we may, from a modern point of view, take his ideals for granted, even in scholarly writing, but they were not as self-evident as they may seem, not in his day. Stylistic simplicity has certainly not been a general hallmark of scholarly prose; not uncommonly it may even seem as if a certain degree of complexity has been deliberately sought, putatively to give the text a special ring of scientific accuracy and credibility. National traditions may vary on this point, and efforts in Sweden during the last half-century or so to make the official language easy to understand probably represent an extreme end of the scale. However, the rather radical linguistic ideal which characterizes modern scholarship in Sweden may indeed be a heritage from the eighteenth century. With his far-reaching authority, Linnaeus was closely followed by his students, and they conveyed his ideals to later generations. It is an important fact that one of his so-called apostles, Carl Peter Thunberg, succeeded Linnaeus's son on his father's chair. From there, the same scientific and linguistic ideals were defended for more than a century. That may be enough to found a tradition.

Neither Linnaeus himself nor his students took an active part in the language debate as such, at least not to a crucial extent. Their contributions are primarily scientific. It can, however, be argued that Linnaeus's published documentation of his own travels through different parts of Sweden in the 1730s and 1740s was taken as a model by his "apostles" when, later on, they reported from their own expeditions all over the world. Thus, a homogeneous type of factual prose for scholarly purposes was developed, characterized by Linnaeus's own stylistic ideals and, of course, by his own manner of detailed reporting. In this way, the Linnaeans did after all contribute actively to the standardization process, particularly in the sense that they used the Swedish language extensively for scientific purposes.

4. Linnaeus's travel reports

Linnaeus's diary from his journey to northern Sweden in 1732 was the first of its kind, but certainly not the last. A later expedition in Sweden took him to the two large islands in the Baltic, as hinted above, and was undertaken nine years later. This time his notebook was revised – or rather, rewritten – and published. In the introduction to his *Öländska och Gothländska Resa*, printed in 1745, we find many formulations worth noting (Linnaeus 1745, p. 1; capitals and italics in the original maintained):

RESAN til ÖLAND

1741 Maji 15.

> STOCKHOLM reste wi ifrån i den behageligaste Wårtiden, kl. 11 förmiddagen. Solen sken klar och Luften war något kulen.
>
> Wåren, som ei bör mätas efter calendarium, utan efter climatet och wärman, war så wida kommen at Lönnen utslagit sina blomor, men ei blad; at Biörken nyligen utspruckit och blommade som bäst; Alens stiplar woro nyligen utbrustne och Granen hade på sina ytterste qwistar små röda smultron-lika Knoppar, som woro thes hanblomster, men ännu ei miölige; Lind, Ek och Asp stodo ännu sofwande i sin winter-dwala.
>
> Blommor woro inga andra än *Draba prima, Caltha, Glechoma, Leontodon, Hepatica, Anemone secunda, Oxalis, Adoxa, Salices*.

Ibland Foglarne hade Giöken nyss begynt gucka och Ladu-Swalan såg man i dag den första.

[Stockholm we left in the most agreeable springtime, at 11 o'clock in the morning. The sun was bright and the air somewhat chilly.

The spring, which is not to be measured by the calendar, but by climate and warmth, was so far gone that the maple was in bloom but had no leaves; that the birch had just burst into leaf and was blooming; the stipules of the alder had just burst, and the pine had on its outermost twigs small red buds, much like wild strawberries, which were its male flowers, but not yet floury; the lime, the oak and the aspen were still in their winter sleep.

Among the flowers there were none but *Draba prima, Caltha, Glechoma, Leontodon, Hepatica, Anemone secunda, Oxalis, Adoxa, Salices*.

As far as the birds are concerned, the cuckoo had recently started to call and the first swallow could be seen today.]

This text is characterized by clarity, precision, and details. To what extent Linnaeus deliberately and consciously trained his disciples in this reporting technique is impossible to tell, but most of them seem to have tried to follow their master's example. Travel reports became a popular text type, having an impact on far more readers than they were primarily aimed at. In the late eighteenth century, travel books in general seem to have been popular, both in Sweden and abroad. The reports of the Linnaeans, though packed with facts and details, were no exception. This is attested by the famous Swedish nineteenth-century writer Erik Gustaf Geijer. In his memoirs he states (Geijer [1834], p. 74):

Det menskliga samhället i alla sina likheter och olikheter under särskilda odlingsgrader och omständigheter har alltid varit mitt älsklings-ämne. Det närvarande, det förflutna, allt är mig i detta hänseende intressant. Jag är derföre äfven en bland de största läsare af resebeskrifningar, och har varit det från min barndom.

[Human society, in all its similarities and varieties due to differing degrees of culture and circumstances, has always been my favourite subject. The present, the past – I find everything in this respect interesting. Therefore I am also one of the greatest readers of travel books, and have been since childhood.]

The reports were intended to be scholarly in nature, but they also served as entertainment. They are likely to have had a stimulating effect on both the reading public and the writing minority, though they have largely been neglected by Swedish historians of literature. Their influence on the development of general Swedish prose is undeniable. As a key source of inspiration, Linnaeus has an obvious role in the history of literary genres in Sweden.

5. Linnaeus's influence on later writers

The interesting question, however, is whether his influence goes beyond that. The introduction just mentioned may be compared to the following one, taken from his *Skånska Resa 1749* (printed in 1751). After some preliminaries it goes like this (Linnaeus 1751, p. 1; capitals, boldface, and italics in the original maintained):

> VÅREN war nu förhanden, sedan wintren så draget sig undan, at han näppeligen hade några snöfläckar qwar i diken och skuggan. **Sommaren** hade ock föga fått fram några få af sina förelöpare, som endast fram tittade på södra sidan af renarna och på bägge sidor af wägen, där Gräsen begynte sticka up med sine gröna blader; men inga örter hade ännu wågat uppsätta något blad utom *Alchemilla Fl. Svec.* 135 och *Potentilla adscendens Fl.* 419, som i dag först wisade sig. Icke hade eller något Trä wågat slå ut sina blader, mer än Krusebärs-busken Fl. 195, som äfwen först i dag öpnade sine knoppar och framsläpte sina blad. Ingen blomma sågs i dag på hela fälten.
>
> [Spring was now here, as winter had withdrawn to such an extent that there were only a few patches of snow left in the ditches and in the shade. The summer had also produced just a few of its forerunners, which peeped out only on the south side of the ditch banks and on both sides of the road, where green blades of grass were beginning to emerge; but no herbs had so far dared to show their leaves, apart from *Alchemilla Fl. Svec.* 135 and *Potentilla adscendens Fl.* 419, which today appeared for the first time. Neither had any tree dared to unfold its leaves, other than the gooseberry bush Fl. 195, which likewise only today opened its buds and let its leaves appear. No flower could be seen today anywhere in the fields.]

It has been argued by literary historians that the pioneer novelists of the early nineteenth century were indeed influenced to some extent by the Linnaeans, in their emphasis on details when describing nature. It is hard to believe that the following account of early spring has been conceived without a knowledge of Linnaeus's reports mentioned above (Cederborgh [1856], p. 30; italics as above):

> Det var en vacker vårdag. – Dagsmejan hade borttinat nästan all snön, endast några fläckar här och der lågo qvar på nordliga sidan om krusbärsbuskarna. Svärdsliljornas första skott och tulpanbladen började redan titta opp ur de fuktiga, ännu med löf och skräp öfverhöljda och ouppgräfda rabatterna. Vattnet av den smälta snön formerade små rännilar i sandgångarna, der en och annan spild ärta redan börjat på att gro och tillkännage att våren med stora steg ingick. Inga löf voro dock ännu utslagna; de tittade endast fram ur frodiga knoppar, och i en vild häck

började en enda videbuske att på naken stjelk, utslå sina luggiga blommor. Den tidiga vårfogeln ännu ensam ropade sitt *Lite hö!* –

[It was a beautiful spring day. – The midday thaw had melted almost all the snow away, only some patches here and there were still lying on the northern side of the gooseberry bushes. The first sprouts of the irises and the tulip leaves were already starting to peep out of the damp flower beds, still heaped with leaves and debris and not yet dug up. The water from the melted snow formed little rivulets in the sandy paths, where the occasional dropped pea had already started to germinate and to announce that spring was advancing with great strides. However, so far no leaves had come out; they were only peeping out from exuberant buds, and a single osier bush in a wild hedge was beginning to bloom on a bare stem. The early springtime bird, still alone, cried its "A little hay!"]

The quotation is taken from Fredric Cederborgh's novel *Ottar Trallings lefnadsmålning*, famous among his contemporaries, which was published in four volumes in 1810–18. We cannot dwell on a close reading here, but the similarities are extremely striking and cannot possibly be coincidental.

It is interesting to note, then, that Cederborgh's text has not been associated specifically with Linnaeus, but with August Strindberg. In 1879, the latter published his novel *Röda rummet* (*The Red Room*). This book not only led to its author's breakthrough as a recognized writer, but symbolizes the modern breakthrough itself in Swedish literature. The very beginning of the novel is one of the most frequently quoted passages in Swedish literature. Here it is, again:

Första Kapitlet.

Stockholm i fogelperspektiv.

Det var en afton i början av Maj. Den lilla trädgården på Mosebacke hade ännu icke blifvit öppnad för allmänheten, och rabatterna voro ej uppgräfda; snödropparne hade arbetat sig upp genom fjolårets löfsamlingar och höllo just på att sluta sin korta verksamhet för att lemna plats åt de ömtåligare saffransblommorna, hvilka tagit skydd under ett ofruktsamt päronträd; syrenerna väntade på sydlig vind för att få gå i blom, men lindarna bjödo ännu kärleksfilter i sina obrustna knoppar åt bofinkarne, som börjat bygga sina lafklädda bon mellan stam och gren; ännu hade ingen menskofot trampat sandgångarne sedan sista vinterns snö gått bort, och derför lefdes ett obesväradt lif derinne af både djur och blommor. Gråsparfvarne höllo på att samla upp skräp, som de sedan gömde under takpannorna på navigationsskolans hus; de drogos om spillror af rakethylsor från sista höstfyrverkeriet, de plockade halmen från unga träd som året förut sluppit ur skolan på Rosendal – och allting sågo de! De hittade barège-lappar i bersåer och kunde mellan stickorna på en bänkfot draga fram hårtappar efter hundar, som icke slagits der sedan Josefina-dagen i fjor. Der var ett lif och ett kif.

[CHAPTER 1

A BIRD'S-EYE VIEW OF STOCKHOLM

> It was an evening in the beginning of May. The little garden on "Moses Height," on the south side of the town had not yet been thrown open to the public, and the flower-beds were still unturned. The snowdrops had worked through the accumulations of last year's dead leaves, and were on the point of closing their short career and making room for the crocuses which had found shelter under a barren pear tree; the elder was waiting for a southerly wind before bursting into bloom, but the tightly closed buds of the limes still offered cover for love-making to the chaffinches, busily employed in building their lichen-covered nests between trunk and branch. No human foot had trod the gravel paths since last winter's snow had melted, and the free and easy life of beasts and flowers was left undisturbed. The sparrows industriously collected all manner of rubbish, and stowed it away under the tiles of the Navigation School. They burdened themselves with scraps of the rocket-cases of last autumn's fireworks, and picked the straw covers off the young trees, transplanted from the nursery in the Deer Park only a year ago – nothing escaped them. They discovered shreds of muslin in the summer arbours; the splintered leg of a seat supplied them with tufts of hair left on the battlefield by dogs which had not been fighting there since Josephine's day. What a life it was! (Translation by Ellie Schleussner, 1913.)]

In spite of the similarities between Strindberg's text just quoted and the extract from Cederborgh given before, it cannot be proved that Strindberg ever read Cederborgh. He was, however, well acquainted with Linnaeus, from his childhood onwards, and that may have been enough. Already in 1882, Strindberg included a long, enthusiastic chapter on Linnaeus in his monumental work on the Swedish people (*Svenska Folket*). This was a quarter of a century before Levertin's essay, which clearly demonstrated the literary ability of Linnaeus. Both Strindberg and Cederborgh may have been inspired by Linnaeus, independently of each other. I have treated Strindberg's relationship to Linnaeus more extensively elsewhere (e.g. Ralph 2007).

There is no space for a careful analysis here, but there is an interesting detail to observe. There is a puzzling wording in Strindberg's text, viz. "lindarna bjödo ännu kärleksfilter i sina obrustna knoppar åt bofinkarne" [the tightly closed buds of the limes still offered cover for love-making to the chaffinches]. Here, Strindberg made a correction in his manuscript, from *afrodisiaka* to *kärleksfilter*. There is, however, no such word as *kärleksfilter* in Swedish, but it has been suggested that Strindberg had *kärleksdryck* in mind, in the sense of 'love potion', 'philtre'; he could have been influenced by Latin *philtrum*, which has this meaning. Linnaeus, too, lingers on the reproduction of the trees in his text. A connection between the two passages is quite possible.

If these connections were to prove valid, Linnaeus's contribution to the history of the Swedish language would not be restricted to his potential importance for the development of Swedish special-purpose language and specialized terminology in the eighteenth century and for the refinement of certain literary genres like travel reports. He may also to some extent have served as a model for the most influential Swedish writer ever, and he would thus – at least indirectly – have played a role in the modern breakthrough, more than a hundred years after his own death.

6. Linnaeus's stylistic inspiration

So, Strindberg did not emerge from nowhere, but what about Linnaeus? Where did he find his inspiration? Strindberg was not particularly impressed by Georg Stiernhielm, at least not in his youth, although Stiernhielm is traditionally referred to as the father of Swedish poetry. Instead, the towering figure of the Swedish Reformation, Olaus Petri, is claimed by Strindberg to have played a much more important role in the history of the language and of literature in Sweden. During the period when he was working on *Röda rummet*, he was almost obsessed by "Master Olof". He had already written one version of the play carrying that name, and he was constantly rewriting another version.

In the early sixteenth century, the reformer of the Swedish church repeatedly expressed his linguistic ideals, to the effect that simplicity and clarity should be the guiding stars. In fact, his formulations come very close to the corresponding declarations made by Linnaeus and Strindberg. However, they are not entirely the reformer's own invention; his ideals have been inherited from Martin Luther himself, who was very outspoken about the importance of the Bible being accessible to all. That was why, in his view, it had to be translated into the respective vernaculars. Luther's ideals are ultimately based on his experience from his practical work, be it preaching, teaching or translation. This line of influence would come naturally to Linnaeus, too, a vicar's son as he was, living in a dogmatic time. Indeed, he has many of the features that have often been suggested to be typically Lutheran. It has even been argued that the Swedes may have falsely been accused of being more Lutheran than Luther himself – it could just as well be the scrupulously ordering Linnaean tradition that is still in command of their minds.

Martin Luther gives priority to the vernacular as opposed to Latin, as far as the language of religion is concerned. The biblical texts must be translated into the vernacular to make them comprehensible to everybody – this is the general idea. Therefore, massive translation is carried out by Martin Luther and by

Olaus Petri, into their respective languages. In this way, they substantially contribute to crucial events in the history of their respective languages. Linnaeus's attitude in these matters is less programmatic. Indirectly, however, he is under the sway of the language of the Reformation Bible, and his works can be fitted into the written production of the period during which Swedish eventually became the principal language of Sweden – in all linguistic domains. His stylistic ideals became decisive to the Swedish tradition of technical and special-purpose language, and he may even have had an impact on the modern breakthrough in Swedish literature. It is hard to imagine a more central position in the development of the Swedish language.

References

Cederborgh, Fredric
[1856] *Ottar Trallings lefnads-målning* [Ottar Tralling's life depicted]. Tredje delen. In: *Valda skrifter af Fr. Cederborgh efter de äldre original-upplagorna utgifna* [Selected writings by Fr. Cederborgh, according to the older original editions]. Stockholm: Axel Hellstens förlag.

Dalin, Olof
1732 *Then Swänska Argus* [The Swedish Argus]. N:o I. Stockholm: Benjamin Gottlieb Schneider.

Geijer, Erik Gustaf
[1834] *Minnen* [Memoirs]. In: *Erik Gustaf Geijers Samlade skrifter* [Erik Gustaf Geijer's complete writings]. Förra afdelningen. Tredje bandet. Stockholm, 1851: P. A. Norstedts & söner.

Hannesdóttir, Anna Helga
2011 From vernacular to national language: Language planning and the discourse of science in eighteenth-century Sweden. In: Britt-Louise Gunnarsson (ed.), *Languages of Science in the Eighteenth Century*, 107–122. Berlin, New York: De Gruyter Mouton.

Kolmodin, Olof
1732 *Biblisk Qwinno-Spegel* [Biblical lady's mirror]. Stockholm: Historiographi Regni Tryckerij (Hartwig Gercken).

Levertin, Oscar
1906 *Carl von Linné. Några kapitel ur ett oafslutadt arbete* [Carl von Linné. Some chapters from an unfinished work]. Stockholm: Albert Bonniers förlag.

Linnaeus, Carl
[1732] *Caroli Linnæi Iter lapponicum Dei gratia institutum 1732* [...]. [Ed. by Magnus von Platen and Carl-Otto von Sydow.] 1975. Stockholm: Wahlström & Widstrand.

Linnaeus, Carl
[1734] *Caroli Linnæi Vulcanus docimasticus Fahlun 1734*. Efter originalmanuskriptet i Linnean Society i London utgivet av G. A. Granström. Uppsala and Stockholm 1925: Almqvist & Wiksells Boktryckeri-A.-B. (Skrifter utgivna av Svenska Linnésällskapet. N:r 2.)

Linnaeus, Carl
1739 *Tal, om märkvärdigheter uti insecterna, hållit för* Kongl. Vetens. Academien *uti Auditorio illustri, Då första* Præsidentskapet *aflades 1739 D. 3 October* [Speech about remarkable things in the insects, given to the Royal Swedish Academy of Sciences, when the first presidency ended on 3 October 1739]. Quoted from the 3rd edition: På Kongl. Vet. Academiens befallning, tredje gången uplagd. Stockholm 1752: Lars Salvius.

Linnaeus, Carl
1745 *Carl Linnæi* [...] *Öländska och Gothländska Resa på Riksens Högloflige Ständers befallning förrättad Åhr 1741* [Linnaeus's Öland and Gotland journey 1741]. Stockholm/Upsala: Gottfried Kiesewetter.

Linnaeus, Carl
1751 *Carl Linnæi* [...] *Skånska Resa, på Höga Öfwerhetens Befallning Förrättad År 1749* [Linnaeus's Scanian journey 1749]. Stockholm: Lars Salvius.

Ralph, Bo
2007 August Strindberg – unik genombrottsman med föregångare [August Strindberg – a unique pioneer with predecessors]. In: Wollin et al. 2007, 229–239.

Strindberg, August
1879 *Röda rummet. Skildringar ur Artist- och Författarlifvet* [The red room]. Stockholm: Jos. Seligmann & C:is förlag.

Strindberg, August
1882 *Svenska Folket* [The Swedish people]. I–II. Stockholm: C. E. Fritze's K. Hofbokhandel.

Teleman, Ulf
2002 *Ära, rikedom och reda. Svensk språkvård och språkpolitik under äldre nyare tid* [Glory, wealth, and order. Swedish language cultivation and language policy in early modern times]. (Skrifter utgivna av Svenska språknämnden, 85.) [Stockholm:] Norstedts Ordbok.

Teleman, Ulf
2011 The Swedish Academy of Sciences: Language policy and language practice. In: Britt-Louise Gunnarsson (ed.), *Languages of Science in the Eighteenth Century*, 63–87. Berlin, New York: De Gruyter Mouton.

Wollin, Lars, Anna Saarukka and Ulla Stroh-Wollin (eds.)
2007 *Det moderna genombrottet – också en språkfråga?* Föredrag vid nionde sammankomsten för svenska språkets historia, i Åbo 19–20 maj 2006 / Studier i svensk språkhistoria 9 [The modern breakthrough – also a matter of language?]. (Skrifter från svenska institutionen vid Åbo Akademi 5/2007.)

Calendar and aphorism: A generic study of Carl Linnaeus's *Fundamenta Botanica* and *Philosophia Botanica*

Han-Liang Chang

1. Introduction

This chapter discusses the classical and Renaissance genre legacy which influenced Carl Linnaeus (Carl von Linné, 1707–1778) in his writing of *Philosophia Botanica* (1751) and *Fundamenta Botanica* (1736). The two key words that constitute the main title, "calendar" and "aphorism", evoke the classical genre tradition, while at the same time suggesting a double denotation of literary genre and biological genus; both can be traced back to classical antiquity, but converged in the writings of Linnaeus. It was Linnaeus who enlightened the Age of Reason by founding modern botany and systematics, and who codified botanical Latin as a tool for scientific inquiry. This paper traces the generic sources of Linnaeus's writings and discusses the theoretical implications of aphorism as genre with reference to contemporary speculations on the linguistic sign in the wake of Ferdinand de Saussure. It is assumed that generic and formal devices, such as aphorism and calendar, are secondary textual constraints superimposed on the primary language code of botanical Latin, which has its own time-honoured stylistic conventions. Because of their rich generic links, the two texts by Linnaeus call for careful intratextual and intertextual readings, as the present author will demonstrate towards the end of the chapter.

2. The two faces of Linnaeus

It is a critical commonplace that Linnaeus has two incongruous faces (Lindroth 1994; Chang 2007). Scholars have extended this self-contradiction to other aspects of his life and career, including the diversity of his research methods and writings. One could, for example, compare Linnaeus's classification method used in *Systema Naturae*, which is encoded in scientific Latin and follows closely the Aristotelian-Porphyrian a-priori logic, and his travel writings, which are written in Swedish and characterized by narrative randomness and occasional bursts into lyricism. Even through English translation, one can still suspect a conspicuous stylistic inconsistency. What one is dealing with in this regard is not only the author's linguistic code as the primary modelling system, consisting of the lexical, syntactic and semantic aspects, but also the many stylistic and generic devices, which are built on top of the base linguistic code to form his secondary modelling system. For such a conceptualization, we subscribe to the discovery and descriptive procedures proposed by the Tartu-Moscow School of Semiotics in the 1970s (Lotman 1977; Lotman et al. 1973).

3. Two generic conventions: aphorism and calendar

A truly exceptional case is Linnaeus's *Philosophia Botanica* (hereafter cited as *PB*), developed from the prototypal *Fundamenta Botanica* (hereafter cited as *FB*), in which the author makes exclusive use of the discursive device of aphorism to complete the preordained empty calendar. Whereas Linnaeus's classification is deductive and systematic, his calendars, along with his many travelogues, are random and sketchy, discursive but inconclusive, characteristic of the essayistic and aphoristic writings of the Renaissance, in particular those by Francis Bacon (1561–1626). Now there are visibly two genres fused in the text of *PB*, namely, calendar and aphorism. Both genres, marginal as they may seem, can be traced back to classical antiquity and were revived, with various semantic and cultural investments, in the Renaissance. Their relationship cannot be one between form and content, because aphorism is a form in itself. What can be said is that without rich semantic investment, a calendar, for all its rigid 365 daily divisions, remains an empty form, and it takes the protean aphorism to materialize. We shall therefore take a quick glance of calendar as genre.

4. Calendar as genre

The *Oxford English Dictionary* gives two definitions of *calendar*:[1]

(1) The system according to which the beginning and length of successive civil years, and the subdivision of the year into its parts, is fixed; as the Babylonian, Jewish, Roman, or Arabic calendar.

---

1. The citations from the *OED* and the *Merriam-Webster*, especially of the definitions of aphorism later in the paper, are aimed at elaborating on the semiosic distinction of dictionary and encyclopaedia outlined by Eco (1976, 1984). To me, no other

(2) A table showing the division of a given year into its months and days, and referring the days of each month to the days of the week; often also including important astronomical data, and indicating ecclesiastical or other festivals, and other events belonging to individual days. Sometimes containing only facts and dates belonging to a particular profession or pursuit, as *Gardener's Calendar*, *Racing Calendar*, etc. Also a series of tables, giving these facts more fully; an almanac.

Compared with the *OED*, the less authoritative Merriam-Webster Dictionary ironically gives definitions more relevant to our purpose:

(1) A system for *fixing* the beginning, length, and divisions of the civil year and *arranging* days and longer divisions of time (as weeks and months) in a definite order;
(2) An orderly list: as a list or schedule of planned events or activities *giving* dates and details. (Emphases mine.)

Only in this sense of non-prescribed acting (i.e., "fixing" and "giving") by individuals can calendar be a kind of writing – literary or otherwise; and only in this sense of textualization can it exemplify the act of genre praxis, of *beinhalten*, or "filling content into form", at any historical moment.

As suggested by the *OED*, there are specialized calendars traceable to classical antiquity but popularized in the Renaissance, e.g., the shepherd's calendar; the farmer's or gardener's calendar; the floral [not the florist's] calendar. Under the influence of classical pastoralism, a number of British poets have produced works of the same title, *The Shepherd's Calendar*, from the sixteenth to the nineteenth centuries, namely, Edmund Spenser (1552–1599), James Hogg (1770–1835), and John Clare (1793–1864). The farmer's/gardener's calendar has perhaps an older history, attributed to Hesiod's (*fl.* 8th century BCE) *Works and Days*. Aristotle's disciple Theophrastos (372–287 BCE), a philosopher-botanist like Linnaeus, contributed to the genre with his *De causis plantarum* and *De historia plantarum*. He was followed by the Roman Vergil (70–19 BCE) in the latter's didactic poetry *Georgics*. Finally, we come to the calendar of flora. Linnaeus's English translator, Benjamin Stillingfleet, identifies the genre in Theophrastos' *De historia plantarum*, to which he adds that by Alexander

example than aphorism can serve more powerfully as a metacommentary to examine the distinction. The current online edition of *Encyclopaedia Britannica* bases its entry on aphorism on the 11th edition, which in turn bases its entry on the *OED*. See n. 3 below.

M. Berger (Linnaeus), dated 1755, and his own of the same year. (Linnaeus [1775]1977; Heller 1983: 215–216.)[2]

It is Stillingfleet who renders justice to calendar by granting the sterile form a prestigious genre status. In an essay entitled *The Calendar of Flora, Swedish and English*, published together with his translations of Linnaeus's miscellaneous tracts in 1755, Stillingfleet defines what he means by the calendar of flora: "…and that time [the right time] may not be according to certain calendar days, but according to a hitherto unobserved calendar, which varies several weeks in different years" (Linnaeus [1791] 1977: 235). He observes that the calendar existed a long time ago, in Hesiod's time, "but when artificial calendars came into vogue the natural calendar seems to have been totally neglected, for I find no traces of it after his [Hesiod's] time" (Linnaeus 1977: 235). Therefore, he calls for a revival of the genre, and illustrates it with his own calendar as follows:

> I have retained the division of months according to budding, leafing, flowering, &c … But I am convinced that this method marks more precisely when we may expect the flowering of any plant, or the any bird, &c. than the bare mention of the day of a common calendar month, and at the same time marks it more universally. (Linnaeus 1977: 244–245)

There is no coincidence that Linnaeus's *A Tour in Lapland* makes use of both kinds of mechanical and natural calendars, the one prescribed, the other composed through observation:

> *May* 12, 1732, old style
>
> I set out alone from the city of Upsal on Friday May 12, 1732, at eleven o'clock …
>
> At this season Nature wore her most cheerful and delightful aspect, and Flora celebrated her nuptials with Phoebus. (Linnaeus [1811] 1971: 2)

Stillingfleet's description shows an important pragmatic function of the calendar of flora. Much as in keeping journals, based on lived experience and quotidian observations, the writer does not aim for self-communication, as is the case when keeping a private diary; rather, he means to communicate what he has observed and recorded to fellow naturalists, often across national borders, for comparative research. A question can be posed in this connection: Are Lin-

2. Apparently Stillingfleet takes the author of the dissertation to be Alexander M. Berger, a student of Linnaeus's, but with historical hindsight, contemporary scholars generally agree to attribute these dissertations supervised by Linnaeus to him rather than to the students who submitted and signed them because of the special supervision system used in Sweden at the time. See Knut Hagberg's (1939) and Stearn's (1957) comments, quoted in Heller (1983: 216, n. 13).

naeus's *Fundamenta Botanica* and *Philosophia Botanica* floral calendars? If not, what are they? On the surface, the two works retain one feature of the mechanical calendar, i.e., its numeral and numerical symbolism of 365, and the author inserts into the given 365 textual spaces his learned and speculative aphorisms, though they are not entirely based on observations of nature, as he insists at the end of *PB* that they should be: *In scientia Naturali / Principia veritatis / Observationibus confirmari debent* (In natural science / the elements of truth / ought to be confirmed by observation) (Linnaeus 2003: 307). At least here, between the two poles of human faculties for nature studies, it is reason rather than experience that prevails.

5. Aphorism as genre

What about the other genre, aphorism? Perhaps there is no coincidence that the *OED* (online 2nd ed. 1989), in its entry *aphorism*, gives a few sample sentences, one of which is by the English essayist Bacon and the last of which refers to Linnaeus:[3]

> **1605** BACON *Adv. Learn.* I. v, Knowledge, while ... in aphorisms and observations ... is in growth.
> **1879** DE QUATREFAGES *Hum. Spec.* 50 The aphorism ... which was formulated by Linnæus in regard to plants.

The second citation belongs to the English translation of Jean Louis Armand de Quatrefages de Bréau's (1810–1892) *L'Espèce humaine* (1877). It may not be able to do justice to Linnaeus because his *PB* and *FB* had already been published for more than a century. But there is apparently a genre tradition linking Bacon to Linnaeus, criss-crossing many other writers on natural history and natural philosophy during the seventeenth and eighteenth centuries, including, as we shall see, Joachim Jung (1587– 1657), Hermann Boerhaave (1668–1738), and Christian Gottlieb Ludwig (1709–1773). This genre convention is a secondary modelling system (or subcode) superimposed on the primary modelling system of New Latin or such vernaculars as English, Dutch and Swedish.

Did, as De Quatrefages said, Linnaeus "formulate" aphorisms in regard to plants? Who are his predecessors? The term was first used in the *Aphorisms* (400 BCE) of Hippocrates (*c.*460–368 BCE), which is a series of propositions

3. Ephraim Chambers' (*c.* 1680–1740) *Cyclopædia enters the term in v*olume 1, dated 1728, but it does not cite Linnaeus for obvious chronological reasons.

concerning the semeiology of disease and the art of curing. The opening aphorism is a most famous quote:

> Life is short, the Art long, opportunity fleeting; experiment treacherous, judgment difficult. The physician must be ready, not only to do his duty himself, but also to secure the co-operation of the patient, of the attendants and of externals. (Hippocrates [1931] 1992, 4: 99)

Another collection of medical aphorisms is Hermann Boerhaave's *Aphorismi de cognoscendis et curandis morbis*, published in Leiden in 1709. Within a few years, it was translated into English as *Boerhaave's Aphorisms, Concerning the Knowledge and Cure of Diseases* and published in London in 1715 (Boerhaave [1715] 1986). Several anecdotes can be evoked to establish the two writers' *rapports de fait*. In June 1735, at the University of Harderwijk, Linnaeus took his degree examination, which included "an exposition of two of Hippocrates' *Aphorisms*" (Blunt 2001: 94). Shortly afterwards, on 5 July 1735, Linnaeus met Boerhaave, nicknamed "Hippocrates Redivivus" (Blunt 2001: 96), from whom Linnaeus learned much "that was to be valuable to him in later life" (Blunt 2001: 97). The English biographer Blunt continues his narration: That winter, Linnaeus stayed with Johannes Burman who "assisted with two books on which he was working at the time – his *Fundamenta Botanica* and *Bibliotheca Botanica*" (Blunt 2001: 100).

One can surmise Linnaeus's indebtedness from his extensive use of Boerhaave in annotating his *PB*. In the opening, he quotes from Boerhaave to annotate his Aphorism 3:

> *Boerh. Hist. 3*. Planta est corpus organicum, alteri cuidam corpori Cohaerens per aliquam partem sui, per quam Nutrimenti & Incrementi & Vitae materiam capit & trahit.
>
> (*Boerhaave, Historia 3*. 'A plant is an organic body, adhering to some other body by some part of itself, through which it draws the matter for nourishment, growth and life.') (Linnaeus 2003: 9)

The same Aphorism 3 is glossed by two other writers' aphorisms. The late William T. Stearn (1966: 30), President of the Linnean Society of London, who supervised many translation and research projects on Linnaeus, comments on an earlier work, Joachim Jung's *Isagoge Phytoscopica* (1678), as such: "His work is thus very formal in character, consisting of aphorisms ...". It is to a definition in this work that Linnaeus compares his own definition of a plant:

> Jung. Isagog. c.1. Planta est corpus vivens [vivum] non sentiens, s. [Sive est] [corpus] certo loco aut certae sedi affixum, unde nutriri, augeri, denique se propagare potest.

(*Jung, Isagoge Ch. 1*. 'A plant is a living body without feeling, fixed in a particular place or abode from which it can receive nourishment and growth and eventually propagate itself.') (Linnaeus 2003: 9)

The third writer of aphorisms is Christian Gottlieb Ludwig, author of a title identifying its generic belonging: *Aphorismi botanici* (1738), a work published only two years later than *Fundamenta Botanica*. There is evidence that Ludwig had read Linnaeus's *Systema Naturae*. But here Linnaeus, in glossing Aphorism 3, refers to another title of Ludwig's:

Ludwig. veget. 3. Corpora naturalia eadem semper Forma & Loco-motivitate praedita appellantur *Animalia*; eadem semper forma, sed locomotivitate destituta *Vegetabilia*; & quae diversam formam obtinent *Mineralia* dicuntur.

(*Ludwig, Vegetabile 3.* 'Natural bodies, that are always provided with the same shape and with locomotion, are called *animals*; those that keep the same shape but are without locomotion, are called vegetables; and those that are subject to variations in shape, *minerals*.') (Linnaeus 2003: 9)

He does refer to *Aphorismi botanici* when defining botany:

Ludwig. aphor. 1. Botanica est scientia vegetabilium, s. cognitio eorum, quae per plantas & in plantis fiunt.

(*Ludwig, Aphorismi 1.* 'Botany is the science of vegetables, to wit, the investigation of the things that are done by plants and in plants.') (Linnaeus 2003: 10)

Linnaeus might have borrowed the form from Boerhaave, but more likely the Swedish physician did not. The classical legacy of Hippocrates being very much around and the Greek writer's work required reading for all physicians, Jung's *Isagoge Phytoscopica* and Ludwig's *Aphorismi botanici* bear good evidence that this genre dominated the medical discourse of the time. A standard joke in my field of comparative literature runs like this: One does not need to read a Petrarchan sonnet to write a Petrarchan sonnet. The reason is self-explanatory. Such a form as aphorism has become a common property, indeed an anterior textual space in which people write with ease, echoing or emulating one another. Such earlier collections serve the triple function of, firstly, alluding to the oldest medical aphorisms attributed to Hippocrates, then evoking, unwittingly perhaps, their theoretical articulation in Bacon, and finally anticipating the calendar writings of Linnaeus who, as we have discussed, adopts another classical genre of natural writings, the floral or botanical calendar which dates back to Theophrastus and has been popular throughout the centuries. Even regarding this calendar genre, critical opinion divides as to whether Linnaeus follows the ancient tradition or a more immediate religious genre of the Lutheran Almanac (Koerner 1999: 40–41).

6. Aphorism reconsidered

What is aphorism? The *OED* gives two definitions.

1. A 'definition' or concise statement of a principle in any science. ex. **1664** POWER *Exp. Philos.* III. 190 The old and uncomfortable Aphorism of our Hippocrates.
2. Any principle or precept expressed in few words; a short pithy sentence containing a truth of general import; a maxim. ex. ***c*1590** MARLOWE *Faustus* i. 19 Is not thy common talk sound aphorisms?

Why has aphorism as genre not received much critical attention? Boerhaave, in his preface to the reader, describes it as "a Stile adapted entirely to the Subject; not with the Elegance of the Augustaean Age" (Boerhaave [1715]1986: A4). As late as 1968, the German-American comparative literature scholar Ulrich Weisstein could still note aphorism's belated reception as a genre concept:

> … the theory of literary genres is also charged with circumscribing and defining such marginal forms as are often excluded from the realm of belles letters. We think, for example, of the essay, the biography, and the autobiography and, among the miniature forms, the maxim, the aperçu, the aphorism, the Character (*caractère*), and perhaps even the Romantic fragment – many of which forms are just beginning to gain a measure of scholarly recognition. (Weisstein [1968] 1973: 109)

In terms of its brevity in expression, basically on the sentential rather than the discursive level, aphorism can better be termed a form than genre, though not so "empty" a form, i.e., to be "in-formed", as calendar. However, this "miniature form" or trivializing form, or, in Roman Jakobson's words, "transitional genre", was historicized as a major formal device in the botanical and medical writings of Renaissance Europe (Jakobson [1935] 1971: 86). What is more important, Weisstein made his comments at the time when post-structuralist thinkers were reviving, as did their German Romantic predecessors, such minor genres as aphorism, fragment, and essay, holding that they could give form to speculative inquiry of the highest order. (Lacoue-Labarthe and Nancy, 1988; Kauffmann 1989.)

I would like to argue that the simple form of aphorism has a number of logical and semiotic functions to perform, running the gamut from definition to argument, and at the same time displaying language semiosis on both the textual and intertextual levels. Among the very few writers who consciously cultivate the logical and rhetorical power of aphorism, Bacon figures most prominently. His *Novum Organum* (1620), with a subtitle of *Aphorisms Concerning the Interpretation of Nature and the Empire of Man*, is written in aphorisms, and as such serves reflexively as a metacommentary on aphorism. The rationale of such style had been propounded fifteen years before in 1605, which saw the publication of *The Advancement of Learning*. The two texts can be therefore read as a pair: whilst the earlier text lays the theoretical foundation, or poetics, so to speak, of aphorism, the later text provides the reader with a large-scale concrete demonstration of the kind of writing the author has advanced.

Bacon introduces aphorism as the proper approach to knowledge in lieu of fixed dogma, or, in his word, "method", because it has the advantage of growth. Where dogma closes a concept prematurely, aphorism opens it and lets it grow. He compares knowledge already set and that still in growth to young men's bodies:

> But as young men, when they knit and shape perfectly, do seldom grow to a further stature; so knowledge, while it is in aphorisms and observations, it is in growth: but when it once is comprehended in exact methods, it may perchance be further polished and illustrate and accommodated for use and practice; but it increaseth no more in bulk and substance (1.5.4). (Bacon 1974: 34)

Aphorism then is very much like the essay form for which Bacon is better known. Like essay, aphorism is tentative and inconclusive; it encourages and invites inquiry and participation on the part of the scientific researcher. But unlike essay, aphorism is noted for its brevity and precision in expression; it can be said that aphorism is the basic or minimal structural unit of the essayistic form.

The English philosopher favours aphorism precisely for its lack of rhetorical trappings and wordiness. It is, as it were, language in minimalism!

> [F]or aphorisms, except they should be ridiculous, cannot be made but of the pith and heart of sciences; for discourse of illustration is cut off: recitals of examples are cut off; discourse of connexion and order is cut off; descriptions of practice are cut off. So there remaineth nothing to fill the aphorisms but some good quantity of observation: and therefore no man can suffice, nor in reason will attempt, to write aphorisms, but he that is sound and grounded (2.17.7). (Bacon 1974: 135–136)

Now a question arises from this anti-rhetorical stance: If aphorism is the kernel of discourse and by the very nature of language it grows, in what does it grow, if not in language, and to what, if not to full-fledged discourse? Maybe Bacon has made a bad move in using a linguistic unit, namely, the sentence, to represent

Renaissance man's inquiring mind. The fact is that whilst he believes he is referring to the linguistic form, he is actually talking about the message transmitted by that form, that is, language's basic representational function. This "short, pithy sentence" is to begin with a grammatical concept, and when reified, it can serve as the vehicle of a statement, a definition or a proposition, depending on how and when language is used, as discourse or in logic. One could compare excerpts from the English translations of Hippocrates, Bacon, and Boerhaave to arrive at the conclusion that the seemingly fragmentary aphorisms, when articulated in discourse, perform all the functions which regular declarative sentences or constative utterances do.[4] In fact, Bacon's aphorisms in *Novum Organum*

4. Perhaps ironically to Bacon, in aphoristic writing all kinds of Aristotelian syllogism are used. I suspect that the rhetorical enthymeme is not uncommon. For a discussion of Hippocratic semiosis and abduction, see Manetti 1993: 47–48.
The following excerpts are taken from the three authors under discussion:
Bacon, *Novum Organum*, 2000 (1620) [The numbers follow those of the original order of aphorisms.]
(1) Man, being the servant and interpreter of Nature, can do and understand so much and so much only as he has observed in fact or in thought of the course of nature. Beyond this he neither knows anything nor can do anything.
(2) Neither the naked hand nor the understanding left to itself can effect much. It is by instruments and helps that the work is done, which are as much wanted for the understanding as for the hand. And as the instruments of the hand either give motion or guide it, so the instruments of the mind supply either suggestions for the understanding or cautions.
(3) Human knowledge and human power meet in one; for where the cause is not known the effect cannot be produced. Nature to be commanded must be obeyed; and that which in contemplation is as the cause is in operation as the rule.Boerhaave, *Practical Aphorisms*, (1715) 1986: 1–2.
(1) Whatever State of the Human Body doth disorder the Vital, the Natural, or even the Animal Functions of the same is call'd a *Disease*.
(2) And that part of the Science or Art of Physick, which teacheth how to find out the Disease actually afflicting the Patient, and how to cure the same, is called *The Practice of Physick*.
(3) Whosoever therefore is unacquainted with what is requisite to perform well all vital, natural and animal Functions; and consequently is ignorant of the Causes of Life and Health; can never discern and know rightly the Defects, that is, the Diseases of them.
(4) A *Cure* therefore is the *Changing* of a Disease into Health: and consequently supposes a Knowledge of what is just now mentioned (3.) and consequently the Knowledge and the Cure of Diseases require the Knowledge of those Doctrines which expound what it is, which is strictly termed

tend to become longer and longer as the author's argument develops, so that if the numeral headings were removed, the work would read just like an extended treatise.

It seems that our exploration into aphorism as a style concept has come to a theoretical and pragmatic impasse. Perhaps there is another kind of aphorism which is less discursive and rhetorical, suited to botanical representation, as exemplified above by the cross-references among Jung, Boerhaave, Ludwig, and Linnaeus, despite the fact they are all definitional. This kind of aphorism is the language used in botanical diagnoses and descriptions (Stearn 1966: 143). Whether a plant's characters are differential or essential, the wording for them must be lexically and syntactically concise and precise. John Lindley's (1799–1865) description of the language used for such purposes can be the kind of alternative "economical" aphorism we have in mind: "differential characters express in the least possible space the distinctions between plants; they should contain nothing superfluous"; and "[t]he *essential* character of a plant expresses … those peculiarities known by experience to be most essential to it; but admits nothing unimportant or superfluous…" (quoted in Stearn 1966: 143; italics original). The epitome of this kind of writing is no doubt Linnaeus who "devoted much thought to the drafting of these diagnostic phrases" and "held that these diagnoses should not exceed twelve words in length" (Stearn 1966: 144).

For a book claiming to be "philosophy", then there must be two types of aphorism: Type A, that for argumentation in the domain of philosophical discourse; Type B, that for definition, diagnosis, and description befitting a manual. The following two aphorisms qualify as Type B.

> 79. Vegetabilium (78) PARTES, primum a Tyrone distinguendae, sunt III: *Radix*, *Herba*, Fructificatio.
> (79. 'The PARTS of vegetables [78], that must be distinguished by the beginner, are Three: the *root*, the *herb*, and the *fruit-body*.) (Linnaeus 2003: 50, 53)

Life and *Health* in Man, *viz.* the Institution of Physick, or the Animal OEconomy, or the Theory of Physick.

Hippocrates, *Aphorisms*, 1992, 4: 105.

(1) Old men endure fasting most easily, then men of middle age, youths very badly, and worst of all children, especially those of a liveliness greater than the ordinary. (Book I, XIII)

(2) Growing creatures have most innate heat, and it is for this reason that they need most food, deprived of which their body pines away. Old men have little innate heat, and for this reason they need but little fuel; much fuels puts it out. For this reason too the fevers of old men are less acute than others, for the body is cold. (Book I, XIV)

127. COMPOSITI (117) flores implentur (121) vel Petalis tubulatis vel planis.
(127. 'COMPOUND flowers [117] are filled up [121] either by tubular petals or by flat ones.') (Linnaeus 2003: 95)

They are in sharp contrast to the polemical propositions of Type A, as follows:

9. PHILOSOPHI (18) Scientiam Botanicam demonstrative ex principiis rationalibus in formam scientiae reduxerunt; ut Oratores (20), Eristici (21), Physiologi (22), Institutores (23).
(19. 'The PHILOSOPHERS [18], by demonstration from rational principles, have reduced botanical knowledge to the form of a science: for instance, *rhetoricians* [20], *controversialists* [21], (*physiologists* [22]) and *legislators* [23].') (Linnaeus 2003: 21. The parenthesized part is missing in Freer's English translation.)

20. ORATORES, (19) quaecunque Scientiam docte ornant, proposuerunt.
(20. 'The RHETORICIANS [19] have expounded all things that are learned Ornaments of science. (Linnaeus 2003: 21)

365. Oeconomicus Usus plantarum generi humano utilissimus est.
Usus plantarum pro Oeconomia universali naturae sedulo indagandus.
In scientia Naturali
Principia veritatis
Observationibus confirmari debent.
(365. 'THE ECONOMIC use of plants is of great utility to the human race.
The use of plants for the *Universal Economy of Nature* should be assiduously researched.
In natural science
the elements of truth
Ought to be confirmed by observation.') (Linnaeus 2003: 207)

7. Conclusion: Textual semiosis and reading

This leads us to our final question. How does Linnaeus structure his text, save the mechanic numeral of 365 and the numbering of aphorisms accordingly? As Aphorisms 19 and 20 indicate, there is the technique of enchainment (*enchaînée*), with 20 referring back to 19 and forward to 21, thus forming an interlocking chain. The immediately interlocked aphorisms often show semantic and syntactic continuity, corresponding to the numerical linearity. Following Saussure's conceptualization of language structure in binary opposition, we may call such sections syntagmata; in fact, each aphorism constituting a syntagm in itself. There is an in-built coherence among the elements forming a syntagm and in their *mise-en-scène*. Aphorism 7 on the Botanist identifies him as the Collector (8) and

Methodizer (18); Aphorism 8 on the Collector refers to him simultaneously as the Father (9), Commentator (10), Illustrator (11), Describer (12), Monographer (13), the Meticulous (14), Adonis (15), Compiler (16) and Traveller (17). When these are introduced, each in his turn, they are likewise mutually referred to. This textual practice on the part of the author provides the reader with guidance to infinitely repeatable synoptic reading.

Furthermore, the instances of 7 and 8 suggest another order of relationships: they point to the "absent" linguistic signs, in Saussure's words, "*in absentia*" in the aphorism under consideration, but "present" ("*in praesentia*") as paradigms within the lexicon of the text (Saussure 1983: 122). The same textual practice goes beyond an individual work, in search of intertextual resonances (cf. Todorov 1982). The annotations of an aphorism and the references to other similar parallels outside the text of *PB*, such as the aphorisms by Jung, Boerhaave, and Ludwig, evoked to illuminate Aphorism 3 (see above), suggest the existence of a larger textuality which inscribes and informs the scientific discourse in the eighteenth century.

It may not be a coincidence that we have referred to Saussure in elucidating the organizing principle of the Linnaean texts. In his fragments, recently republished with the newly discovered Orangery manuscripts (Saussure 2002; 2006), the founding father of linguistic semiology comes to the defence of aphorism:

> Quelques vérités qui se retrouvent [ ]. – (Saussure 1974 tome 2, fascicule 4: 42) Ne parlons ni d'*axiomes*, ni de principes, ni de thèses. Ce sont simplement et au pur sens étymologique des aphorismes, des *délimitations*. – [ ] mais des limites entre lesquelles se retrouve constamment la vérité, d'où que l'on parte [ ] (Saussure 2002: 123)
>
> Let us not speak of *axioms* or principles or theses. These are simply, and in the pure etymological sense, aphorisms, definitions. – ... but limits between which truth is constantly to be found, whatever the starting point... (Saussure 2006: 82)

Aphorism has been traditionally used in medical and scientific treatises precisely because of its function in definitional logic. Now definition raises a special situation of one kind of iconic sign. At first sight, the two parts bridged by the copula, i.e., the term (*definiendum*) and its definition (*definiens*), are, logically speaking, analytical and tautological, but, in semiotic terms, mutually iconic. However, the relationship between *definiendum* and *definiens* is non-reversible. This necessitates the terms that constitute the *definiens* to be further defined *ad infinitum*. This is the logic which witnesses the continued syntagmatization of paradigmatic elements, the passage from iconic sign to indexical sign, and the transformation of the semiosis of dictionary to that of encyclopedia (Eco 1976; 1984). In this sense, Linnaeus has provided a semiotic text well ahead of his time.

References

Bacon, Francis
1974 The Advancement of Learning and New Atlantis. Edited by Arthur Johnston. Oxford: Clarendon Press.

Bacon, Francis
2000 The New Organon [Novum Organum: Aphorisms concerning the Interpretation of Nature and the Kingdom of Man]. Cambridge: Cambridge University Press. First published London [1620].

Bacon, Francis
1994 The Tvvoo Bookes of Francis Bacon Of the proficience and aduancement of Learning, diuine and humane. London: Henrie Tomes. Reprint. Bristol: Thoemmes Press. First published London [1605].

Blunt, Wilfrid
2001 *Linnaeus: The Compleat Naturalist*. PrincetonPrinceton University Press. First published London [1971].

Boerhaave, Herman
1986 *Boerhaave's Aphorisms: Concerning the Knowledge and Cure of Diseases*. Translated by J. Delacoste. London: B. Cowse and W. Innys. Reprint. Birmingham, AL.: The Classics of Medicine Library. First published London: B. Cowse [1715].

Chang, Han-liang
2007 Natural history or natural system? – Encoding the textual sign. In: Marcello Barbieri (ed.), *Biosemiotics: Information, Codes and Signs in Living Systems*, 165–178. New York: Nova Science Publishers

Eco, Umberto
1976 *A Theory of Semiotics*. Bloomington: Indiana University Press.

Eco, Umberto
1984 *Semiotics and the Philosophy of Language*. London: Macmillan.

Heller, John Lewis
1983 *Studies in Linnaean Method and Nomenclature*. (Marburger Schriften zur Medizingeschichte 7.) Frankfurt am Main: Peter Lang.

Hippocrates
1992 Aphorisms. In: *Hippocrates in Six Volumes*, Volume 4. Trans. W. H. S. Jones. (Loeb Classical Library 147). Cambridge, MA: Harvard University Press. First published Cambridge, MA [1931].

Jakobson, Roman
1971 The Dominant. *Readings in Russian Poetics: Formalist and Structuralist Views.* Edited by Ladislav Matejka and Krystyna Pomorska. Massachusetts: MIT Press. 82–87.

Jungius, Joachim
1678 *Isagoge Phytoscopica*. Hamburgi: Pfeifferus. (Collection Leibnizressourcen digital 1) http://diglib.hab.de/drucke/xb-4827/start.htm

Kauffmann, R. Lane
1989 The skewed path: Essaying as unmethodical method. In: Butrym, Alexander J. (ed.) *Essays on the Essay: Redefining the Genre*. Athens: University of Georgia Press. 221–40.

Koerner, Lisbet
1999 *Linnaeus: Nature and Nation*. Cambridge, MA: Harvard University Press.

Lacoue-Labarthe, Philippe and Jean-Luc Nancy
1988 *The Literary Absolute: The Theory of Literature in German Romanticism*. Translated by Philip Barnard and Cheryl Lester. Albany: State University of New York Press.

Lindroth, Sten
1994 The two faces of Linnaeus. In: Tore Frängsmyr (ed.) *Linnaeus: The Man and His Work*. (Uppsala Studies in the History of Science 18.) Rev. ed. Canton, MA: Science History Publications.

Linnaeus, Carl
2003 *Linnaeus' Philosophia Botanica*. Translated by Stephen Freer. Oxford: Oxford University Press.

Linnaeus, Carl
1740 *Systema Naturae*. Editio secunda. Stockholmiae: Apud Gottfr. Kiesewetter.

Linnaeus, Carl
1740 *Fundamenta Botanica*. Editio secunda. Stockholmiae: Gottfr. Kiesewetter.

Linnaeus, Carl
1751 *Philosophia Botanica*. Stockholmiae: Apud Godofr. Kiesewetter.

Linnaeus, Carl
1787 *Philosophia Botanica in qua Explicantur Fundamenta Botanica*. Editio Quarta. Mark Garland/botanicallatin.org./10 November 2000.

Linnaeus, Carl
1977 *Miscellaneous Tracts Relating to Natural History, Husbandry, and Physick*. Trans. Benjamin Stillingfleet. Reprint. New York: Arno Press. First published London: J. Dodsley [1775].

Linnaeus, Carl
1971 *Lachesis Lapponica or a Tour in Lapland*. London: White and Cochrane. First published London [1811].

Lotman, Jurj
1977 Primary and secondary communication modeling systems. In: Daniel P. Lucid (trans. and ed.), *Soviet Semiotics: An Anthology*. Baltimore: Johns Hopkins University Press. 95–98.

Lotman, J. M., B. A. Uspenskij, V. V. Ivanov, V. N. Toporov and A. M. Pjatigorskij
1973 Theses on the semiotic study of culture (As applied to Slavic texts). In: Jan van der Eng and Mojmír Grygar (eds.) *Structure of Texts and Semiotics of Culture*. Paris/The Hague: Mouton. 1–28.
Manetti, Giovanni
1993 *Theories of the Sign in Classical Antiquity*. Translated by Christine Richarson. Bloomington: Indiana University Press.
Mautner, Franz H.
1967 "Maxim(es), sentences, fragmente, aphorismen." *Actes du IVe Congrès de l'AILC, Fribourg 1964*. Edited by François Jost. The Hague: Mouton, 1967. 2 volumes. Volume 2: 812–9.
Merriam-Webster Online
The Oxford English Dictionary. 2nd Online edition. 1989.
Saussure, Ferdinand de
1974 *Cours de linguistique générale: Édition critique par Rudolf Engler*. Tome 2, fascicule 4. Wiesbaden: Otto Harrassowitz, 1974.
Saussure, Ferdinand de
1983 *Course in General Linguistics*. Translated by Roy Harris. London: Duckworth.
Saussure, Ferdinand de
2002 *Écrits de linguistique générale*. Établis et édités par Simon Bouquet et Rudolf Engler. Paris: Gallimard.
Saussure, Ferdinand de
2006 *Writings in General Linguistics*. Translated by Carol Sanders and Matthew Pires. Oxford: Oxford University Press.
Stearn, William T.
1966 *Botanical Latin: History, Grammar, Syntax, Terminology and Vocabulary*. London: Thomas Nelson.
Todorov, Tzvetan
[1978] 1982 *Symbolism and Interpretation*. Translated by Catherine Porter. Ithaca: Cornell University Press.
Weisstein, Ultrich
[1968] 1973 *Comparative Literature and Literary Theory: Survey and Introduction*. Bloomington: Indiana University Press.

The reflective cultivator? Model readers in eighteenth-century Swedish garden literature

Andreas Nord

The boundaries between the popular and academic scientific domains were vague in eighteenth-century Sweden – science was supposed to be useful, and almost everything useful could be labelled *wet(t)enskap* 'science' (Lindroth 1967, Kukkonen 1989, Gren-Eklund in this volume). It is not surprising, therefore, to find that gardening was also referred to as a "science", especially since it was considered an area of importance. As the dominant ideology prescribed self-sufficiency, the yield from gardening was seen as economically important on a national level, and there was hope of improved, expanded gardening in Sweden as a result of technological progress, the spread of new knowledge and the introduction of new crops.

The present study encompasses an analysis of aspects of knowledge in the Swedish popular instructive discourse on gardening of the eighteenth century. This is a relevant subject, as it can illustrate the practices of and views on written mediation of knowledge in one of the non-academic fields of "science", as well as the construction of reader identities in the field, thus demonstrating the possible role of one type of handbook. For this reason, the emphasis is on aspects related to practical cultivation and techniques. The material studied consists of excerpts from eight popular handbooks on gardening that address laymen (private garden owners) from different socio-economic classes or practising professionals (gardeners or garden apprentices).[1]

The theoretical and methodological framework of the study is social semiotics (e.g. Halliday and Matthiessen 2004; Martin and White 2005), and the key concept is "model reader". This is the reader construed within the text, i.e. it is not a physical, empirical reader, but a text-internal representation, an aspect of

1. Nevertheless, they were considered parts of the "scientific" discourse. In the books, several references to gardening as "this science" etc. can be found, and in three of them approvals from the Royal Swedish Academy of Sciences are cited, e.g. in Lundberg 1754, where it is stated that the book "contains the most detailed information in this Science which has so far been printed in our mother tongue".

the "addressivity" of the text (cf. Bakhtin 1986). The model reader is, in a way, a representation of the "ideal" reader – the reader most loyal to the stance taken in the text. This concept originates in the reception semiotics of Umberto Eco (e.g. 1984), and has been discussed and developed, for example, by Tønnesson (2003) and Björkvall (2007). Eco, using the concept for the analysis of fictional prose, introduces it in the following way:

> To make his text communicative, the author has to assume that the ensemble of codes he relies upon is the same as that shared by his possible reader. The author has thus to foresee a model of the possible reader (hereafter Model Reader) supposedly able to deal interpretatively with the expressions in the same way as the author deals generatively with them. (Eco 1984: 7)

Thus, the construction of the model reader of the text is the author's tool for, so to speak, creating his own reader. This is done intentionally, but not necessarily consciously (as noted by Tønnesson 2003) – all choices made in forming a text contribute to constructing the model reader (although some choices may be considered more significant than others). The "reader-in-the-text", however, is not just an aspect of the text – it also has a normative function in that it offers a model for identification for the physical, empirical reader (cf. Björkvall 2007 on the normative potential of the model reader).

In other studies in social semiotics, there are theoretical counterparts. Gunther Kress explains how a text construes an ideal reader by offering "a certain 'reading position' from where the text seems unproblematic and 'natural'" (1989: 36). Kress stresses the manipulative potential of that reading position: "The text attempts to coerce the reader, by its 'obviousness' and 'naturalness', to become its ideal reader, to step into the reading position constructed for the reader in the text" (1989: 36). J. R. Martin and P. R. R. White (2005) write in their introduction to the social semiotic Appraisal Theory model about the "construed reader", the "putative reader", the "putative addressee" etc. of a text in much the same sense. They underline how, through the interaction of appraisal, texts "construct a model of the putative addressee and position the author with respect to that addressee" (2005: 210). The main difference between, on the one hand, Eco's model reader concept and, on the other, Kress's and Martin and White's concepts, is that Eco's version of the concept is used to describe aspects of a whole text or at least large portions of a text, while those of Kress and Martin and White are applied more locally, to single propositions or smaller portions of a text.

As the purpose of the current study is to characterize whole texts, or at least to generalize between different parts of texts, the model reader concept is preferable, although with the acknowledgement that the overall model reader is constructed through an aggregation of the local "construed readers" distributed across the text (cf. Tønnesson's [2003] discussion about the possibility of several model readers of a text). One theoretical purpose of the study is to connect the model reader concept to certain approaches in social semiotics, and to evaluate its potential as an interpretive concept in a historical study. In particular, the concept may make possible a systematic discussion, in particular, of the interpersonal side of the text in relation to its context (as described, for example, by Halliday and Matthiessen 2004): its *tenor*, the extratextual relationship between the participants (writer, reader), as the model reader can be related both to the (possible) empirical readers and the target group presented.

1. Action orientation or knowledge orientation?

One key aspect of the model reader is his relationship and attitude to knowledge and practice, which are particularly relevant in this study of knowledge-mediating texts. The study focuses on this aspect of the model reader with respect to questions about methods and techniques for the cultivation of plants and fruit trees. Because of the handbook character of the books, the model reader may be expected to be relatively action-oriented. The question is to what extent a book is directly oriented towards action, or towards building up the reader's general competence. The distinction can be related to Britt-Louise Gunnarsson's (2009: 102) typology of different reading purposes, where a distinction is made between directly action-oriented comprehension and indirectly action-oriented comprehension (integration in world conception). In the first case, the reading purpose is to find out how, according to the text, to act in a given situation – the reading of the text has the aim of influencing action in the present time. In the second case, the reading purpose is to integrate the text with previously acquired knowledge and experiences – the reading provides a readiness to act in the future. In relation to this, the main question is whether the book is designed for a directly or indirectly action-oriented reading purpose – for example, is the model reader supposed to be interested only in *what* to do, or also in knowing *why* to do it that way, or in different views on the subject matter?

The more specific aspects of the model reader under study are therefore related to knowledge orientation vs. action orientation, and the construal of the model reader's relationship to knowledge and practice. The analysis includes the connection of the model reader to a certain process type and participant role, and the construal of the reader role through engagement, viz. the dialogistic patterning and the presence or absence of certain types of propositions.

1.1. Process type and participant role

Of particular interest is whether the model reader is construed as active or passive. The analysis follows the description of transitivity (process types and participant roles) of systemic functional grammar (Halliday and Matthiessen 2004; Holmberg and Karlsson 2006), and focuses on the transitive material process type. The latter realizes actions, and is associated with the participant role of Actor. Is the model reader thus construed as an Actor in a transitive material process (acting) or as passive (other roles) – or are there no "blanks" in the text that can be identified with the model reader?

1.2. Engagement

The analysis of dialogical engagement follows Martin and White's (2005) Appraisal Theory model. Engagement involves "those meanings which in various ways construe for the text a heteroglossic backdrop of prior utterances, alternative viewpoints and anticipated responses" in key propositions (Martin and White 2005: 97). The approach is informed by the dialogism of Bakhtin and Voloshinov. The focus here is on the anticipatory aspect of the text, "the signals speakers/writers provide as to how they expect those they address to respond to the current proposition and the value position it advances" (2005: 93).

There is a basic distinction between *monogloss* and *heterogloss*. With monogloss (bare, categorical assertions), a value position can be presented as one which can be taken for granted. This is connected to an accepting reader position. Some types of heterogloss are also connected to an accepting reader, viz. the ones where the authorial voice challenges, fends off or restricts the scope of dialogically alternative positions and voices ("dialogistic contraction"). The difference compared with monogloss is that there is still a signal of the possibility of different viewpoints. As is the case with the *denial* sub-type (typically marked by negation), it can be seen as "a resource for introducing the alternative positive position into the dialogue, and hence acknowledging it, so as to reject it" (2005: 118), and the *countering* type has a similar function, representing the proposition as replacing a proposition which would have been expected in its place, marked, for example, by *men* 'but'. (For other sub-types, see Martin and White 2005.)

An interesting aspect of the countering and denial types concerns the way the heteroglossic backdrop challenged by the authorial voice is coded. The disaligned viewpoint can be construed as a (potential) viewpoint of the reader, or from the context as that of a third party or as one generated as an inference of a proposition in the text, e.g.: "Some claim that before they [the carrots] are put down [in the winter preservation pit], they must be completely dry [...]. *But this makes no difference* [...]".[2] Here, the disaligned viewpoints are ascribed to "some [people]", not the reader. "Never dry the carrots", on the other hand, would construe the disaligned action as possibly connected to the reader. Martin and White note that a countering of the reader's (potential) viewpoint can be seen as "corrective" but not necessarily confrontational (2005: 120).[3]

Other types of heteroglossic propositions, the dialogically expansive ones, actively make allowances for dialogically alternative positions and voices. Here, the value position is presented as one which is likely to be questioned, resisted or rejected. This is connected to a critical reader, one interested in valuing the sources of knowledge. The expanding heterogloss types are *attribution*, whereby a proposition is marked as dissociated from the authorial voice, either distancing (marked for example by *påstå* 'claim') or acknowledging, and *entertainment*, whereby the authorial voice allows for dialogic alternatives (typically marked by modals). The entertainment type has subtypes depending on what kind of modal meaning is marked. The main type is connected to modality of the probability type (such as *kan* 'may' and *möjligen* 'possibly'), and its function is to "make allowances for, and hence to make space for, alternative voices and value positions in the ongoing colloquy within which the text is located" (2005: 108). Modality of the permission and obligation types (the counterpart of the traditional category of "deontic") may have a somewhat different effect, e.g. in a directive like "The cutting must be performed almost every day".[4] The dialogistic relationship here can be described as "relationships of control and compliance/resistance rather than of the offering of information and viewpoints" (2005: 110–111), more closely related to a monoglossic imperative counterpart: "Perform the cutting almost every day!"[5] There is a certain difference in that the deontic modal propositions are still heteroglossic, allowing alternatives into the dialogic space, but in instructive discourses they are, I would suggest, functionally more closely related to the contractive types of heterogloss, as the authorial voice certainly allows other viewpoints, but also pronounces a certain alternative to be more or less valid or preferable.

2. The example is taken from one of the books in the study: "Några påstå, at för än de nedläggas, skola de wara ganska torra, [...]. Men det första gör ej til saken, [...]" (Fleischer 1795: 138).
3. Countering may also be aimed at too rigid inferences drawn internally from the discourse, e.g. when a recommendation is nuanced in a following sentence. This type is less significant in mapping the positioning of the reader.
4. The example is taken from one of the books in the study: "Utbrytningen måste skee nästan hwar dag, [...]" (Ahlich 1722: 119).
5. The imperative is the most relevant monoglossic counterpart, as it is equivalent in speech act terms.

To summarize, heterogloss of the expansive types – with the exception of the type involving deontic modals – can be connected to a critical, reflective model reader, one interested in the views of different sources, while monogloss and contractive types of heterogloss (including the deontic modal type) suggest a more directly action-oriented model reader.

1.3. Types of propositions

The engagement analysis sheds light on the forming of the model reader by dialogic positioning and naturalized response. As a complement, the occurrence of certain functional proposition types is mapped. This is relevant, as the type of discourse contributes to construing the model reader through the type of response that is natural to it. The aspects focused on are the presence or absence of explanations, which construe the model reader as oriented towards critical evaluation and reflection, and the (potential) organization of discourse for instruction, which is the maximally action-oriented discourse type and thus construes the model reader as action-oriented.

The analysis is based on Jean-Michel Adam's (1992) typology for the sequential organization of propositions in texts, adapted for Swedish by Ledin (2000) and Nord (2008), an approach easily integrated in a social semiotics context, and with fundamental similarities to the Sydney School's descriptions of *genre* (Martin and Rose 2008). Two of the five main types of sequences discussed by Adam are relevant, explication and description. Description can however, according to earlier conceptions of the typology (Adam 1987, 1990), be split into description proper and instruction, and it is instruction that is in focus here. The sequences are established by relations between the propositions forming them. The explicative sequence is the sequence forming a full explanation. It consists of four propositions:

Introduction. A problem area is presented

Problem, based on the questions *why* or *how*.

Explanation, answering the question *because*, *by means of*...

Comment, rounding off the explanation.

The instructive sequence can be described as consisting of four propositions, one of which may be realized either initially or finally:[6]

6. The description given is a modification of Adam's earlier draft of the system, proposed by Nord (2008: 105) as more appropriate for the description of instructions generally, since earlier conceptions seem to have been too specifically designed for the description of recipes.

Result product. The intended final product is mentioned.

Condition(s). Different material or situational circumstances are mentioned, so that the reader/hearer knows the conditions for creating the object in question.

Transformation(s). The measures to be taken for the creation of the object are mentioned. These are one or more actions the reader/hearer is to carry out in order to reach the goal. The actions are usually ordered chronologically.

Result situation. Material or situational circumstances after the transformation are mentioned. They may form Conditions for further Transformation–Result situation pairs.

Result product. The intended final product is mentioned.

Both sequences can be realized in a fragmented or elliptical form, or have the propositions in inverted order, but the central propositions, Explanation and Transformation, cannot be omitted, and they must be anchored in some other part of the sequence, as the sequence is established through the interrelations of the propositions. A proposition may be given a double function, e.g. a Result doubling as a Problem in relation to an Explanation.

The linguistic form of an Explanation typically includes lexis for cause, consequence or hypothesis (e.g. *depend on*, *result in*, *for that reason*, *because* and *since*) and the proposition Problem may be realized through interrogative sentences (Ledin 2000: 75–76). According to Ledin, a linguistic indication of an instruction (Transformation proposition) in Swedish is a material process with the reader/hearer as implicit or explicit Actor (2000: 77).

2. Gardening in eighteenth-century Sweden – possible readers

The model reader analysis must be evaluated in relation to the target group (the intended reader), which may be explicitly stated in the text, as well as in relation to the possible empirical readers (the *de facto* readers) (cf. Tønnesson 2003). We know very little about the actual reading of garden literature in eighteenth-century Sweden, but the possible empirical readers (possible target groups) can be identified. (The stated target groups of the books will be described in the next section.)

In eighteenth-century Sweden, gardening was the concern of a growing number of people. As in the 17th century, formal gardens (later, English parks as well) were being laid out at palaces and manors, but they increased in number as a new, rapidly growing class of manufacturers and merchants acquired manors in the countryside, and town dwellers built summer villas for utility and pleasure. In the countryside, however, the common people had normally not yet started practising gardening of a more advanced kind.

The discourse of usefulness as well as the boom in botany and "oeconomia" (the science of national economy and production) – the academic sciences closest to gardening – were associated with a desire to disseminate new ideas, techniques and crops, although in practice this seems to have had only a marginal effect. One example is the failure to establish a corps of professionals (*provincialschäfrar*, literally 'provincial shepherds') with the mission of promoting the use of new crops, techniques and ideas for livestock breeding and agriculture, including gardening, among smallholders in the countryside. There was no popular education of any significance until the second part of the nineteenth century, nor was there any professional education. The first formal education started at the end of the eighteenth century but was of no real significance. Still, a move towards a professionalization of gardening did take place, albeit from a very low level initially. A gardeners' guild was established in Stockholm in 1735, partly as an effect of the ongoing debate on the role of immigrant gardeners, who were better educated and were regarded as a threat to their native Swedish counterparts.

The potential target groups of popular written discourse on gardening can be restricted to professional or semi-professional gardeners, reading to improve their own skills, garden apprentices, and private garden owners, reading either to improve their own skills or to acquire a theoretical education, or to oversee their gardener in a more informed way. Towards the end of the century, gardening school students formed another potential target group, although marginal in number.

3. The material – excerpts from eight books

Eighteenth-century Swedish garden literature was relatively rich: Halldor (1935) lists more than 100 titles, including several smaller "tracts" and similar prints, as well as dissertations and articles in the transactions of the Royal Swedish Academy of Sciences. There were about a dozen monographs (handbooks), some of them in several editions. Here, the first editions of the eight most important are studied.[7] They will be referred to by the name of the author and the year of publication. Eight themes from the book excerpts studied are chosen: the

7. Excluded from the study are three less important books (Dahlman 1728, 1732 and Brehmer 1738), which are mostly compiled from older sources, and a book by Lundberg (1775), which is a supplement to one of the books included in the study.

cultivation of carrots, cucumbers, melons, peonies and tulips (or flower bulbs generally), pruning and grafting (or inoculation) of fruit trees, and principles for watering. These themes are intended to represent the core content of the books: advice and instructive discourse on practical and relatively hands-on matters that represent the ordinary as well as the more exclusive, the simpler as well as the technologically more complicated. The excerpts are listed in Table 1.[8]

Table 1. Excerpts (number of words)

Book	*Carrots*	*Cucumbers*	*Melons*	*Peonies*	*Tulips*	*Pruning*	*Grafting inoculation*	*Watering*	*Total number*
Ahlich 1722	271	191	595	143	2,438	2,323	494	–	6,455
Kammecker 1731	138	179	1,909	–	–	1,899	712	450	5,287
Lundberg 1754	235[a]	83	1,046[b]	–	239	1,968	1,290	–	4,861
Lissander 1768	–	–	–	–	–	363	779	–	1,142
Broberg 1775	–	–	–	–	–		2,601[c]	–	2,601
Jörlin 1784–1794	238	282	740	(–)	362	40	1,237	–	2,899
Fleischer 1795	2,379	1 595	243	–	1,885	2,519	3,391	–	12,012
Vothmann 1796	586	1,204	1,333	–	–	–	–	158	3,281

[a]) Includes parsley root.

[b]) Includes hotbeds.

[c]) The excerpt consists of one paragraph, describing the sowing, planting, grafting and pruning of apple trees.

In Table 2, data about the books are presented. They vary in significant ways. The size of the books ranges from some 90 to 653 pages. One of the books, Fleischer 1795, is a translation from Danish, another, Vothmann 1796, was translated from German (although the author was Danish) and adapted to the Swedish climate. Most were granted royal privilege, prohibiting unauthorized later printings, and three of them cite approvals from the Royal Swedish Academy of Sciences.

Four of the authors were gardeners: Ahlich (in Stockholm), Lundberg (in Stockholm; also chairman of the Gardeners' Guild), Broberg (gardener at the Botanic Garden in Uppsala) and Vothmann (gardener in Sønderborg in southern Denmark). Only one of them (Jörlin) was an academically trained botanist, but both the translators were academics. In the book by Vothmann (1796), which was an adapted translation, not only the status of the author but also that of the translator/adaptor is foregrounded in metatexts. The translator, Olof Swartz, was a renowned botanist, a member of the Royal Swedish Academy of Sciences and professor at its school of gardening. Other books were written by laymen.

8. Not all of these themes are covered in every book; Lissander 1768 and Broberg 1775 mainly deal with fruit trees, while Kammecker 1731 and Vothmann 1796 concentrate on the kitchen garden and orchard.

Table 2. Data about the books under study

Author (translator)	*Year of publication (year of reprint/new edition), no. of pages in first edition*	*Author's (translator's) background*	*Source of knowledge presented*	*Target group (according to preface etc.)*	*Privileges/ approvals*
Johann Ahlich (*c.* 1680–1743)	1722 (revised edition 1744), 192 pp.	gardener in Stockholm of German origin	own experience	amateur gardeners, gardener employers, housewives, gardeners?	royal privilege
Johan Kammecker (1701–41)	1731, 317 pp.	clergyman in London and Stockholm	"good English books", personal contact	upper classes, including the clergy	royal privilege
Peter Lundberg (1711–1787)	1754 (1763, 1773, 1780, 1798), 222 pp.	gardener, chairman of Gardeners' Guild in Stockholm	own experience	wealthy garden owners, farmers, gardener employers, gardeners	approval by Academy of Sciences, royal privilege
Anders Lissander (1704?–1786)	1768, 351 pp.	director of the Parliament's board of manufacturing industry	own experience	amateur gardeners, gardener employers, garden apprentices	approval by Academy of Sciences
Lars Broberg (1725–1795)	1775, 88 pp.	gardener at Botanic Garden in Uppsala	–	common people in the countryside	approval by Academy of Sciences
Engelbert Jörlin (1733–1810)	Part 1: 1784 (1788, 1793), 44 pp.; 2: 1786 (1790, 1794), 23 pp.; 3: 1794, 14 pp. (in sum 91 pp.); all parts in joint edition 1796	botanist, scholar	books and scientific articles	garden owners without gardener	royal privilege (1796 edition)
Esaias Fleischer (1732–1804), translation: Carl Fredrik Fallén (1764–1830)	1795 (1797, 1805), 653 pp.	Danish county governor, experienced layman (translator: scholar, entomologist)	own systematic experiments	"less wealthy" garden owners without a gardener	royal privilege
Johann Georg Vothmann, translation/ adaptation: Olof Swartz (1760–1818)	1796, 222 pp.	gardener in Sønderborg in southern Denmark (translator: botanist, gardening school professor)	(translator's authority stressed on title page)	garden owners in the countryside, gardening school students	

Kammecker was a clergyman. Both Lissander and Fleischer were civil servants and amateur gardeners. The basis for the knowledge of several authors – both gardeners (Ahlich, Lundberg) and laymen (Lissander, Fleischer) – is in prefaces or other metatexts said to consist simply of their own rich experience. Others refer to written or oral sources (Kammecker, Jörlin). In the case of Broberg and Vothmann/Swartz, sources of knowledge are not explicitly discussed.

Most books contain statements about the intended target group. The four oldest books (Ahlich 1722, Kammecker 1731, parts of Lundberg 1754, Lissander 1768) address amateur gardeners and people who employ a gardener. Some of them also explicitly address professional gardeners (parts of Lundberg 1754, and Ahlich 1722) or garden apprentices (Lissander 1768). In addition, some parts of Lundberg 1754 address "Landtmannen" 'the man in the countryside'. The four more recent books address less well-to-do garden owners, viz. those not employing a gardener. According to the subtitle of Broberg 1775, the target group is the common people of the countryside, *Allmogen*, although in practice it does not seem to address the poorest section of the rural population. One book, Vothmann 1796, is, according to the translator's preface, also intended for students at the gardening school run by the Royal Swedish Academy of Sciences.

4. Action-oriented model reader: examples

From the analysis, two relatively distinct patterns emerge. In order to demonstrate how the analysis has been carried out, two relatively clear-cut examples of the two main patterns will be described in some detail. The first type can be connected to an action-oriented model reader, the second to a more critical, and possibly also more reflective, reader.

The first example is taken from Ahlich 1722, and describes the cultivation of melons. The first part of the highly detailed exposition reads:[9]

(1) Thet 68. *Capitlet.* **Om *Meloner,* huru the skola rätt vpbringas, skötas och bijbehållas.**

THet första och förnämsta härwid är, at man sig winlägger om godt frö och thet af åtskilliga arter, så många man kan öfwerkomma.

9. English translations are intended to reflect the Swedish wordings as accurately as possible. Thus, the awkwardness of the translations does not reflect any particular awkwardness in the Swedish originals. – Italics and boldface in examples always reflect features in the original text: boldface reflects Schwabacher typeface and italics Serif typefaces (bold italics Serif typeface italics). Regular typeface reflects black letter (Fraktur).

När man nu fått sådant frö, giör man wid paß mit uti *Februario,* också mit uti *Martio,* i full Måna en dyng-säng för ett eller tu fönster, som man kallar *Melon*-Scholan, på ett ställe, som ligger rätt emot Middags Solen. Then fyller man med god, lång, warm hästedynga vp. När hon legat i några dagar och sunckit, så trampar man henne wäl i hop och lägger, om så nödigt är, mera warm dyngia theri, kastar sedan, til goda 2. händers högd, wäl wattnad Dynge-Jord theröfwer, och låter henne så ligga i 2. eller 3. Dygn, och sticker då och då fingret i Jorden, til at känna, om dyngan något bränner. (Ahlich 1722: 119)

'*Chapter* the 68th. On Melons, how they should be properly propagated, raised and maintained.

Of primary and chief importance in this matter is that one strives for good seeds, and from several species, as many as one can procure.

When one has procured such seeds, one sets up – approximately in the middle of February, as well as the middle of March, under a full Moon – a dung bed for one or two windows, which one calls the Melon Nursery, on a site situated straight in the Midday Sun. One fills this with good, long, warm horse dung. When it has been lying for some days and has subsided, one tramples it down well and adds more warm dung thereto, if such is necessary, then throws well-watered Dung-Earth – to a height of a good 2 hands – over it, and lets it lie for 2 or 3 Days, and puts a finger now and then into the Soil to feel if the dung is somewhat heated.'

The transitivity pattern is straightforward: A recurrent *man* 'one' functions as a reader representation, and is generally identified with the Actor of transitive material processes. In some cases, the Actor is suppressed by use of the passive voice (e.g. *watnas* 'be watered'), but can still be associated with the model reader.

The propositions in the example are all monoglossic, "taken-for-granted" advice. In other parts of the melon article, a few heteroglossic propositions of the deontic modal type occur (e.g. the recommendation "The cutting must be performed almost every day").

This is a typical instructive sequence (although we do not reach the result in the example above). In the example given, there are no traces of other sequence types, but later there are two minimal explanations embedded in the instruction, e.g.:

sequence	embedded sequence	text	
Transformation		tå the ock sedan icke mera måste watnas, än i högsta nödfallet, och då allenast jorden wid roten,	then they must not be watered any more, unless it is an absolute emergency, and then solely on the soil at the root,
	Problem	utan at komma wid blommorna och frukterna;	without touching the flowers and fruit;
	Explanation	emedan watnet är Melon-Grenarne alldeles emot.	because the water is detrimental to the Melon Branches.

The last part of the Transformation can be read as the Problem of a very elliptical explicative sequence. There are no more detailed explicative sequences.

In other parts of Ahlich 1722, this takes a slightly different form. In the exposition in the chapter on pruning, heteroglosses dominate. The main type, however, is the deontic modal. A typical organization of utterances is the following:

heterogloss: deontic modal	Nu måste man wäl laga så, at träet får lufft mitt uti eller på sidan, til at emottaga solskenet:	Now one must carefully make sure that the tree gets space in the middle or on the side, to receive sunshine:
heterogloss: countering (intratextual) + entertaining (deontic modal)	men man får likwäl icke beskära träet alt förmycket; [...]	but one still must not prune the tree too much; [...]

The first piece of advice is given in the form of a directive. It is followed by a second piece of advice adding nuance to the first, which is marked as countering an inference of the first proposition, and also as a deontic modal heterogloss. The dialogistic load is fairly light.

The pattern in Ahlich 1722 could be summarized as

- sequence/proposition types: recurrent instructive sequences with elaborate transformations, few explications;
- transitivity: model reader generally construed as Actor of a transitive material process;
- engagement: monogloss or heterogloss of the entertainment type with deontic modal markers.

The interpretation here is that the model reader is clearly action-oriented, ready to accept the authorial voice's recommendations without question and without asking for explicitness about other views on how to do something or reasons for the recommendations. The writer-reader relationship of the text thus resembles a classical master-apprentice relation where the writer/master is the unquestioned and unquestionable authority on matters of knowledge and the reader/apprentice is just a compliant follower of the instruction given.

5. Critical or reflective model reader: example

My example of a text that creates a critical or reflective model reader is taken from Fleischer 1795. It consists of one of the paragraphs in the lengthy exposition about the cultivation of carrots:

(2) §. 5.

Jag sade at såningen skal ske i första hälften af December. Orsaken är den, at man därefter kan wänta frost, så at man icke kan komma ned i jorden – men så länge man ser at kölden uteblifwer, låter man bli at så. När sängen är tillagad, så är hon snart besådd, om man märker, at luften ser ut til frost. Oförfarna torde kan hända le åt detta, och tänka at man ju borde så innan frost kom, så hade man ej af nöden, at så noga passa på wäderleken, och låta kölden öfwerrumpla oß. Men oärfarne tänka blott så. Jag har sjelf sedt Trägårdsmästare så Morötter, i blid wäderlek, midt i Nowember; men följden blef den, at de fingo aldeles inga om Wåren, och däraf dömde de, som almänneligen sker, när försöket mißlyckas, at sjelfwa föreskriften icke duger – De och deras Läromästare "Förstodo saken bättre – Man skulle allenast blifwa wid „det gamla – o. s. w.„ Men detta Folket obserwerade icke, at orsaken til det mißlyckade låg uti den tidiga såningen, och uti det at ingen frost följde snart derpå; ty så länge fröet ligger i jorden, tål det frost, men grodden, när han komer up, tål den icke. Sår man nu omedelbarligen innan frost kommer, blötes fröet imedlertid uti jorden, och börjar möjeligtwis at slå skott til rot, samt drifwer tidigt om wåren up och wäxer; men infaller mildt wäder sedan man sådt, innan frosten kommer, så begynner fröet at skjuta up ur jorden, och när då kölden kommer på, lider sådden skada. (Fleischer 1795: 135)

'§ 5.

I said that sowing should take place in the first half of December. The reason is that thereafter one can expect frost, so that one cannot penetrate the soil – but as long as one sees that the cold is not coming, one refrains from sowing. When the bed is prepared, it is readily sown, if one notices it looks like frost in the air. The inexperienced might perhaps smile at this, and think that one should of course sow before the frost comes, then one would not have

> to attend to the weather, and let the cold surprise us. But only the inexperienced think this way. I myself have seen Gardeners sowing Carrots in mild weather in the middle of November; but the result was that they did not get any at all in Spring, and from this they judged – as is generally done when the experiment fails – that the very prescription is of no use – they and their masters "Understood the matter better" – one was simply to keep to "the old way – and so on". But those People did not notice that the reason for the failure was the early sowing, and the fact that no frost followed soon after it; because as long as the seed is in the soil, it can endure frost, but the sprout, when it emerges, cannot withstand it. If one sows immediately before the frost comes, the seed, however, is moistened in the soil, and possibly starts to sprout roots, and comes up in early spring and grows; but if mild weather occurs after one has sown, before the frost comes, the seed begins to put forth shoots from the earth, and when the cold comes, the seed is damaged.'

The process type patterns are less straightforward than in example (1). The author takes a place in the text in the form of a repeated *jag* 'I', and the reader is occasionally represented as *man* 'one', but *man* can also refer to other participants, as in the reported thoughts of "the inexperienced" ("The inexperienced might perhaps smile at this, and think that *one* should of course sow before the frost comes"). The process type primarily connected with the reader is material (*sår* 'sows', *såningen* 'sowing').

The engagement pattern is very complex: almost all the propositions are heteroglossic, and both contractive and expansive heterogloss types can be found, although denials and counterings dominate. It is to be noted, however, that the authority of the author is never challenged. Other voices are explicitly let in, but only eventually to be dismissed by the authorial voice. The most interesting feature is how the argument for a certain sowing time for carrots is heteroglossically staged as a debate between the author ('I') and two groups of others: the group called *oförfarna/oärfarne* 'the inexperienced', and gardeners who jump to incorrect conclusions from a failed experiment. By ascribing those faulty views to a third party (through attribution), another voice is let in, and it can be dismissed without being potentially ascribed to the model reader. On the contrary, the authorial voice allows the reader to show solidarity with his view on the best time for sowing carrots.

The sequence type pattern is also complex, as different sequence types are intertwined. Transformations from the instructive sequence may be discerned, but so may Explanations from the explicative sequence, as well as parts of an argumentative sequence with several arguments and a narrative sequence (the related chain of events).

On some points, it is not easy to summarize the pattern in this example, but based on the analysis of the passage given and the other excerpts from the book, the main characteristics are as follows:
- sequence/proposition types: instructive sequences as well as diverse other sequences; explications occur somewhat more frequently and are occasionally more elaborate;
- transitivity: model reader mainly construed as Actor of a transitive material process;
- engagement: heterogloss of diverse types, including both the contractive and expansive types, letting other voices in, although construing the authorial voice as the ultimate authority.

Despite being rather heterogeneous, this pattern construes a model reader interested in valuing different views on the practices of gardening (albeit not a reader who is at all free, as the authorial voice indicates which of the different existing views is to be preferred). The transitivity pattern, however, still suggests a relatively action-oriented model reader (albeit less action-oriented than in the example discussed in the previous section).

The same pattern, with some differences, is found in instructions in Lissander 1768, as in the following example:

(3) Man kan ympa hela Kärn-Scholan, om man wil, i andra eller tredje året, bwilket [read: hwilket] sker på följande sätt. Jag tager näst föregående årets skott af goda Fruktträn som hafwa full Sol; lika mycket om skottet sitter högre eller lägre, eller på hwilken sidan det wuxit, allenast det är fylligt, friskt och har goda knoppar: Skär derutaf ympeqwistar, hwardera med 2 à 3 ögon. Sedan anställes ympningen: 1.o Jag skär utaf unga stammar 1/2 aln från jorden, med 1/2 3/4 eller hel tums snedd, alt som de äro tjocka til; på samma sätt, eller med lika snedd, skäras ympqwistarnas nedra ändar: [...] (Lissander 1768: 54)

'One can graft the whole Tree-School, if one likes, during the second or third year, which is done in the following way. I take the previous year's branches from good Fruit Trees, which are exposed to full Sun; regardless of whether the branch is located higher or lower, or on what side [of the tree] it has grown, only that it is strong, healthy and has good buds: Cut from it grafting twigs, each with 2 to 3 eyes. Then the grafting is performed: 1. I cut young stems 1/2 an ell from the ground, 1/2, 3/4 or a whole inch on a diagonal, according to their thickness; in the same way, or with the same diagonal, the bottom parts of the grafting twigs are cut: [...]'

The exposition quoted is instructive, but the potential representation of the model reader as Actor of a transitive material process (*man* 'one' in the first sentence, the implied Actor of the imperative *Skär* 'cut' in the third sentence) is replaced in some strategic propositions with *I*. This turns it into a description of the author's practices, with the implicit function of positioning the advice (Transformations) as subjective. The function is similar to that of contractive heterogloss, although the heteroglossic framing is implicit: by describing how "I" do it, the authorial voice allows other "I's" to do it in other ways, thus acknowledging other points of view. The use of the first person in this way is a feature outside Martin and White's engagement system, but in this example it proves to bear functional similarities to heterogloss.

6. Two patterns – similarities and distribution

As noted, there are no major differences in transitivity between the patterns, and although the explications are more elaborate in the second pattern, the main difference between them turns out to be one of engagement: the absence or (explicit) presence of other voices in the text, and the explicit valuing of knowledge. The distribution of the patterns is summarized in Table 3. The first pattern predominates in Ahlich 1722, Lundberg 1754 and Vothmann 1796.[10] The second pattern clearly predominates in Fleischer 1795 and Lissander 1768 (with some specific features). In one book, Broberg 1775, there are tendencies towards a strategic shift: the first pattern is dominant, but in some strategic passages (e.g. about the planting of fruit trees) a shift towards the second can be discerned. In another book, Kammecker 1731, different parts of the book are connected with different patterns, some with the first and others with the second. To some degree, the distribution is related to the complexity of the subject. The cultivation of melons and pruning of trees (second pattern) are undoubtedly more complicated than the cultivation of carrots and cucumbers (first pattern), and might therefore be arenas for more diverse views, thus demanding more elaboration in accounting for the advice given.

In Jörlin 1784–1794 the first pattern predominates, but in the excerpt on grafting a shift towards the second can be seen (through heterogloss of the attribution and countering types), possibly also as an effect of the difficulty of the subject matter. Thus, in most cases, there is a relatively consistent model reader of one of two types, in itself a notable result. Kammecker 1731 is more heterogeneous.

10. The similarity between Ahlich 1722 and Lundberg 1754 is enhanced by the fact that a large portion of the excerpt on the pruning of fruit trees in Lundberg 1754 is obviously copied from Ahlich 1722.

Table 3. Distribution of the patterns in the books

Book	*Excerpt*	Pattern 1	Pattern 2
Ahlich 1722		×	
Kammecker 1731	excerpt on melons and pruning		×
	excerpts on carrots, cucumbers and grafting	×	
Lundberg 1754		×	
Lissander 1768			×
Broberg 1775		×	(×)
Jörlin 1784–1794	excerpt on grafting	(×)	×
	other excerpts	×	
Fleischer 1795			×
Vothmann 1796		×	

As is evident from the table, no clear tendency for the distribution of patterns to change over time can be detected. It may not be a coincidence, though, that the two books most clearly construing a critical, reflective model reader (Lissander 1768 and Fleischer 1795) are from the later part of the century – after all, it would be surprising if no change at all had occurred, as the eighteenth century was a period of great change in Sweden as a knowledge society. For instance, the material conditions for written communication improved and the academic sciences closest to gardening, such as botany and economy ('oeconomia'), were booming, and these factors, together with the general spread of gardening, would presumably make it natural to expect a more informed and possibly reflective reader during the later part of the century.

One tendency towards a change over time, however, was obvious from the very start: the change in target groups (as indicated by explicit statements on the subject), with a clear tendency towards the end of the century (after about 1770) to address those who were "less wealthy", instead of more well-to-do garden owners. The target groups do not, however, seem to influence the form of the books – a (possible) adaptation to different target groups cannot explain the distribution of patterns. Ahlich 1722, Kammecker 1731, Lundberg 1754 and Lissander 1768 address roughly the same categories but still construe at least partly different model readers, and this also holds for Broberg 1775, Jörlin 1784–1794, Fleischer 1795 and Vothmann 1796, which all address the less privileged garden owner. The results of the study may, on the contrary, be used to contest the explicit statements about addressees given in the books, as they are not in fact always consistent with the contents of the book as reflected in the model reader analysis. For example, Ahlich 1722 could be considered less appropriate for "noble people" interested in gardening as a hobby and in overseeing their gardener, whom the book explicitly addresses, as this use of the text would demand a somewhat deeper understanding of matters than this very action-oriented book provides. Fleischer 1795, on the other hand, can of course be useful for the "less wealthy" garden owner it explicitly addresses, but the very detailed advice it gives would make it equally (or even more) suitable for professional gardeners and scientifically oriented experimenters.

Furthermore, the background of the authors can be taken into consideration, and this turns out to be more relevant. The tendency is for books by gardeners (Ahlich 1722; Lundberg 1754; Broberg 1775; Vothmann 1796), as well as those connected with academic botanists (Jörlin 1784–1794; Vothmann 1796 with the renowned botanist Swartz as adaptor), to construe a clearly action-oriented and less critical model reader, while model readers of books written by laymen (Kammecker 1731; Lissander 1768; Fleischer 1795) tend to be relatively action-oriented but more critical. The overall pattern may at first seem paradoxical, but can be related to how the authority of the writer generally is established. In all the books the author is central; in most of the books there is an *I* present in the text, and the role and background of the author are stressed in the subtitle or preface. With an author presented as a reliable – and satisfactory – source of knowledge, the action-oriented, less critical, apprentice-like model reader is natural. The laymen authors, on the other hand, are not authorities per se in the same way, even though their wealth of experience may be stressed (Lissander 1768; Fleischer 1795). This makes it more natural to choose to construe a more critical, though still action-oriented, model reader. If the normative aspect of the model reader is considered, the two different types are not just natural, but can be seen as a tool for (re)establishing a certain relationship of authority through the book.

This, in turn, makes the construed model reader not only a manifestation of the author's position in the knowledge production system, but also a normative offer of (ideal) readership. Seen from this perspective, some of the laymen authors at least (Lissander, and particularly Fleischer) seem to be active proponents of a project to re-form the readership of this type of non-fictional prose. While the professional gardeners strive to reproduce a more authoritarian and traditional view of knowledge, the 'competent layman authors' invite the reader to be like themselves: reflective and with an interest in the sources of knowledge.[11]

11. This assumption is strengthened in part by quite lengthy references in the prefaces of these two books, where sources of knowledge are discussed and even surveys of previous written sources of knowledge on the subject matter are given.

7. Conclusion

Eighteenth-century Sweden is characterized by a very optimistic view of knowledge, a fact that is sometimes mirrored even in the most hands-on literature, such as handbooks on gardening. The study has revealed the occurrence of two different patterns, or model readers: one action-oriented and one more critical or reflective. These model readers can be related to slightly different reading goals and views on the purpose of reading (cf. Gunnarsson 2009). If model readers are seen as normative for the reading positions taken by actual readers, they convey a message of somewhat different ideal readerships and views on knowledge. This, it may be claimed, can be interpreted as part of an endeavour to educate readers for a society based on useful knowledge.

This result also demonstrates the potential usefulness of the methodological approach of the study. One theoretical purpose was to connect the model reader concept to proposition-based approaches to reader positioning within social semiotics. This proved relatively successful, although a (not at all surprising) result may be an observation that texts can be heterogeneous, seemingly offering different model readers in different sections of one book. If this can be accepted, the model reader concept may indeed function as a concept that makes possible the step from the reader positioning of individual propositions to the discourse level.

Another theoretical purpose was to evaluate the potential of the model reader concept as an interpretive concept in a historical study. The concept turns out to be a useful tool for connecting patterns in the text to features in its context by allowing a comparison between the model reader of a text, revealed by careful, theoretically guided micro-linguistic analysis (focusing e.g. on the *tenor* of the text), and its extratextual relations (e.g. the background of the author, the relationship between author and possible readers). It also provides a tool for discussing the interpersonal normative role of texts in, for example, the construal of readership. Furthermore, it enables a systematic comparison between texts from different domains or times at a more holistic level. Similarities as well as differences between different domains or communities can be pinpointed, as can structural changes in the ideological context and the social relations as acted out through text.

A touchstone for the analytical usefulness of a theoretical concept is if it enables observations to be made that go beyond the obvious. This, it can be claimed, is fulfilled by the model reader concept as demonstrated by this study. The handbooks in question, for example, have for a long time been an object of interest to garden historians, who have generally focused on the prefaces and other paratexts, which are sometimes rhetorically elaborate, sometimes quarrelsome towards rivals. But the dimension of (re)construing readership uncovered in this analysis has not been acknowledged, although this may prove to be a valuable clue to the historical role of some of the texts – and the role of some of the non-fictional prose generally – within the eighteenth-century semi-scientific discourse.

Primary sources

Ahlich, Johan
1722 *Den Swenske Lust- Örte- och Trä-gården, Beprydd med Blomster/ Köks-Krydder och Örter Samt Fruchtbärande Trän* [The Swedish flower garden, herb garden and orchard, decorated with flowers, spices and herbs as well as fruit-bearing trees]. Stockholm: J. L. Horrn.

Broberg, Lars
1775 *Några korta Anmärkningar och Rön Angående Trä-Planteringen Samt Köks-Trägårds-Skötseln* [Some brief notes and findings about tree planting as well as kitchen garden cultivation]. Stockholm: Johan Georg Lange.

Fleischer, Esaias
1795 *Conference-Råd Esaias Fleischers Trägårds-Bok* [The garden book of *Conference-Råd* Esaias Fleischer]. Lund: Prof. Joh. Lundblad.

Jörlin, Engelbert
1784 *Flora Macelli Hortensis. Swenska Köks- och Krydde-Gården* [The flora of the vegetable market. The Swedish kitchen and herb garden]. Lund.

Jörlin, Engelbert
1786 *Swenska FrucktTrägården. Eller Konsten at Så, Plantera och Flytta allahanda Fruckt-Trän, samt Oculera och Ympa* [The Swedish orchard. Or the art of sowing, planting and moving various types of fruit trees, as well as inoculating and grafting]. Lund: Magister Doc. Joh. Lundblad.

Jörlin, Engelbert
1794 *Swenska Trägårdens Tredje Del, Om Blomstergården* [Third part of the Swedish garden. About the flower garden]. Lund: Prof. Joh. Lundblad.

Kammecker, J. L.
1731 *En Til sitt kära Fäderneslandz Tienst och Nytta utgifwen Trägårdz-Man* [A garden man published for the service and benefit of his beloved fatherland]. Stockholm: Benjamin G. Schneider.

Lissander, Anders
1768 *Anmärkningar Wid Swenska Trägårds-Skötslen* [Notes on Swedish gardening]. Stockholm: C. Stolpe.

Lundberg, Peter
1754 *Then rätta Swenska Trägårds-Praxis, Eller Kort Underrättelse huru Kiöks- Trä- Örte- och Lust-gårdar Tillika med Orangerier samt Humle-gårdar böra anläggas, skiötas och conserveras* [The proper Swedish garden practice, or brief advice on how kitchen gardens, orchards, herb gardens and flower gardens as well as orangeries and hop gardens should be arranged, cultivated and preserved]. Westerås: Joh. Laur. Horrn.

Vothmann, J. G.
1796 *J. G. Vothmans Trägårds Cateches, Förswänskad och tilökt af Olof Swartz, Professor wid Bergianska Trägårds-Skolan och Medl. Af flere in- och utländske Lärde Societ. Första Delen* [J. G. Vothman's garden catechism, turned into Swedish and augmented by Olof Swartz, professor of the Bergian School of Gardening and member of several national and international learned societies, Part I]. Upsala: Joh. Fr. Edman.

References

Adam, Jean-Michel
1987 Types de séquences textuells élémentaires. *Pratiques* 56: 54–79.

Adam, Jean-Michel
1990 *Éléments de linguistique textuelle. Théorie et pratique de l'analyse textuelle*. Liège: Mardaga.

Adam, Jean-Michel
1992 *Les textes: types et prototypes. Récit, description, argumentation, explication et dialogue*. Paris: Nathan.

Bakhtin, Mikhail Mikhajlovich
1986 The Problem of Speech Genres. In: *Speech Genres and Other Late Essays*, 60–102. Austin: University of Texas Press.

Björkvall, Anders
2007 Social identities offered in Swedish magazine advertising. In: José María Bernardo Paniagua, Guillermo López García, Pelegrí Sancho Cremades and E. Serra Alegre (eds.), *Critical discourse analysis of the mass media*, 234–244. Eco, Umberto
1984 *The Role of the Reader: Explorations in the Semiotics of Text*. Bloomington: Indiana University Press.

Gunnarsson, Britt-Louise
2009 *Professional Discourse*. London: Continuum.

Halliday, M. A. K. and Christian M. I. M. Matthiessen
2004 *An Introduction to Functional Grammar*. 3rd ed. Revised by Christian M. I. M. Matthiessen. London: Arnold.

Holmberg, Per and Anna-Malin Karlsson
2006 *Grammatik med betydelse. En introduktion till funktionell grammatik* [Grammar with meaning. An introduction to functional grammar]. Uppsala: Hallgren & Fallgren.

Kress, Gunther
1989 *Linguistic Processes in Sociocultural Practice*. 2nd ed. Oxford: Oxford University Press.

Kukkonen, Pirjo
1989 *Från konst till vetenskap: Begreppet* vetenskap *och dess språkliga uttryck i svenskan under 400 år* [From art to science: The concept of *science* and its linguistic expressions in Swedish during the last 400 years]. Helsinki.

Ledin, Per
2000 *Veckopressens historia: del II* [The history of the weekly press, part II]. Lund: Institutionen för nordiska språk, Lunds universitet.

Lindroth, Sten
1967 *Kungl. Svenska Vetenskapsakademiens Historia 1739–1818* [The history of the Royal Swedish Academy of Sciences 1739–1818]. Uppsala: Almqvist & Wiksells.

Martin, J. R. and P. R. R. White
2005 *The Language of Evaluation: Appraisal in English*. Basingstoke, Hampshire/New York, New York: Palgrave Macmillan.

Martin, J. R. and David Rose
2008 *Genre Relations: Mapping Culture*. London/Oakville: Equinox.

Nord, Andreas
2008 *Trädgårdsboken som text 1643–2005* [The garden book as text 1643–2005]. With an English Summary. (Acta Universitatis Stockholmiensis. Stockholm Studies in Scandinavian Philology. New Series 47.) Stockholm: Institutionen för nordiska språk, Stockholms universitet.

Tønnesson, Johan
2003 Model readers and other textual strategies in Norwegian historian Professor Finn Olstad's paper "New perspectives on the rise and fall of the working class" (2000). In: Kjersti Fløttum and François Rastier (eds.), *Academic discourse: Multidisciplinary approaches*, 158–180. Oslo: Novus Press.

The linguistic construction of scientificality in early Swedish medical texts

Britt-Louise Gunnarsson

1. Introduction

In this chapter I will present and discuss an analysis of the linguistic construction of scientificality in twelve medical articles published in the Transactions of the Royal Swedish Academy of Sciences between 1750 and 1769. In the history of Swedish medical science, the eighteenth century is an important period in several ways. Medical practice and medical science are generally said to have experienced a breakthrough in Sweden at that time. The number of doctors increased fivefold and a regular system of general and mental hospitals was created. Efforts to organize the Swedish medical profession, begun in the late seventeenth century through the *Collegium medicum* (1661) and the *Societas chirugica* (1686), continued in the eighteenth. In the middle of the eighteenth century, moreover, there was a flourishing of medical teaching with the chairs in medicine at Uppsala University held by the two prominent scientists Carl Linnaeus and Nils Rosén von Rosenstein.[1]

Medicine was one of the useful sciences by which great store was set in eighteenth-century Sweden, and when the Royal Swedish Academy of Sciences was created in 1739 that body also proved crucial to the country's medical profession. Carl Linnaeus was one of the founders, but other prominent medical men also figured among its members: Nils Rosén, Peter Jonas Bergius, David Schultz, Olof Acrel, Joh. Lor. Odhelius and J. G. Wahlbom (cf. section 3). And it was in the Academy's quarterly Transactions that such scientists presented and discussed their medical findings.[2] The Academy's choice of Swedish rather than Latin as the language of these publications was certainly an important factor in the emergence of an indigenous language of science (cf. Teleman, this

1. For a more detailed account of the history of medical science, see Lindroth (1967 and 1978).
2. An overview of eighteenth-century scientific and popular medical literature in Swedish can be found in Gunnarsson (1988b).

volume). Sweden's scientists were forced to create a Swedish scientific language and text genre that could present their theories and data in a trustworthy way to the readers of the Transactions.

By means of a detailed analysis of some early medical texts in Swedish, I intend in this chapter to explore how eighteenth-century medical professionals textually created scientificality. My analytical framework can be described as socio-constructivist, which means that I regard "scientificality" as a construction at different levels, or within different layers, of texts: a cognitive, a social and a societal layer. In the next section, "The construction of professional discourse", I outline this theoretical framework, which I then refer back to when discussing my results. The description in section 3 of the corpus studied is followed by three results sections. In the first, section 4, the analysis focuses on the cognitive content of the texts and how it is structured, that is, on textual features that can primarily be linked to the cognitive layer. In the second, section 5, I consider a number of linguistic phenomena (references, names and pronouns) that can be related to the social layer; here, the analysis is concerned with how the scientific social network is constructed in the texts, i.e. how authors and their colleagues are explicitly mentioned there. In the third, section 6, I examine how, in their texts, authors constructed a scientific role in society for the group to which they belonged. Here the link is to the societal layer, that is to say, I analyse how – by using terms, figures, names etc. – the author-scientist-expert can be said to have positioned himself in relation to other groups in society. In the concluding section (7), finally, I apply a developmental perspective to the eighteenth-century texts studied, relating my findings to different stages in the development of medical science.

2. The construction of professional discourse

As an introduction to my analysis of the texts, let me begin with an outline of my socio-constructivist framework.[3] Scientific language and discourse emerge in a cooperative and competitive struggle among scientists to create the knowledge base of their field, to establish themselves in relation to other scientists and other professional groups, and to gain influence and control over political and

3. Studies that have applied a socio-constructivist approach to scientific English include those of Bazerman (1988), Atkinson (1999) and Valle (1999). Gross, Harmon, and Reidy (2002) present analyses of English, German and French texts from the seventeenth century to the present. Taavitsainen and Pahta (2000) discuss the medical case report from the late nineteenth century to the present day.

socio-economic means. Historically, language has played a central role in the creation and development of different professions and academic disciplines. By means of their texts, professionals have attempted to carve out a space for their field within society and to establish themselves in contact and competition with others within their group. Their knowledge base and its linguistic forms have thus been created in a societal and social framework.

Beginning with the *cognitive* layer of texts, we find that the medical profession had a certain way of viewing reality. Socialization into this profession involved learning how to discern the relevant facts and how to view the relations between different factors. Language, texts and spoken discourse were part of this construction process, that is to say, medical terminology, medical text structures and medical discourse patterns developed as a means of dealing with reality in a manner that was appropriate for medical purposes. The way in which language was used was related to existing knowledge within the field and also to conceptions about what constituted knowledge and the attitude that should be adopted to it. Consequently, what is important is not only what medical scientists knew in the eighteenth century, but also what knowledge they believed to be relevant and how they considered that data should be collected, observed and analysed.

Secondly, regarding the *social* layer, every professional group is also formed by the establishment of an internal group structure, a group identity, group attitudes and group norms. The need for a professional identity, i.e. for a professional sense of "us-ness", has of course played an important role in the construction of a medical group language. The use of medical scientific language during a given period is thus related to the type and level of the professional community (the social group), its size, structure, degree of professionalization, degree of internationalization, extent and nature of mutual contacts, existence of publications, etc.

Thirdly, as regards the *societal* layer, every professional group also stands in a particular relationship to the society in which it operates. The members of a profession play a role in relation to other actors in society; they play – or do not play – a role on the political scene, within the business world, in the education system, in relation to the media, and so on. It is through language that professional groups perform their societal function. If they are to play a role in society, they have to construct their communicative behaviour, i.e. their texts, in a way that is appropriate for that purpose. Professionals thus adapt to established genres, but are also involved in developing new ones.

The three layers described above are strongly related to the emergence and continuous re-creation of medical language and discourse. The cognitive establishment of the medical field took place at the same time as the scientists concerned were fighting for their place in society and to strengthen their group in relation to others.

On the basis of the three layers, we can also describe different scientific stages. Looking at how medical science has evolved over the last few hundred years, we can see that medical knowledge and practice took a major step forward in the seventeenth and in particular in the eighteenth century. Only gradually, though, did medicine develop into a science in the modern sense. Since the eighteenth century, all societies have undergone radical change. Likewise, there have been changes within the medical scientific community:

1. Medical knowledge has grown enormously.
2. Science in general and the philosophy of science have changed. Statistics and empirical methods have developed. In many sciences, positivism has become the only accepted view.
3. The medical profession has gradually become increasingly established and recognized. Today, doctors are highly valued professionals, and medical scientists and medical research are considered of great importance to society.
4. The medical scientific community has become much larger. The number of doctors, medical scientists and students of medicine has increased, as has the number of medical journals and conferences.

Important changes have thus occurred over the last three centuries in medical science, science in general, the medical profession and the medical scientific community. One claim that I will make in this chapter is that language and discourse are essential elements in the construction of science and in shaping the scientific community; another, that changes in language and text patterns can be described in relation to three scientific stages: the pre-establishment stage, the establishing stage and the specialized stage. For each layer, the three stages can be summarized on a developmental axis (cf. Gunnarsson 2005: 711, also 2009:62):

Cognitive layer

• Individual findings — Accumulation of findings — Theoretical integration

Social layer

• Isolated researchers — Academic grouping — Developed scientific community

Societal layer

• Scientists act within society — Scientists act within society and academic groupings — Scientists act within scientific community

In my conclusions, I will return to these developmental axes.

3. Medical articles analysed

I will now go on to describe the corpus of texts studied.[4] My aim, as already indicated, has been to analyse how scientificality was constructed in the earliest medical articles in Swedish. In that context, I have not applied a strict definition of science – since relatively few Swedish eighteenth-century texts would presumably then have qualified for inclusion – but have used "scientific" to refer, rather, to a text written by a university-educated scientist, with other professionals – in this case doctors – as its intended readers. My interest here is in original Swedish works, i.e. papers presenting observations and investigations by scientists working in Sweden. (Translations or presentations of foreign books and articles, in other words, are not included.)

As my source I chose the Transactions of the Royal Swedish Academy of Sciences (*Kongl. Vetenskapsakademiens Handlingar*), that is, the forum in which the doctors and medical professors of the day could present and discuss diseases, their progress, and different methods of treatment. From these publications, I selected all the texts from the period 1750–1769 dealing with (a) smallpox and (b) cataract and other eye conditions. The 20-year period in question seemed appropriate for two reasons: first, publication of the Transactions (begun in 1739) had by then been going for a number of years, presumably allowing time for certain text conventions to emerge; and second, the period is described in the literature as a heyday of the Transactions (Lindroth 1967: 6) and, in particular, of the medical articles appearing there.

Papers on smallpox and cataract were chosen, on the one hand, because smallpox and cataract were common diseases in eighteenth-century Sweden and methods of curing them were considered to have been found; and on the other, because one of these two areas, smallpox, represented a medical and the other, cataract, a surgical problem.

Specialization, it is true, was far less pronounced in the eighteenth century than it is today. In earlier centuries however, and continuing into the eighteenth, a distinction had been made between physicians, who treated internal disorders, and surgeons (*fältskärer* 'field surgeons'), who attended to external conditions. This distinction admittedly seems to have become less marked during the Age of Liberty (1718–1772) as far as education and professional status were concerned (Lindroth 1978: 418), but as a difference in professional practice and, it may be assumed, theory formation, it persisted. Furthermore, the subject indexes to the different volumes of the Transactions have separate headings for surgery, under which cataract and other eye conditions appear, and medicine, which includes

4. An earlier analysis of this corpus is presented in Gunnarsson (1988a).

smallpox and smallpox inoculation. As it cannot be ruled out that different text patterns may have become established among physicians on the one hand and surgeons on the other, texts from both these specialties were chosen.

All the original articles from the period dealing with the two areas referred to were included in the study. The list below gives the author, volume (and quarterly issue, where there are several articles from the same year by the same author), title and page numbers of each paper in the Transactions.

Articles studied, published in the Transactions of the Royal Swedish Academy of Sciences, 1750–1769

Medicine: On smallpox and smallpox inoculation

1 Rosén, N., 1751, Sätt at förekomma elaka koppor [Ways of preventing confluent smallpox]. Pp. 31–33.

2 Bergius, P. J., 1756:1, Rön om en Koppo-smitta, som liknar ympning, jämte skäl till koppo-ympningens antagande här i riket [Finding concerning a smallpox infection resembling inoculation, together with reasons for the adoption of smallpox inoculation in this country]. Pp. 40–45.

3 Bergius, P. J., 1756:2, Rön, om frossors och ältans uphäfvande genom koppor [Finding concerning the supersession of ague and childhood fever by smallpox]. Pp. 131–134.

4 Schultz, D., 1757, Rön vid kopp-ympningen [Finding made in connection with smallpox inoculation]. Pp. 197–199.

5 Bergius, P. J., 1766:1, Rön och anmärkningar om en midt uti en koppympning inkommen mässlings-smitta. Förra stycket [Finding and observations concerning a measles infection arising in the middle of a smallpox inoculation. First part]. Pp. 68–88.

6 Bergius, P. J., 1766:3, Rön och anmärkningar om en midt uti en koppympning inkommen mässlings-smitta. Senare stycket [Finding and observations concerning a measles infection arising in the middle of a smallpox inoculation. Final part]. Pp. 199–211.

Surgery: On cataract, cataract surgery, and other eye conditions

1 Wahlbom, J. G., 1756, Anmärkningar om starr-stickning [Observations on cataract surgery]. Pp. 261–270.

2 Acrel, O., 1757, Jämförelse emellan de förmoner och olägenheter, som åtfölja hvarje starrstickningssätt, i stöd af egna rön och anmärkningar [Comparison of the advantages and disadvantages accompanying each method of cataract surgery, based on the author's own findings and observations]. Pp. 94–117.

3 Odhelius, J. L., 1764, Beskrifning på en besynnerlig ögonskada [Description of a strange eye injury]. Pp. 290–292.

4 Odhelius, J. L., 1765, En ganska sällsynt ögonskada [A rather rare eye injury]. Pp. 290–292.

5 Odhelius, J. L., 1767, En ganska sällsynt ögonskada [A rather rare eye injury]. Pp. 145–146.

6 Acrel, O., 1767, Anmärkningar vid föregående rön [Observations on earlier findings]. Pp. 147–148.

4. Constructing scientificality at the content level

The first part of my analysis focuses on the cognitive content of the texts (4.1) and the structuring of that content (4.2). My claim will be that the results of this analysis can be related to the stage reached by the domain of medical science, in terms of degree and type of scientificality.

4.1. Cognitive content

The cognitive content of the twelve articles was categorized in relation to five *cognitive worlds*: a scientific world, a practical world, an object world, a private world and an external world. Each item of information was assigned to one of these five worlds, and within these to different *aspects*. Within the *scientific world*, which I will focus on in this chapter, the aspects identified were theory, classification and experiment. For the analysis of different parts of a text, I also found it relevant to distinguish between two *content types*: state-descriptive and action-descriptive, and within these, between different (temporal) *dimensions*: cause, phenomenon, process and change/result (cf. Gunnarsson 2009: 30–38).

Figure 1a shows the scientific world and its related aspects (theory, classification and experiment) and dimensions (cause, phenomenon, process and result) for a *state-descriptive* text. Figure 1b sets out the corresponding categories for an *action-descriptive* text: here, the dimension of cause is related to preventive measures, that of process to treatment etc.

The cognitive analysis has in a number of studies proved a useful tool for describing the differences between popular science and scientific texts, and also for describing how text patterns change from one period to another.[5] Within each cognitive world, it is possible to distinguish certain abstract categories

5. The method of analysis applied is described in detail in Gunnarsson (2009).

	Cause	**Phenomenon**	**Process**	**Change**
Theory	Theory of causes of disease	Theory of disease	Theory of disease process	Theory of change in disease
Classification	Classification of causes of disease	Classification of diseases	Classification of disease processes	Classification of changes in disease
Experiment	Observation of causes of disease	Observation of disease	Observation of disease process	Observation of change in disease

Figure 1a. Scientific world for a medical text. State-descriptive parts of text

	Cause	**Phenomenon**	**Process**	**Change**
Theory	Theory of preventive measures	Theory of disease	Theory of treatment methods	Theory of result of treatment
Classification	Classification of preventive measures	Classification of diseases	Classification of treatment methods	Classification of results of treatment
Experiment	Experiment with preventive measures	Observation of disease	Experiment with treatment methods	Observation of results of treatment

Figure 1b. Scientific world for a medical text. Action-descriptive parts of text

that are common to different texts. These refer, at one level, to differing aspects that may be in focus in the text. Within the scientific world, as noted, the aspects of *theory*, *classification* and *experiment* can be distinguished (forming the rows in the diagrams above). At another level (forming the columns), the categories refer to different (temporal) dimensions between which the text can move. For state-descriptive parts of a medical text, these dimensions are *cause of disease*, *disease*, *disease process* and *change in disease*. For action-descriptive parts of the text, the corresponding categories are *preventive measures, disease, treatment* and *result of treatment.*

The cognitive content of the twelve papers studied was categorized in relation to all five cognitive worlds (see above). Table 1 shows what proportion of each text was taken up by each of these worlds. My categorization of the content was relatively coarse, as percentages were calculated on the basis of entire lines of text. Headings, as well as any tables, diagrams, captions or notes in the articles, were not included in the calculations.

Table 1. Cognitive worlds in eighteenth-century articles on smallpox and cataract (Percentages refer to share of body text assigned to each world)

	Cognitive worlds				
	Scientific	**Practical**	**Object**	**Private**	**External**
Article	%	%	%	%	%
Smallpox					
1 Rosén 1751	0	0	58	42	0
2 Bergius 1756:1	57	3	25	8	6
3 Bergius 1756:2	20	7	73	0	0
4 Schultz 1757	26	67	0	7	0
5 Bergius 1766:1	8	89	3	0	1
6 Bergius 1766:3	56	44	0	0	0
Cataract					
1 Wahlbom 1756	61	30	9	0	0
2 Acrel 1757	64	36	0	0	0
3 Odhelius 1764	0	18	82	0	0
4 Odhelius 1765	0	17	65	19	0
5 Odhelius 1767	0	85	15	0	0
6 Acrel 1767	86	14	0	0	0

To give a broader picture of the "scientificality" of these eighteenth-century texts, three main categories can be distinguished, based on the percentages in the "Scientific" column of the table:

No scientific world

The articles placed in this category either give advice to the public, Rosén (1751), or describe specific cases, Odhelius (1764, 1765 and 1767). The latter type, i.e. case descriptions, can be said to be common in the Transactions, a key aim of which was to bring together empirical findings (see further below). Four of the articles, then, belong in this category: one medical (Rosén 1751) and three surgical (Odhelius 1764, 1765 and 1767).

Small proportion of scientific world

In this category I have included articles in which a small proportion, up to about 25%, of the text was assigned to the scientific world. These are articles containing both a scientific discussion and accounts of cases. The category comprises three medical articles: Bergius (1756:2), Schultz (1757) and Bergius (1766:1).

Large proportion of scientific world

This category comprises articles in which a large proportion of the text was assigned to the scientific world. Bergius (1766:3) and Acrel (1767) are comments on earlier descriptions of cases, while Bergius (1756:1), Wahlbom (1756) and Acrel (1757) consist of independent contributions to the scientific debate. Bergius's 1756 article advocates the introduction of smallpox inoculation in Sweden, while Wahlbom and Acrel call for the use of different methods of cataract surgery. In all, then, five articles belong in this category: two relating to smallpox and smallpox inoculation, and three dealing with cataract and other eye conditions.

My analysis of the cognitive content of the eighteenth-century texts studied thus shows that the majority of them include passages that can be assigned to the scientific world; only four of the twelve articles do not move within that world at all. Five of the articles have a large proportion of their text within the scientific world, i.e. they contain a significant element of classification and theorizing.

4.2. Structuring of the content

The next step in the text analysis focusing on the cognitive layer looks at how the cognitive content is structured in the texts. As noted earlier, only papers presenting the authors' own observations or investigations were included in the study. The writers' chief aims may therefore be assumed to have been relatively similar, i.e. they presumably wished to give their presentations a form that would be recognized as scientifically credible. In this part of the analysis too, then, the focus is on "scientificality", in this case on how it was constructed through the chosen text structure.

To begin with, it may be noted that it is not possible to discern a uniform or common pattern to the content structuring of the texts, even when the two subgroups, i.e. the medical and surgical articles, are considered separately. Rather, these texts exhibit what to a present-day reader is a striking variation, in terms both of how they begin and how they are structured from that point on. Above all, as a modern reader one is struck by the prominent role given to the description of specific cases. Careful observation of individual cases was of course fundamental to the exploration of nature, and hence also of diseases, in the eighteenth century. And in the texts studied, case descriptions in fact emerge as the common denominator, with one or more of them to be found in all the articles except one (Rosén 1751). Where they are placed, however, varies.

To present the results of this part of the analysis, I have chosen below to divide the articles into three main categories, based on how the author begins the text: by describing (1) a case, (2) a (treatment) problem or (3) the current state of knowledge. Within each main category I then distinguish two or three subgroups, denoted by the letters *a*, *b* and *c*. The subgroups assigned the letter *a* comprise structurally straightforward, unproblematizing articles in which the case (1a), the treatment problem (2a) or a brief account of current knowledge + a case (3a) are followed by a conclusion. Subgroups 1b, 2b and 3b consist of papers that include a discussion section. Subgroup 3c, finally, comprises articles describing a gap in existing research.

1. Texts that begin by describing a case

Five articles open with a description of a specific case. This is how Wahlbom (1776), for example, begins:

> En bonde-piga från Öland, 25 år gammal, hade på bägge ögonen den grå starren. För 6 eller 7 år sedan hade hon under sit arbete fått lika som et skott eller slag i det vänstra ögat, hvarpå följde värk och inflammation, och synen försvan inom kårt tid. Et par år därefter kände hon också stickningar, med värk i det högra ögat, hvarefter synen på detta försvan, så at hon blef aldeles blind. Var eljest flussig och plussig till constitution.
> Då hon kom till mig, den 16 Junii detta år, fant jag sjelfva uvea på inre sidan af det väntra ögat vara remnad, och således syntes pupullen helt sned och innanföre sjelfva Starren., hvilken betäktes …
>
> 'A peasant's maid from Öland, 25 years old, had cataract in both eyes. 6 or 7 years ago, while at work, she received some sort of projectile or blow to her left eye, whereupon pain and inflammation ensued, and within a short time sight was lost. A couple of years later she also felt stinging and pain in her right eye, after which the sight in that eye was lost, leaving her completely blind. Otherwise she was plump and chubby in constitution.
> When she came to see me, on the 16th of June this year, I found the actual uvea on the inside of the left eye to be torn, and thus the pupil appeared to be completely askew and within the actual Cataract, which was covered …'

Following the opening descriptions of cases, however, are sections of differing types in the different articles. In Odhelius (1764 and 1765) what follows is a short conclusion, while other authors use the case as the basis for a scientific argument, with explanations and a discussion (Wahlbom 1756, Bergius 1766:3 and Acrel 1767). We thus find two subgroups:

1a Case – Conclusion
Odhelius (1764 and 1765)

1b Case – Result – Explanation – Discussion
Wahlbom (1756), Bergius (1766:3), Acrel (1767)

2. Texts that begin with a (treatment) problem

Two of the articles relating to smallpox inoculation (Rosén 1751 and Schultz 1757) begin by describing a problem arising in connection with treatment. This is the introduction to Schultz's paper:

> Många hafva funnit, at personer ej smittats efter första inskårningen, men ändock fått kopporna efter förnyad ympning. Denna omständighet har varit oläglig för Läkaren, och obehagelig för de inoculerade, som antingen länge måste vänta på en besvärlig, om icke farlig sjukdom, eller ock stanna uti osäkerhet, om de icke en eller annan gång torde smittas af naturlige koppor. Brist uppå naturlige Koppor, af hvilka jag kunnat tilreda ympningstråd, har gifvit mig anledning til ett rön, som i framtiden lärer kunna fria ympningen ifrån denna olägenhet. Jag var försedd med en årsgammal tråd, men som D:r HADOW och Herr GUYOT försäkra, at en gammal tråd åstadkommer senare och svagare utslag, så förskref jag Kopp-tråd både från Åbo och Upsala. Ympningen anstäldes uppå trenne unga Fruentimmer, hos hvilka uti …
>
> 'Many have found that persons were not infected after the first incision, and yet did contract the small-pox after a further inoculation. This circumstance was inconvenient for the Physician, and unpleasant for the inoculated, who had either to wait a long time for a troublesome, if not dangerous, disease, or else to remain in uncertainty about whether they might at some time become infected naturally with the small-pox. The lack of natural Small-pox from which I might prepare thread for inoculation has led me to a discovery which could probably in future free inoculation of this inconvenience. I was provided with a year-old thread, but as Dr. HADOW and Mr. GUYOT assure us that an old thread causes a later and weaker eruption, I prescribed Small-pox thread from both Åbo and Upsala. Inoculation was carried out on three young Women, in whom …'

The articles taking their starting point in a treatment problem also vary in the way they continue. In Rosén (1751), the solution to the problem is described in an authoritative manner, in the form of instructions for action, while Schultz

(1757) formulates the solution as a thesis, for which case descriptions subsequently serve as proof.

2a Problem – Solution
Rosén (1751)

2b Problem – Thesis regarding solution – Case(s) (= proof) – Discussion
Schultz (1757)

3. Texts that begin by describing the current state of knowledge

Five of the articles start off with a description of the existing knowledge. The introduction to Bergius (1756) can serve as an example:

> Ibland de väsendteliga förmoner, som den senare tiden tilskyndat oss framför våra förfäder, kan man med nog skäl räkna den nyttiga kunskap om koppors ympande, som under nu varande tide-hvarf uti Europa inkommit. En kunskap, hvilken så mycket högre förtjenar at värderas, som han tjenar at mota en of människo-slägtets grymmaste fiender, och at märkeligen hämma en landfördärfvelig folkbrist. Vi se med förundran af Herr De la Condemine och flera, med hvad besynnerlig framgång koppo-ympningen i Turkiet, Engeland och flerstädes blifvit fortsatt, så at derigenom många tusende människor blifva vid lif och utan lyte, då deremot på de andra ställen halfva människo-hopen antingen blifvit förödd eller ock på något sätt vanskapad.
>
> 'Among the considerable advantages that have accrued to us in recent times, compared with our forefathers, there is good cause to include the useful knowledge concerning the inoculation of the small-pox that has arrived in Europe during the present epoch. Knowledge which deserves to be valued all the more highly because it serves to combat one of the human race's cruellest enemies, and to appreciably check a nationally destructive loss of population. We see with amazement from Mr. De la Condemine and others the extraordinary success with which inoculation against the small-pox has been pursued in Turkey, England and elsewhere, with the result that many thousands of people remain alive and without defect, whereas in other places half the population has either been destroyed or else become in some way disfigured.'

The opening account of the current state of knowledge varies both in length and in the degree of scientificality. In Odhelius (1767), a brief outline of existing knowledge is followed by a description of a case and a conclusion (category 3a), while in other articles the account of current knowledge leads into a wider-ranging discussion and presentation of evidence (3b). I also identify a further category (3c), comprising the two articles in which the author describes a knowledge gap, drawing attention to the incomplete state of existing knowledge in a way that is typical of present-day scientific texts.

3a Brief description of current knowledge – Case – Conclusion
Odhelius (1767)

3b Current knowledge – Problem – Thesis regarding solution – Case(s) – Discussion – Conclusion
Bergius (1756:1), Acrel (1757)

3c Current knowledge – Knowledge gap – Case(s)
Bergius (1756:2 and 1766:1)

4.3. Discussion

Linking the above grouping of content structure to the earlier grouping of the texts according to their cognitive content, we find the following associations:

Cognitive content	**Text structure**	
No scientific world	1a	Case – Conclusion
	2a	Problem – Solution
	3a	Brief description of current knowledge – Case – Conclusion
Small proportion of scientific world	2b	Problem – Thesis regarding solution – Case(s) – Discussion
	3c	Current knowledge – Knowledge gap – Case(s)
Large proportion of scientific world	1b	Case – Current knowledge/– Explanation – Discussion – Conclusion
	3b	Current knowledge – Problem – Thesis regarding solution – Case(s) – Discussion – Conclusion

As this overview shows, articles with a large proportion of their text assigned to the scientific world may begin either with a case description or with an account of current knowledge. What distinguishes this group of articles, rather, is that they include a discussion of results and of solutions to problems.

There is no common structural pattern, then, to the papers in the Transactions,[6] or to either of the two subgroups, i.e. the texts on smallpox and smallpox

6. One of the findings of my earlier analysis (Gunnarsson 1988a: 172–173) was that external marking of text structure is rarely to be found in these eighteenth-century texts.

inoculation or those on cataract and other eye conditions. For the individual authors, on the other hand, a degree of consistency in the way they structure their articles may be noted. Odhelius's three contributions are of types 1a and 3a, while Bergius's four papers are of types 3b and 3c. The third author with more than one article in the corpus is Acrel. However, his two papers cannot be compared, as one of them (Acrel 1767) is a comment on another article. We thus find indications of individual text patterns, but little sign of any clearly developed genre conventions.

To sum up, the content and content structuring of the texts studied can – despite the elements of classification, theorizing and description of current knowledge – be said to suggest a pre-establishment stage of science. One indication of this, of course, is the considerable weight given in the articles to particular cases involving specific individuals. Another is the lack of any marked uniformity of text structure; that is to say, individual authors were free to choose their own forms of presentation. Although the medical case report as such is not a genre which is exclusively characteristic of an early stage of medical science, analyses of medical literature from the nineteenth and twentieth centuries show case reports of another kind. For one thing the case report has a less central function in recent literature, for the other its role is present unusual cases of common diseases or well-documented cases of rare diseases (Taavitsainen and Pahta (2000:61). When the modern type of statistically based research became important, the descriptions of single cases were marginalized.

5. Constructing a scientific network

In this section I will discuss a number of linguistic phenomena that can primarily be linked to the social layer of the texts. By looking at how the author-scientist himself appears in his text and how he refers to his colleagues, I wish to provide a picture of the construction in the text of a scientific network. In the eighteenth century, too, scientificality was formed at a group level, which meant that, to have his findings accepted as truths, the author-scientist needed to textually construct a relationship between himself and other members of the scientific community.

In modern scientific writing, this relationship is crucial, although it is often disguised in the form of references to colleagues' work and to earlier studies by the author. My eighteenth-century corpus, however, contains very few references of a modern type. To explore how scientific relationships were constructed during that period, therefore, I chose to include in my analysis both references to other works and names of colleagues (5.1), and in addition explicit mentions of the author himself, i.e. the frequency of the pronoun *jag* 'I' in the texts (5.2). My claim will be that the results of this analysis of the textual construction of a scientific network can be related to the stage reached by the medical scientific community, i.e. to its size and its degree of establishment.

5.1. References and names of colleagues

To gain a picture of how eighteenth-century authors related to their colleagues' work, then, I began by noting the use of references to other literature in the articles studied. It should be pointed out first of all that none of the articles include a list of references at the end. A couple of the papers (Bergius 1766:3 and Acrel 1757) do, however, contain references in notes. In the others, such references are to be found in the main body of the article. The figures for references in Table 2 cover all references, including those appearing in notes.

Other doctors and scientists are frequently mentioned by name in the texts. It is stated, for example, that Drs M and N were present at the smallpox inoculation or the operation, or that Dr O has made a similar finding. Such mentions are not references in the modern sense, but their purpose, I would argue, is essentially the same. They anchor the author's own finding within the scientific community and thus link it to existing knowledge. To get an idea of the frequency of such "semi-references", I counted names of other doctors and scientists in the articles (names of patients were not included). Each name was counted only once in any given paper. Names of authors of articles that were counted as references were not counted as instances of naming.

Table 2 shows how many references and names of colleagues are to be found in the texts. The length of each article, in standard pages of 3,000 characters, is also given as a basis for comparison.

As the table makes clear, the numbers of references and names of colleagues in these eighteenth-century texts are fairly modest; the difference in this respect compared with present-day medical articles is considerable. Only in five of the twelve papers are there any references at all to other literature, and only four of them contain more than two names. An interesting question, then, is what the distinguishing features are of the articles in which names and references do occur. What purpose did references and naming of colleagues serve for the authors of papers published in the Transactions?

To begin with, we can note the comparatively large number of references, and also of colleagues' names, in Acrel (1757). Although the fact is not stated in the text, Acrel's paper was a counter-attack on Wahlbom's propaganda (Wahlbom 1756) for a particular method of cataract surgery. These two articles marked the start of a heated polemic between their authors over which method was to

Table 2. References and names of colleagues in eighteenth-century articles on smallpox and cataract

Article	Length in pages	References Number	Names Number
Smallpox			
1 Rosén 1751	1.1	0	0
2 Bergius 1756:1	2.9	0	1
3 Bergius 1756:2	1.5	0	1
4 Schultz 1757	1.2	0	11
5 Bergius 1766:1	11.4	6	6
6 Bergius 1766:3	5.6	3	1
Cataract			
1 Wahlbom 1756	4.4	3	9
2 Acrel 1757	10.4	8	18
3 Odhelius 1764	0.5	0	1
4 Odhelius 1765	0.8	0	0
5 Odhelius 1767	0.6	0	2
6 Acrel 1767	0.8	4	0

be preferred, the St Yves method, advocated by Acrel, or that of Ferrein, which Wahlbom favoured (Lindroth 1978: 468). An important surgical problem was thus at stake. As we see, Wahlbom cites a number of sources (3 references) and colleagues (9 names) in support of his argument, but Acrel goes even further, invoking 8 sources and 18 colleagues. It should also be noted, though, that Acrel includes references even in his short article in the Transactions in 1767.

Of the medical articles, only the two published by Bergius in 1766 (Bergius 1766:1 and 1766:3) contain references; there are thus none to be found in his papers from 1756. The 1766 articles describe in detail the discovery which the author made when he was able to observe what happened when patients contracted smallpox and measles at the same time. This finding must have been judged to be of great interest to the scientific community at the time, as so much space was devoted to it in the Transactions (17.2 standard pages in all), and the references and naming of colleagues in these two articles were presumably designed to lend credibility and weight to the author's discovery; it may be mentioned that four of the individuals named were present at the event and are thus invoked as witnesses.

Also to be noted is the large number of names to be found in Schultz (1757). Schultz, too, feels that he has made an important discovery. He claims that it is one that could in future free inoculation of the inconvenience of occasional shortages of natural smallpox for the preparation of thread for use in the procedure. Once again, it may be assumed that the purpose of mentioning colleagues by name was to enhance the credibility of this useful finding.

To sum up, then, we can note the modest number of references and names of colleagues in the eighteenth-century texts studied. It does not appear to have been common practice, as it is now, to refer in articles to the works of others. It could of course be objected that this is only to be expected, as there was little literature to refer to at that time, at least not literature published in Sweden. Although some medical publications did begin to appear in Swedish in the eighteenth century, the number involved was fairly limited. In other European countries, however, medical matters had long been written about in journals and books, and the well-travelled and internationally trained Swedish doctors of the eighteenth century were no doubt familiar with this literature (Lindroth 1978: 437). It may be noted, though, that papers published in the Transactions made very little reference to foreign authorities. When authors did choose to refer to other sources – native or foreign – they seem to have done so in order to lend weight to their arguments (Wahlbom, Acrel) or credibility to their findings (Bergius, Schultz).

5.2. The personal pronoun "I" in the texts

As many studies of scientific language have shown, one of its characteristic features is its impersonality, i.e. the fact that the author chooses not to appear in the "I" form, but rather as a "we" or, better still, indirectly through facts (see, for example, Gunnarsson 2001). To get an idea of how the eighteenth-century articles examined here rate on an imaginary scale from personal to impersonal, I went through the texts, counting occurrences of the pronoun *jag* 'I'. Table 3 shows the frequency of this pronoun, overall and per standard page of 3,000 characters, in each of the twelve texts.

As we see from the table, the first-person singular pronoun occurs quite frequently in the eighteenth-century texts studied. The author appears in the "I" form in all the longer texts, and in eight of the twelve papers he does so, on average, several times per page. All of the articles with no use of "I" are on surgical subjects; it should be noted, however, that these are short texts, less than 1 standard page in length (0.5, 0.8 and 0.8 pages). Considering the six medical texts as one group and the six surgical texts as another, we find a small difference in the frequency of the pronoun between the two. The medical texts, i.e. those on smallpox and smallpox inoculation, have more occurrences of "I" per page on average than the surgical texts, 4.5 compared with 3.6. Overall, though, the style

Table 3. Frequency of *jag* 'I' in eighteenth-century articles on smallpox and cataract

	Frequency of *jag* 'I'	
	Overall	Per page
Smallpox		
1 Rosén 1751	9	8.2
2 Bergius 1756:1	14	4.8
3 Bergius 1756:2	4	2.7
4 Schultz 1757	9	7.5
5 Bergius 1766:1	11	1.0
6 Bergius 1766:3	14	2.5
Cataract		
1 Wahlbom 1756	23	5.2
2 Acrel 1757	32	3.1
3 Odhelius 1764	0	0
4 Odhelius 1765	0	0
5 Odhelius 1767	8	13.3
6 Acrel 1767	0	0

of these eighteenth-century texts is not impersonal at the superficial linguistic level that is reflected in pronoun choice.

5.3. Discussion

As this analysis of the construction of a scientific network in the texts makes clear, the author-scientist is present as an individual, an "I", in most of the articles. It is common for authors to explicitly mention by name other members of the scientific community. The textual construction of a scientific social relationship is thus a central feature of the eighteenth-century medical article. What is particularly striking is that this relationship is primarily established as one between individuals, i.e. between the author, present as an "I", and other members of the medical community, who are mentioned by their personal names. These results, too, can be linked to a pre-establishment stage of science, i.e. to a very small scientific community in which it is important to manifest the connection between one individual and another.

6. Constructing a scientific role in society

In this results section I will examine how, in their texts, authors constructed a scientific role in society for the group to which they belonged. Like other scientific texts, the eighteenth-century articles studied here manifest a relationship not only between the individual author-scientist and his close colleagues (the other members of the scientific network), but also between the author's own group, medical scientists, and other members of society: lay people and other professionals. My focus in this section, in other words, will be on how these texts construct a societal role for medical scientists. The questions I will be seeking to answer are: Does the author-scientist place himself at a great distance from other groups in society, i.e. does he, through his text, construct the role of a scientific expert? And does the author-scientist-doctor place himself at a great distance from the people he examines, his patients? To gain an impression of the degree of distance in the first respect, that is, between the author-expert and "others", i.e. non-experts, I examined the occurrence of terms and figures (numbers) in the texts (6.1). To get a picture of the distance created in the second respect, between the author-scientist-doctor and the patient, I looked at how cases were introduced in the texts (6.2).

My claim will be that the results of this analysis, too, can be related to the stage reached by the medical scientific community at the time, in this case with respect to the scientific role played by the medical profession in society.

6.1. Terms and figures in the texts

Text researchers attempting to characterize scientific usage have often mentioned the use of terms and figures (i.e. numbers) as markers of distance.[7] Here, of course, it is a matter of distance in relation to "others", i.e. to those who are not experts in the field.

To obtain a picture of how distance is marked in my eighteenth-century corpus, I counted occurrences of terms and figures in the texts. Conscious that it is not easy to decide what constitutes a term, I decided, for the sake of simplicity, to count as a term any word which, to me as a native speaker of Swedish, seemed foreign; that is to say, chiefly words of Latin, French or Greek origin. Compounds including a foreign element, such as *incisions-såret* 'the incision wound', were also regarded as terms. A prescription consisting of a sequence of several abbreviations was counted as a single term. As figures, I counted all numbers referring to frequency, quantity, age or the like. Years,

7. See for instance Bungarten (1986) and Adams-Smith (1990).

other dates, and ordinals (*1sta* '1st') were not included. Table 4 shows how many terms and figures are to be found in the texts, overall and per standard page. It should be noted that, as far as terms are concerned, the numbers in the table do not refer to the number of *different* terms, but to the aggregate occurrence of terms.

Table 4. Occurrence of terms and figures in eighteenth-century articles on smallpox and cataract

	No. of terms		No. of figures	
	Overall	Per page	Overall	Per page
Article				
Smallpox				
1 Rosén 1751	13	11.8	17	15.5
2 Bergius 1756:1	2	0.7	2	0.7
3 Bergius 1756:2	4	2.7	11	7.3
4 Schultz 1757	6	5.0	2	1.7
5 Bergius 1766:1	336	29.5	15	1.3
6 Bergius 1766:3	60	10.7	7	1.3
Cataract				
1 Wahlbom 1756	215	48.9	3	0.7
2 Acrel 1757	464	44.6	32	3.1
3 Odhelius 1764	20	40.0	0	0
4 Odhelius 1765	18	22.5	0	0
5 Odhelius 1767	17	28.3	2	3.3
6 Acrel 1767	42	52.5	0	0

What can be noted, first, is that terms are considerably more frequent in the surgical texts (39.5 terms per page on average) than in the medical ones (10.1 on average). Although I am unable to quote any figures for comparison, I would claim that the surgical texts are rich in terms even by modern-day standards. They include terms referring, for example, to parts of the eye and to various instruments.

In the medical texts, the density of terms varies from one article to another. The highest density is found in Bergius (1766:1), and more specifically in the part of that article that is a summary, or perhaps a direct transcription, of the author's case notes on the smallpox inoculation of seven children. The language there differs from that found in the rest of the article, and also from that of other papers by the same author. It may be assumed that Bergius normally attempted to write a more everyday language, but that here he left the text as it was, perhaps to lend a note of authenticity to the important medical event described. It may be observed, though, that the term density of this article is lower than the average for the six surgical papers (29.5, as against 39.5). Furthermore, the average for the other five articles on smallpox (i.e. excluding Bergius 1766:1) is only 6.2, which can again be compared with the average of 39.5 for the surgical texts.

The difference in the density of terms between the medical and surgical papers can presumably be explained in part by the differing character of their subject matter. Knowledge about the eye was perhaps more precise and term-bound than knowledge relating to smallpox. It seems not unreasonable to assume, though, that part of the explanation for the much higher term density of the surgical texts, compared with the medical ones, is to be found in the differing external circumstances of physicians and surgeons. The former were the established doctors, who controlled Swedish medicine through the *Collegium medicum*. The latter, by contrast, were engaged in a fierce battle at this time for equal status. They were heirs to the old occupation of the field surgeon, which had previously been a craft occupation rather than an academic profession like medicine (Lindroth 1975: 378 f.). It may be supposed that, to secure the same status as the physicians, surgeons now had to demonstrate their academic credentials. Perhaps the density of terms could be seen as a product of their endeavour to become, like physicians, an established group in the existing scientific community.

Turning next to the occurrence of figures in the texts, we see that it is very low. In the surgical texts there are on average 1.2 figures per standard page, and in the medical articles, 4.6. Among the medical texts, Rosén (1751) stands out with its comparatively high frequency of figures. If that article is excluded, the average for the medical papers is just 2.5 per page.

Generally, then, we find a low frequency of figures in the texts. The information they convey relates to the age of patients, time between visits etc. In the Rosén article just mentioned, the text with the most figures per page, they refer for example to how many times a child of a certain age should be given a stated number of pills. It is not, in other words, a matter of statistics in the modern sense.

In this context it may also be mentioned that tables are only to be found in one of the twelve articles: Bergius (1766:3, p. 203), where two tables are provided. One of them contains data from a Scottish study, including four columns of dates, together with a column showing the children's ages and another giving the age of what was inoculated. The other table is a presentation of Bergius's own data, and here there are no figures of the type counted: one column gives the names of the children and three columns the dates when different phases in the medical sequence of events occurred. In neither case, then, do we have an ordinary numerical table.

To sum up, it can be said that, through his use of terms, the author-scientist constructs the role of a scientific expert, which entails a marking of distance from other groups in society. The use of figures, on the other hand, is more difficult to interpret. In general, the eighteenth-century texts analysed describe one or two individual cases, rather than major studies, and when a somewhat larger investigation is presented, as in Bergius's 1766 papers, descriptions of particular cases remain the focus of attention. In these texts, numbers serve a different function from what we find in modern-day texts: they refer to the patient's age, time between visits etc., i.e. they are incorporated into verbally rich accounts of individual cases.

6.2. Introduction of cases

As the analysis of text structure made clear, descriptions of cases are a recurring theme in the eighteenth-century texts studied. In this section, I will approach these descriptions from a different angle, i.e. with a focus on the function they serve in the construction of the relationship between the author-scientist-doctor and the patient examined.

To get an idea of how that relationship is constructed in the twelve articles, I have chosen to concentrate on how the authors introduce the individual cases, and patients, in their texts. The quotations below show how they go about this:

Article	**Introduction of cases in articles on smallpox and cataract**
Smallpox	
1 Rosén 1751	"43 försök" '43 attempts'
2 Bergius 1756:1	"en fattig gesälls hustru boende på Södermalm" 'a poor journeyman's wife living in Södermalm'
3 Bergius 1756:2	"Pigan Brita Jertsdotter, 20 år gammal"; "Sjö-mannen Gladers barn, af 6 års ålder"; "Tvenne barn uti et hus, det ena om 9 år, och det andra om 8" 'The maid Brita Jertsdotter, 20 years old'; 'The seaman Glader's child, 6 years of age'; 'Two children in a house, one 9 years old, the other 8'
4 Schultz 1757	"trenne unga Fruentimmer" 'three young Women'

5 Bergius 1766:1	"skedde Koppympningen på Frimurare-Barnhuset här i Stockholm, på 7 barn, ifrån 5 til 8 års ålder, hvaraf 6 voro flicke-barn, och heto Maria, Johanna, Stina, Lisa, Ulrica och Anna Maja, samt en gosse vid namn Jan Peter" 'inoculation against the Small-pox was carried out at the Freemasons' Orphanage here in Stockholm, on 7 children, from 5 to 8 years of age, 6 of whom were girls, called Maria, Johanna, Stina, Lisa, Ulrica and Anna Maja, and one a boy, by the name of Jan Peter'
6 Bergius 1766:3	"Maria"; "Johanna"; "Stina"; "Lisa"; "Ulrica"; "Anna Maja"; "Jan Peter"

Cataract

1 Wahlbom 1756	"En bonde-piga från Öland, 25 år gammal" 'A peasant's maid from Öland, 25 years old'
2 Acrel 1757	"tvänne starrblinda från Östergötland"; "En dalkarl, som förrättade bärgsprängning"; "Äfven en DalkarlsGåsse af 15 år"; "en Dalbonde från Lexa Församling med 2 :ne dess Söner"; "kommo 4 syskon, bonde-barn ifrån Nerike, en dräng och 3 pigor" 'two persons blind from cataract, from Östergötland'; 'A Dalecarlian, who carried out blasting work'; 'Also, a Dalecarlian Boy of 15 years'; 'a Dalecarlian peasant from Lexa Parish with 2 of his Sons'; 'four siblings came, peasant children from Nerike, a boy and 3 girls'
3 Odhelius 1764	"En man blef så träffad af en pisksnärt" 'A man was struck in such a way by the lash of a whip'
4 Odhelius 1765	"en Flicka om 3 år" 'a Girl of 3 years'
5 Odhelius 1767	"Bonden Anders Andersson, ifrån Sundskiär och Frötuna Socken i Roslagen" 'The peasant Anders Andersson, from Sundskiär and Frötuna Parish in Roslagen'
6 Acrel 1767	"Det tilämnade försöket, med Iridis öpning för ljusstrålarnas insläppande til inögat" 'The intended attempt to open the Iris in order to admit the rays of light into the inner eye'

As we see from these quotations, patients are presented in very tangible terms, with details of their age, occupation and, in many cases, their name and place of origin. Only in Acrel (1767), which is a comment on Odhelius (1767), do we find an abstract description of a case of the type common in modern scientific texts. In an earlier article (Acrel 1757), however, the same author is extremely concrete in his descriptions of cases, so abstractness does not appear to be a constant feature of his writing. As for Rosén (1751), it should perhaps be recalled that this article consists mainly of advice to parents, and that the advice is expressed in a fairly tangible way.

In summary, it can be said that the eighteenth-century texts studied do not describe patients as unknown, abstracts objects of investigation, but as recognizable individuals. In other words, no great distance is created between the author-scientist-doctor and his patients, who are ascribed attributes that bring out their individuality.

6.3. Discussion

Returning to the two questions asked by way of introduction, concerning the distance between the author-scientist and other groups in society, we thus find that the use of terms – especially in the surgical texts – can be said to contribute to constructing the role of a scientific expert. The tangible and detailed description of individual cases on the other hand, in which figures represent ages and points in time rather than statistics, suggests proximity, rather than distance, between the author-scientist-doctor and the patients he examines. My conclusion, therefore, is that the societal role constructed in these texts represents medical scientists as an expert group, but one that acts primarily within society, i.e. in close contact with other citizens and societal groups.

7. Conclusions

The purpose of the study presented in this chapter has been to shed light on how "scientificality" was constructed in Swedish medical articles from the eighteenth century. My analysis has been concerned not with what their scientist-authors *thought about* science (cf. Sörman and Selosse, this volume), but with what a text analysis can tell us about how they constructed "scientificality" in texts written for other members of the scientific community. "Scientificality" is of course a complex concept, and my analysis has therefore explored linguistic features linked to different layers of the texts: cognitive, social and societal. In this concluding section, I will begin with a summary of some of my main findings, viewed in relation to what we know about modern scientific articles in the medical field. From earlier studies of a large body of Swedish texts, we have a good understanding of how today's Swedish medical scientists write and structure their articles in scientific journals.[8] In Gunnarsson (1998, 2001, 2005, and 2009: 55–71), I present the results of studies relating specifically to articles in modern-day medical journals, more particularly from the period around 1980 (1975–1985). Accordingly, my comparison below sets the results obtained concerning the twelve eighteenth-century articles against findings on medical articles from around 1980.

Beginning with the similarities, I would draw attention to the following:

- The eighteenth-century papers classify and theorize, as do modern medical texts.
- Several of the eighteenth-century texts describe the current state of knowledge, in some cases in relatively great detail, as do modern medical texts.
- Some of the eighteenth-century texts describe a gap in existing knowledge or formulate a thesis regarding a solution, a feature that we recognize from modern articles.
- Much use is made of terms in the eighteenth-century material, a characteristic that also strikes the reader of a medical science text from the late twentieth century.

As for differences between modern scientific texts and those analysed here, I would put particular emphasis on the following:

- As we have seen, there is little uniformity of text structure in the papers from the eighteenth century. In that respect, they are very different from medical

8. In the framework of two major projects that I have headed at the Department of Scandinavian Languages in Uppsala, an extensive body of scientific and popular science texts has been studied. The project *Facktexter under 1900-talet* [LSP texts in the twentieth century], which was text linguistic in approach, studied 90 scientific and popular science articles in medicine, technology and economics, from three periods of the twentieth century: 1895–1905, 1935–1945 and 1975–1985 (for detailed presentations of the results, see Melander 1991 and Näslund 1991). In the second project, *Fackspråkens framväxt* [The emergence of languages for specific purposes in Sweden], a quantitative analysis was made of the vocabulary and terminology of a corpus comprising 360 scientific and popular science articles from six periods: 1750–1799, 1800–1849, 1850–1880, 1895–1901, 1935–1945 and 1975–1985. The two projects have generated a large number of publications. In this context, mention may be made of Gunnarsson (1998, 2001 and 2005), where the focus is on the language of medical science over three centuries, from the eighteenth to the twentieth.

articles published around 1980, which are uniformly structured, as regards both the text as a whole and how it begins. In the modern articles, moreover, the structure is marked by subheadings, which were not used in the eighteenth-century texts studied.
- Descriptions of cases play a significant, and prominent, role in the eighteenth-century articles. This gives the older texts a distinctive quality compared with their modern counterparts, which present large studies and statistically analysed results. As mentioned earlier, a similar trend has been found in studies on English medical literature (Atkinson 1992, Taavitsainen and Pahta 2000).
- There are very few references in the eighteenth-century corpus, and when they do occur they are incorporated into the main body of the text or appear in notes. This is a clear difference compared with modern medical texts.
- Another marked difference concerns the presence of the author. As we have seen, in the eighteenth-century articles the author appears as an "I". In the medical texts from around 1980, by contrast, there are no occurrences of the first-person singular pronoun and, it should be added, this is the case even in articles with a single author. Using the distinction between author-centered vs object-centered rhetoric, Atkinson (1992) describes a similar shift for English medical science.
- The eighteenth-century texts are associated with individuals in other respects as well. Colleagues are mentioned by name, as are patients. This is of course a major difference between the eighteenth-century and the modern article. In the eighteenth century the scientific network was small and, what is more, the distance from the layperson-patient was not as great as it is in today's scientific community.

Finally, this comparison can be related to the developmental axes mentioned in section 3. As Table 5 makes clear, the eighteenth-century articles end up to the left on all the axes, while modern medical texts end up to the right.

Table 6. Text and context in texts from the eighteenth and twentieth centuries

Layer	Period **1750–1769**	**1975–1985**
Cognitive	Individual findings	Theoretical integration
Social	Isolated researchers	Developed scientific community
Societal	Scientists act within society	Scientists act within scientific community

As regards cognitive content, the eighteenth-century text describes individual findings, whereas its modern counterpart seeks to achieve theoretical integration. Regarding social networks, the medical scientists of the eighteenth century seem to have been fairly isolated, i.e. the individual was significant in a way that we do not find in today's well-developed medical science community, where the emphasis is on the research group, the institute and the university. The medical scientific community of the eighteenth century was thus at a pre-establishment stage, with individual scientists who had observed particular cases presenting them in detail to other known individual scientists. In articles from the eighteenth century, therefore, we encounter a number of different individuals – the author himself, his colleagues and his patients – and their experiences and judgements are described. As I found in another analysis of medical articles (Gunnarsson 2001, 2005, 2009:55–71), the eighteenth-century text is full of explicit, strongly worded and assured evaluations concerning the subject of the study – the illness and the method of treating it – and also concerning advocates of the method, its naive practitioners. In the way he writes, the author treats individual findings as if they exist per se, he describes himself and his colleagues as isolated researchers, and he seems to act within different sectors of society, rather than within a specialized, fully established scientific community.

References

Adams-Smith, Diana E.
1990 Source and Derived Discourse. In: M. A. K Halliday, John Gibbons and Howard Nicholas (eds.), *Learning, Keeping and Using Language*, Vol. II, 415–433. Amsterdam, Philadelphia: John Benjamins.

Atkinson, Dwight
1999 *Scientific Discourse in Sociohistorical Context. The Philosophical Transactions of the Royal Society of London, 1675–1975*. Mahwah, New Jersey & London: Lawrence Erlbaum.

Bazerman, Charles
1988 *Shaping Written Knowledge. The Genre and Activity of the Experimental Article in Science*. Madison, Wisconsin: The University of Wisconsin Press.

Bungarten, Theo
1986 „Sprachliche Entfremdung" in der Wissenschaft. In: Theo Bungarten (ed.), *Wissenschaftssprache und Gesellschaft*, 22–43. Hamburg: Edition Akademion.

Gross, Alan G., Joseph E. Harmon and Michael Reidy
2002 *Communicating Science. The Scientific Article from the seventeenth Century to the Present*. Oxford and New York: Oxford University Press.

Gunnarsson, Britt-Louise
1988a Textmönster i vår äldsta vetenskapliga tidskriftsprosa. En analys av medicinska rön i Vetenskapsakademiens Handlingar 1750–1769 [Text patterns in the prose of our oldest scientific journals. An analysis of medical articles in the Transactions of the Swedish Academy of Sciences 1750–1769]. In: *Nysvenska studier* 67/1987 [printed in 1988]: 155–180.

Gunnarsson, Britt-Louise
1988b Medicinsk facktext i 1700-talets Sverige [Medical texts in eighteenth-century Sweden]. In: *Ingemar Olsson 25 augusti 1988*, 125–139. (MINS 28.) Department of Scandinavian Languages, Stockholm University.

Gunnarsson, Britt-Louise
1998 Academic discourse in changing context frames: The construction and development of a genre. In: Paula Evangelisti Allori (ed.), *Academic Discourse in Europe. Thought Processes and Linguistic Realisations*, 19–42. Rome: Bulzoni.

Gunnarsson, Britt-Louise
2001 Expressing criticism and evaluation during three centuries. *Journal of Historical Pragmatics*, vol. 2(1): 115–139.

Gunnarsson, Britt-Louise
2005 Medical Discourse: Sociohistorical Construction. In: Keith Brown (ed.), *Encyclopedia of Language and Linguistics* (2nd ed.), Vol. 7, Article 2360, 709–716. Oxford: Elsevier.

Gunnarsson, Britt-Louise
2009 *Professional Discourse*. (Continuum Discourse Series.) London and New York: Continuum.

Lindroth, Sten
1967 *Kungl. Svenska Vetenskapsakademiens historia 1739–1818. Tiden intill Wargentins död (1783)* [The history of the Royal Swedish Academy of Sciences, 1739–1818. The period up to the death of Wargentin (1783)]. Uppsala: Almqvist & Wiksell.

Lindroth, Sten
1978 *Svensk lärdomshistoria. Frihetstiden* [The history of science in Sweden. The Age of Liberty]. Stockholm: Norstedt & Söner.

Melander, Björn
1991 *Innehållsmönster i svenska facktexter* [Content patterns in Swedish LSP texts]. Skrifter utgivna av institutionen för nordiska språk vid Uppsala universitet 28. Uppsala universitet.

Näslund, Harry
1991 *Referens och koherens i svenska facktexter* [Reference and coherence in Swedish LSP texts]. Skrifter utgivna av institutionen för nordiska språk vid Uppsala universitet 29. Uppsala universitet.

Sörman, Richard
2011 Science and natural language in the eighteenth century: Buffon and Linnaeus. In: Britt-Louise Gunnarsson (ed.), *Languages of Science in the Eighteenth Century*, 141–155. Berlin, New York: De Gruyter Mouton.

Selosse, Philippe
2011 The Linnaean botanical nomenclature(s) as "a point of view on the world". From theory of ideas to theory of succedaneum. In: Britt-Louise Gunnarsson (ed.), *Languages of Science in the Eighteenth Century*, 157–168. Berlin: De Gruyter Mouton.

Taavitsainen, Irma and Päivi Pahta
2000 Conventions of professional writing. The medical case report in a historical perspective. *Journal of English Linguistics*, vol 28 (1): 60–76.

Teleman, Ulf
2011 The Swedish Academy of Sciences: Language policy and language practice. In: Britt-Louise Gunnarsson (ed.), *Languages of Science in the Eighteenth Century*, 63–87. Berlin, New York: De Gruyter Mouton.

Valle, Ellen
1999 *A Collective Intelligence. The Life Sciences in the Royal Society as a Scientific Discourse Community 1665–1965*. Anglicana Turkensia No. 17. Department of English, University of Turku, Finland.

Eighteenth-century English medical texts and discourses on reproduction

Päivi Pahta

1. Introduction

In eighteenth-century science, reproduction was a key topic, an object of ardent investigation and passionate debate.[1] Groundbreaking discoveries concerning the origin and development of a new creature were made using new scientific methods and tools. Several historians (e.g. Harvey 2002 and Cody 2005) have pointed out that ideas related to reproduction, including sexuality and gender difference, affected the whole scientific paradigm:

> The scientific revolution, ethnographic and medical disciplines, for example, were partly built upon – and reinforced – notions of sexual difference and gender roles, and discussions of the body and its cultural significance have been integrated into works which demonstrate the gendered character of new forms of knowledge and practices. (Harvey 2002: 900)

The word *reproduction* itself was new, beginning to replace the earlier term, *generation*, towards the end of the eighteenth century. According to the *Oxford English Dictionary* (*OED*), the word *reproduction* in the sense of a "process of producing new individuals of the same species by some form of generation" was first used with reference to human beings by the French naturalist Georges-Louis Leclerc, Count Buffon (1707–1788; see Sörman in this volume). This usage spread to English in discussions and translations of his work in the early 1780s. Like so many issues concerning the notion, the word, too, was controversial, as it was seen as bringing man, the crown of creation, down to the same level as other living beings. This is illustrated by the earliest *OED* quotation

1. This research is part of a joint project on Scientific Thought-styles: The Evolution of English Medical Writing, hosted by the Research Unit for Variation, Contacts and Change in English at the University of Helsinki, funded under the Centre of Excellence programme of the Academy of Finland for the period 2006–2011 (see http://www.helsinki.fi/varieng).

from John Wesley (1703–1791), discussing Count Buffon's innovation in an issue of his religious *Arminian Magazine* from October 1782:

> He [*sc*. Buffon] substitutes for the plain word *Generation*, a quaint word of his own, *Reproduction*, in order to level man not only with the beasts that perish, but with nettles or onions. (*OED*, s.v. *reproduction*).

Reproduction was also at the heart of Carl Linnaeus's innovations in the natural sciences. For Linnaeus, too, the origin of life was an intriguing and inspiring mystery, which he addressed in several of his books and dissertations (Müller-Wille 2002). His scientific taxonomy for describing and organizing nature relied profoundly on the reproduction of species. For him, the ability to reproduce was the most salient feature of all life forms, the component that fundamentally distinguished the living from minerals and inanimate objects (Cody 2005: 240). The sexual system for classifying plants introduced in the *Species plantarum* (1753), and subsequently in the tenth edition of *Systema naturae* (1759), was based on numbers of their reproductive organs, stamens and pistils. In comparison with earlier natural philosophers and other taxonomers, Linnaeus is said to have taken the centrality of reproduction to the utmost limit, often transforming descriptions of plants and animals into human terms. He divided creatures into two sexes, and envisioned the story of their courtship and family life as the basis for sorting plants into central categories (see Cody 2005). Linnaeus referred to flowers as "beds" and stamens and pistils as "husbands" and "wives" (see Müller-Wille 2007). For example, he described the marigold (*Calendula officinalis*) as a plant practising necessary polygamy, with "many marriages, in which the married females are barren and the concubines fertile". The tulip (*Hexandria monogynia*), with its six stamens and one pistil, had "six husbands in the same marriage", whereas the fuchsia (*Octandria monogynia*) had "eight men in the same bridal suite with one woman". He also compared plant structure to human anatomy. William Smellie, a Scottish physician and Linnaeus's near-contemporary, wrote that for Linnaeus,

> the calix represents the marriage *bed*; the corolla the *curtains*; the filaments the *spermatic vessels*; the antherae the *testes*; the pollen the *male semen*; the stigma the extremity of the *female organ*; the stylus the *vagina*; the germen the *ovarium*; the pericarpium the *impregnated ovarium*; and the seeds the *eggs*. (Smellie, *The Philosophy of Natural History*, 1790, p. 248)

In this study, I examine the language of eighteenth-century science in a field which today is the concern of reproductive medicine and reproductive or developmental biology. My aim is to provide a bird's-eye view of the texts and discourses constructing and disseminating knowledge on reproduction in this period. Applying the framework of variation analysis, I discuss a number of linguistic characteristics in the discourses of different types of English texts on reproduction, highlighting some distinctive features that can be discerned in them in relation to each other, on the one hand, and to earlier texts on the same topic, on the other. My particular interest is in variation in the discourses of knowledge, i.e. in the concepts by which the writers of the texts operate, and in the modes of knowing, including the sources of knowledge and the levels of certainty. I examine these discourses in their social and disciplinary embedding, specifically in relation to the epistemology of science, the intended target audience, and the dual, to some extent gendered, character of reproductive medicine as theory and praxis.

2. Variationist approach to the discourses of knowledge in medical texts on reproduction

The example of Linnaeus's plant taxonomy in the introduction illustrates my use of the terms *text* and *discourse*, inspired by recent research in (critical) discourse analysis (e.g. Blommaert 2005) and by constructionist approaches to the language of science and its history (e.g. Bazerman 1988, Hyland 2000, Gunnarsson 2006). Here the two terms have distinct meanings: discourse is seen as social interaction, a process of semiosis or meaningful symbolic behaviour, and texts as historically situated cultural products. The production of scientific texts, then, is a process of situated rhetorical action, in which writers have an active role, constructing scientific knowledge through texts with particular reading audiences in mind, and at the same time positioning themselves and their commitments and ideological debts within the field of knowledge they represent, within the broader scientific community, and within society at large. Discourse in this sense is always realized through texts, and texts consist of discourses that create meanings in social embedding, never in a vacuum. These meanings are realized in texts through linguistic or non-linguistic signs, actual words or visual images, and their various combinations.

The realization of discourse in actual linguistic features and their combinations in texts can vary considerably depending on the contextual variables conditioning and guiding text production in the communicative situation at hand. Thus, for example, scientific texts written in different periods display variation in their repertoires of linguistic features. One of the language-external sociocultural contextual factors contributing to this variability is the changing epistemology of science, i.e. the styles of scientific thinking and the ways of doing science that change, and have changed, over time (see e.g. Crombie 1994, Taavitsainen and Pahta 1997, Pahta and Taavitsainen 2011). With changes in the prevailing philosophy of science, epistemological commitments and domain-specific explanatory principles, discourse perspectives also change, causing changes on the surface level of linguistic features. Other contextual factors contributing to varying discourse perspectives and repertoires of linguistic features include, for example, the author's educational background, the target audience, or the tradition or genre of writing (see Taavitsainen and Pahta 2004, 2010, Taavitsainen et al. 2002, Pahta 2006).

According to Gunnarsson (1997, 2006, this volume), it is possible to distinguish three relevant layers of discourses or meaning-making processes in scientific texts, i.e. the cognitive, social and societal layers, and variation in discourse patterns and linguistic repertoires can occur on any of them. In this study, I focus on the cognitive layer of discourse, i.e. the knowledge base of the discipline – its objects of enquiry, facts, methods and tools, the ways in which professionals operating within the field view reality, what constitutes knowledge, and how it is constructed. Analysing data from different types of eighteenth-century English medical texts, I will highlight some distinctive traits in their discourses of knowledge, paying attention to the concepts by which the writers of the texts construct knowledge on reproduction, and to features indicating evidentiality and modality, i.e. the sources of knowing and the certainty of the claims. I will examine texts from three subsets: academic texts on reproductive theory written by and for scientists, texts on reproductive knowledge for a general audience, and books on midwifery, which provide an interesting site for investigating contemporary discourses on reproduction at the crossroads of theory and practice, but also in relation to some important social and societal developments. Most texts referred to are part of the *Corpus of Early English Medical Writing*, a computerized collection of medical texts from 1375–1800 (see e.g. Taavitsainen and Pahta 1997, 2010, Taavitsainen et al. 2005). My analysis of the texts is mainly qualitative and contextualized, but the discourse phenomena highlighted in the study have been identified and verified using basic corpus-linguistic techniques like concordances, frequency lists, key word analysis and dispersion plots (see Baker 2006).

3. Overview of eighteenth-century texts on reproduction

Eighteenth-century medical texts on reproduction are a heterogeneous group of writings, which can be classified in various ways on the basis of different language-external criteria. They include texts written by authors from various countries, in Latin and several other European languages, for example, German,

French, English, Italian, and Dutch. The texts represent a wide range of genres, including scientific monographs, handbooks, collections of case studies and observations, and collections of remedies, but also research articles, experimental reports, letters and book reviews published in scientific journals, established as a new medium for communicating science in the second half of the seventeenth century, notably *Le Journal des Sçavans* (later *Le Journal des Savants*) and the *Philosophical Transactions of the Royal Society*, founded in 1665 in Paris and London respectively (see Kronick 1976 and Valle 1999). Most of the important scientific monographs addressing reproductive physiology and anatomy were originally published in Latin, but translations were often made into several European languages, usually published within a few years of the original. On the other hand, the texts published in the early scientific journals were mainly in the local languages, and only occasionally in Latin; sometimes the reports and letters that appeared in the journals were translated for publication by the editors from texts originally written in other languages. Texts on reproduction were also written by authors with varying social and educational backgrounds, including scientists and a range of medical professionals with or without a formal education, mostly men. Texts were written for different target audiences: scientists, various kinds of medical practitioners, midwives, and students, or non-professionals, ordinary men and women, married and unmarried. Some texts were highly specialized, focusing on one specific detail of the reproductive process, while others aimed to provide a more comprehensive account of reproduction. Some texts were new creations, whereas others had been in circulation for centuries.

In the disciplinary context, two main traditions of writing can be distinguished, corresponding to the two complementary aspects of the reproductive process that have remained at the centre of enquiry throughout the history of science and medicine. One set of texts is theory-oriented, focusing on embryology, and aiming to explain the origin and development of a new living being. In the eighteenth century, a large number of texts in this category were originally written in languages other than English, but many were translated into English shortly after the publication of the original. The other set of texts, for which there are plenty of English originals, is concerned with midwifery, obstetrics and gynaecology, concentrating on the more practical side of the reproductive process. These two sets of texts reflect the dual nature of medicine as an area of knowledge, a science, on the one hand, and as the application of knowledge to medical praxis, an art, on the other. Interestingly, the division also corresponds to the traditional distribution of labour in the medical profession, as it was men who were concerned with the theory of the origin and development of new human beings, while women were in charge of the praxis, attending to the women carrying and delivering these new human beings. To some extent the two traditions of writing had remained separate from one another, but in the eighteenth century some significant social changes were taking place that brought them into closer contact with each other.

4. Discourses of knowledge in learned texts on reproductive theory

In the eighteenth century, reproductive biology was a new emergent field. Until the seventeenth century, reproductive knowledge largely rested on the theories of generation that had been formulated in ancient Greece by Aristotle, Galen, and the writers of the Hippocratic texts, and transmitted to the Latin West via the Islamic world (see Dunstan 1990). Medieval scholastic philosophers and professors of medicine also wrote numerous Latin accounts of generation, but they did not make any significant contributions to the knowledge base.

The core objects of interest from the very beginning included male and female reproductive anatomy, the origin and constitution of the seed, the mechanism of conception, the respective roles of the male and female in it, the principles of sex determination and heredity, the development and growth of the new creature, various types of reproductive anomalies, and multiple births (see Pahta 1998). Other disciplines apart from medicine also shared an interest in the area. The connection between body and soul received much attention in philosophical and theological writings, and until the late Middle Ages astrological influences on the reproductive process were a major concern among scientists. On many of these questions, the opinions of the ancient authors differed quite radically, and the ramifications of their debates can still be found in some eighteenth-century texts.

There was a practical limit to what ancient and medieval scientists could find out about the principles of human reproduction. While it was possible to observe the physical signs of pregnancy in the expectant mother, most of the reproductive process, at all stages, took place inside the woman's body and was thus beyond the means of observation available at the time. Most early embryologists experimented on hens' eggs, opening them at different stages, and these observations also formed the basis for the theory of human embryology (see Needham 1959: 18–25). The model of female reproductive anatomy, on the other hand, relied on observations made in the dissection of pigs. In many cases, however, the answers provided to the core questions were not based on evidence, but on reasoning or pure speculation. In the scholastic period, the prevailing scientific ideology created an obstacle to the advancement of knowledge. Scholastic embryology, like many other branches of scholastic science, was characterized by a respect for ancient theories, a wish to understand and interpret them. There was reluctance to criticize a systematic theory, even on the basis of contradictory evidence obtained from the observation of individual cases. Whatever the actual evidence for making claims was, they were often presented as axioms or universal truths.

This method of constructing knowledge is seen in fifteenth- and sixteenth-century texts on reproduction, illustrated in examples 1a, containing an account of conception from a medieval anatomical and surgical text, and 1b, from one of the few medieval specialized texts in English dealing with embryology. Texts of this period often refer to the diverging views of the authoritative ancient writers, but they tend not to admit to any actual uncertainties or problems in the knowledge base. The cognitive layer of discourse is characterized by a logocentric or quotative mode, with frequent reference to and citation of ancient authors like Galen or Avicenna (example 1a) or Hippocrates (1b), and a high frequency of speech act verbs, including *tell* (1a) or *say* (1b). Expressions indicating certainty and obligation are frequent, and prescriptive or imperative uses of cognitive verbs are common, like *understand that* (1a) and *it is to be known that* or *it is to be noted that* (1b). Modality markers expressing any kind of uncertainty or hesitation are rare.

(1a) I wole telle þe generacioun of embrioun: þat is to seie, how a child is I-gete in þe modir **Galion** & **auicen tellen**, þat of boþe þe spermes of man & of womman, worchinge & suffrynge togideris, so þat ech of hem worche in oþir & suffre in oþir, embrioun is bigete [...] **Vndirstonde þat** þe fleisch & þe fatnes is mad of menstrue blood [...] as **auicen & oþere auctouris tellen** (Lanfrank, *Chirurgia magna*, 15th c., pp. 20–21)[2]

(1b) **It is to be wist and knowen that** bi 3 monethis past in þe man and iiij in the womman bigynne the veynes to issue to the constitucioun and makyng of the nailes. [...] **It is to be noted that** a womman hath vij cellis or receptacles wher sperme is received. [...] **Ipocras saith** that if [...] (*De spermate*, 15th c., pp. 169, 173, 227).

The gradual epistemological shift from scholasticism to empiricism, with its new methods and tools of observation, opened up new avenues for enquiry. Starting from the sixteenth century, the new science of anatomy provided more accurate descriptions of male and female reproductive organs. In embryology, however, there was no immediate impact. Embryologists of the sixteenth century mostly focused on questions and debates that owed little to observation. The methods and writings of the most important writers of this period, such as

2. The relevant linguistic features are highlighted in bold type in the examples.

Volcher Coiter (b. 1534) and Fabricius ad Aquapendente (b. 1537), repeat to a great extent those of their predecessors (Adelmann 1966, 2: 752 ff.). The most widely read specialized text among sixteenth-century embryologists was Aristotle's *De generatione animalium*, whereas, for example, Galen's specialized treatises on the topic were still largely unknown (Nutton 1990: 137–141).

In the second half of the seventeenth century the traditional theory of generation was challenged in its entirety. The revolution began with William Harvey (1578–1657), whose *Exercitationes de generatione animalium* formed the basis for modern embryology. It was published in Latin in 1651 and an English translation, *Anatomical Exercitations on the Generation of Animals*, came out in 1653. Harvey was the first embryologist to carry out a systematic examination of the developing fetus of the mammal. His observations proved several earlier theories false, but with the tools of enquiry at his disposal (e.g. no microscope yet) he was left in the dark on many points, and many of his views on the reproductive process are quite traditional. What is revolutionary in comparison with the traditional theory of generation is Harvey's discourse of knowledge, illustrated in example 2. Unlike his predecessors, Harvey admits that he is not able to explain what really happens in conception. He refers to it as "a dark, obscure business", proposing to offer some suggestions that "seem probable in such dark matters, until [...] they can be convinced of falsity or errour". These few words serve to illustrate the striking epistemological shift in the scientific discourse on reproduction. The writer himself is an active agent in the construction of knowledge; his engagement in the process is reflected, for example, in the frequent use of first person singular pronouns. The whole attitude to knowledge is different: observation of individual cases counts, and uncertainty is an option.

(2) We meet with more **things worthy our wonder** concerning the *Conception*. It is indeed **a dark, obscure business**; however we shall **adventure to propose something in a *problematical* way, in such sort that it shall appeare we doe not onely goe about to subvert other mens opinions, but also to disclose our owne**. And yet whatsoever falleth from me concerning this subject, **I desire may not be so taken, as if I conceived them pronounced by an *Oracle***; [...] **I may offer those things as true, which seem probable in such dark matters, until such time as they can be convinced of falsity or errour.** (Harvey, *Generation of Animals*, 1653, p. 539)

After Harvey's opening, reproduction became one of the most active areas of scientific enquiry. The core questions for late seventeenth- and eighteenth-century scientists were largely the same as for their predecessors. The new methods of experiment and observation yielded new empirical results, but there was also room for speculation: there were still many things that could not be observed, and observations that could not be explained. Embryology was also increasingly affected by the general interest in mechanical physiology, which was gaining the attention of medical researchers throughout Europe (Keller 2000: 324).

One of the most heated controversies concerned the notion of development, partly connected with another animated debate regarding the balance between the male and female biological contributions in the future offspring. There were two main rival schools of thought, both further divided over various offshoot theories and ideological conflicts (Cody 2005: 98–113). The first principal school of thought supported the notion of epigenesis, an explanation that animal life began when the male seed mixed with the female seed. The two seeds were blended, bubbled together in the womb for the first few weeks, and began to grow into a recognizable shape. According to this theory, both parents contributed to the future offspring, and each embryo was newly produced through sequential development from unorganized material (Roe 1981: 1, Keller 2000: 330–331). The second school of thought supported the notion of preformation. Its exponents argued for the idea of pre-existence, claiming that the entire creature was already formed before conception, lay dormant in miniature in the male or female seed until conception, and then began to grow.

In 1672, a Dutch scientist, Regnier de Graaf (1641–1673), announced that he had located eggs in the ovaries of several mammals, including humans. Another Dutchman, Jan Swammerdam (1638–1680), suggested that the female ovarian egg contained a perfectly formed miniature animal that would begin to expand once the egg had been penetrated by the male spirit. Soon afterwards the Dutchman Antoni van Leeuwenhoek (1632–1723) proclaimed that he had discovered tiny particles in male seed with the help of the microscope. He was convinced that these microscopic *animalculae* rather than the female eggs were the germs of animal life (Brockliss 1990: 158–159). This first discovery of spermatozoa was published in *Le Journal des Sçavans* in 1678 (Clarke 2006), and in the same year in English in the *Philosophical Transactions*.

(3) These **animalculae** were **smaller than** the corpuscles which impart a red colour to blood, so that **I judge** a million of them would not **equal in size** a large grain of sand. Their **bodies**, which were **round**, were **blunt in front** and **ran to a point behind**. They were furnished **with a thin tail, about five or six times as long as** the body [...] so that **I can best liken** them to a small 'earth-nut'. They **moved forward** owing to the motion of their tails **like that of a snake or an eel** swimming in water (Leeuwenhoek, *Philosophical Transactions*, 1678).

The short extract from Leeuwenhoek's report illustrates the discourse of observation that became the standard in this field of enquiry. The author is actively engaged in the process of constructing knowledge: his observations and judgements, rather than earlier writings, are the source of knowledge. Observations of what was seen through the microscope are reported in the first person. Speech act verbs are rare, whereas expressions referring to cognitive processes or sensory perception are frequent. There is an emphasis on detail and structure. Physical and spatial descriptions consisting of expressions indicating shape, size, location, and direction are frequent. The descriptions of new phenomena are anchored to existing knowledge by comparisons to phenomena that are already familiar.

The discovery of large numbers of microscopic and motile creatures in the male seed created a great deal of debate concerning their origin and function; it was not at all obvious that the *animalculae* might be the instruments of conception. Throughout the eighteenth century several scientists suggested that these particles were parasites. (Clarke 2006.) Their role in reproduction was also questioned by the most influential figure in eighteenth-century reproductive physiology, Albrecht von Haller (1708–1777), a Swiss polymath, biologist and professor of medicine, anatomy, surgery and botany at the University of Göttingen, whose encyclopaedic book on the physiological elements of the human body (*Elementa physiologiae corporis humani*, 8 vols., 1757–1766) is a landmark in eighteenth-century medical history. Examples 4a, from a contemporary English translation of von Haller's *opus magnum*, and 4b, a present-day English translation from a shorter treatise in French, illustrate von Haller's discourse on embryonic development. He openly acknowledges the problems in studying the origin and development of a new creature, calling it "a very arduous investigation", where "few things are ascertained … difficultly". Here, knowledge rests on "things which experience shows to be true", which must be supplemented by "hypotheses" of things that "are not evident from the subject itself" (4a). Thus, in the absence of actual eye contact with the development of the early embryo, von Haller adopted the prevailing Newtonian mode of explanation, where conclusions concerning phenomena that could not be observed were based on the study of effects and the laws of those effects (Porter 1996: 165). In consequence, von Haller's conviction in favour of preformation was based on what "appear[ed] very probable to [him]" (4b). Even if knowledge here is not based on observation and hard evidence, the author is still actively engaged in the construction of knowledge, drawing on his own experience and cognitive processes.

(4a) This is **a very arduous investigation**, as **we propose to discover the changes which take place in the inward parts of woman** [...] **We shall relate**, in the first place, therefore, **those things which experience shows to be true**; and then **add those hypotheses by which learned men have endeavoured to supply such things as are not evident from the subject itself**. How **few things are ascertained** on this subject, and how **difficultly** they are ascertained, **I have learned** by too much experience. (von Haller, *First Lines of Physiology*, 1786, p. 197)

(4b) It **appears very probable to me** that the essential parts of the fetus exist formed at all times; not it is true in the way that they appear in the adult animal: [...] certain prepared causes [...] form in the end an animal that is very different from the embryo, and yet in which there is no part that did not exist essentially in the embryo. **It is thus that I explain development.** (von Haller, *Sur la formation du coeur dans le poulet*, 1758, vol. 2, p. 186; tr. from French into present-day English in Roe 1981)

Example 4c also illustrates another discourse trait in von Haller's scientific texts. This is the discourse of religion, an influential undercurrent in enquiries and debates on reproduction down the centuries.

(4c) **To me, indeed, experience seems to agree with those things which the mind foresees will follow their own causes**. Namely, that this most beautiful frame of animals is so various [...] that the ends have been foreseen in the eye, in the ear, and the hand, so that to these ends everything is most evidently accommodated: **it appears, therefore, certain to me, that no cause can be assigned for it below the infinite wisdom of the Creator himself**. (von Haller, *First Lines of Physiology*, 1786, p. 207)

It was largely for religious reasons that von Haller rejected the theory of development proposed by a young contemporary German scholar Caspar Friedrich Wolff (1734–1794). Wolff's dissertation, *Theoria generationis* (University of Halle, 1759), "the most refined embryological study" before the nineteenth century (Porter 1996: 169), ranks among the most important essays ever written on reproductive biology. In his thesis and subsequent studies, Wolff revived the doctrine of epigenesis, arguing that all individuals, plants and animals, developed through a gradual building up of structures from unorganized matter in a process of secretion and solidification that is repeated in cycles. He supported his claims with careful and detailed observational evidence which should have dealt a death blow to the theory of preformation. But the time was not ripe: von Haller and some other influential eighteenth-century physiologists severely criticized Wolff's work, claiming that his ideas were against the divine order. Thus, as a result of fundamentally different theological and epistemological commitments, Wolff became a controversial figure, and his work was never accepted among his most distinguished contemporaries (see Schellenberg in this volume).

5. The knowledge base in texts for a general audience

The new discoveries and learned disagreements of the scientists had little immediate impact on the texts on reproduction that enjoyed immense popularity among non-professional eighteenth-century readers. As Porter and Hall (1995: 33) point out, all kinds of reproduction guides, or sex advice literature, started rolling off the presses in western Europe (England, France, the Netherlands and the German-speaking areas) from the seventeenth century onwards. These included books for women, books of "secrets" and midwifery manuals, and works of astrological medicine and almanacs providing advice on fertility, pregnancy and other issues of sexual etiquette and practice, or even childbirth.

The examples under 5 illustrate the knowledge base and discourse characteristics of texts aimed at general eighteenth-century English audiences. Example 5a is from *Aristotle's Masterpiece*, one of the most popular sex manuals ever published. The text ran into multiple revised editions from the seventeenth to the turn of the twentieth century, but is in fact based on materials that were already circulating in manuscript form in Latin and various vernacular languages in the Middle Ages. Containing a layered hotchpotch of materials from various sources, it gives advice on reproductive concerns that were of interest to ordinary people. Later versions added names of the most famous latest theorists or invoked the Royal Society here and there, but in general the text communicated a knowledge base that combined earlier medieval and ancient theories of generation with folklore (Cody 2005: 99; see Porter and Hall 1995: 33 ff.).

(5a) **It is now a received truth**, that a perfect Child may be born in the 7 month, **by the Authority of the learned Hypocrates**. And therefore **we must believe**, that a Child born at the end of the 7th month, in lawful Matrimony, may be lawfully begotten. **Gallen, in the 6th Chapter of his third Book, handleth** this Argument [...] And **from the Authority of Pliny**, who makes mention of a Woman that went 13 months with Child (*Aristotle's Masterpiece*, 1776, p. 70)

Both the notions by which the author operates and the discourse features echo fifteenth-century texts, as illustrated in the examples under 1. No mention is made of the concepts of the new embryology, eggs or *animalculae*. In example 5b the concept of eggs does appear, but the new science is placed, or rather misplaced, within the frame of the old theories.

(5b) But against all this, **our modern Writers affirm, that the Ancients were very erroneous**; forasmuch as the Testicles in women do not afford Seed, but are two Eggs, like those in Fowls, or other Creatures; neither have they any such office as those of men, but are indeed an *Ovarium*, a Receptacle for Eggs [...] **the Truth of this, say they, is so plain**, that if you boil them, their Liquor will have the same taste, Colour, and Consistency, with the Taste of Birds Eggs (*Aristotle's Last Legacy*, 1766, pp. 31–33)

As these brief examples suggest, the discourse of reproductive knowledge in the text layers that were most widespread among non-professional readers continued in the thought-styles of past periods on both the conceptual and the discoursal level, with few affinities to the discourse of observation, experiment and enlightened speculation among the professional scientists of the time.

6. The discourses of knowledge in midwifery texts

There was one area of reproduction, midwifery, where the advances made in the knowledge base during the seventeenth and eighteenth centuries gave rise to dramatic changes both in the social organization of the medical profession and in disciplinary practices, contributing to the birth of a new discipline, obstetrics (see Wilson 1995, Cody 2005 and Lieske 2007). Until the seventeenth century, assisting childbirth was primarily a woman's job: midwives advised and attended the expectant mother; female "gossips", friends and relatives would arrive for the birth. Although men had written most texts on midwifery, they were only admitted to the actual birth scene in the case of an emergency (Cody 2005: 3, Green 2008).

The existence of all-female birth communities is reflected in images depicting birth chambers and in textual evidence. For example, the prologue of a fifteenth-century English gynaecological and obstetrical treatise suggests that the work was intended specifically for female readers, so that women, who out of shame would not admit their diseases to disdainful men, could benefit from the treatise and help themselves and other women (Green 1992: 77). A similar remark is contained in the preface to *The Byrth of Mankind*, the first printed English midwifery handbook (1540), which was a translation of Eucharius Rösslin's *Der Swangern Frauwen und Hebammen Rosegarten* [*A rose garden for pregnant women and midwives*] (1513), the first printed book focussing exclusively on midwifery. The book is based on classical sources, but it also incorporates advice the author had obtained from midwives and women. According to the preface, the knowledge disseminated in the book

> is nowe so plainly set forrth that the simplest Mydwyfe which can reade, may both understand for her better instruction, and also other women that have need of her be receaved and practysed of Mydwyves and all other Matrones with no less successe, then it is with good wyll and desire wrytten to proffite and to do good to other ... by the occasion of this booke to frequent & haunt wemen in theyr labours, carienge with them this booke in theyr handes, and causyge suche part of it as doth cheifly concerne the same pourpose, to be red before the mydwife, and the rest of the wemen being present. (Rösslin, *The Byrth of Mankind*, 1540, fos. Bv–Cviii).

The whole procedure surrounding childbirth had remained roughly similar for centuries. In most depictions of birth scenes, the mother is shown in bed, recovering after the event, while the midwives wash the child. In fact mothers would rarely lie down for birth; they gave birth crouching or on a birthing stool, a chair with back support and a hollowed-out seat. The midwife would normally sit in front ready to catch the baby (Wiesner 1986: 99–105). Male medical practitioners had no place in normal childbirth; they were only called upon in emergencies, usually requiring surgical intervention (Wilson 1995: 1; see also Green 2008 *passim*).

This traditional gendered division of labour began to change towards the end of the seventeenth century in a dramatic shift which has been called a "revolution in obstetrics" (Wilson 1995: 1). A new kind of practitioner appeared on the scene, first of all in France. This was the man-midwife, the *accoucheur*, a male practitioner who assisted not only in emergencies arising in difficult births, but also in normal births. Soon the scene began to change in Britain as well, and by the second quarter of the eighteenth century men-midwives had set up full-time obstetric practices in urban centres like London, Edinburgh and Dublin. Despite severe criticism from several corners, including serious doubts concerning the morals of the men entering such a delicate area of women's lives, by 1800 male attendants had almost displaced female midwives among aristocratic and urban middle-class families in Britain, and were common in more rural areas as well (see Wilson 1995).

The reasons for this social shift have been examined by several historians. According to Cody (2005), it can in many ways be seen as a direct consequence of the changes that had taken place in the epistemology of science, and subsequently in the knowledge base on human reproduction and in the discourses constructing that knowledge. The shift is also related to larger societal issues concerning medical education and gender hierarchies that largely prevented women from gaining access to the new knowledge base. In the eighteenth century, men had access to institutional education and the latest international developments in the field. Female midwives were largely trained into the profession by being apprenticed to an experienced older midwife, and were licensed by the Church (see Evenden 2000: 25–28). Men-midwives, on the other hand, were well-trained professional doctors, most of them practising surgeons, and they could persuade their way into the birth chamber by emphasizing the superiority of their knowledge base over that of their female colleagues, who, conversely, came under severe criticism in this new situation for their lack of knowledge and skills. The most successful men-midwives also began to teach the profession, thus contributing to a new form of institutionalization of midwifery training. William Smellie (1697–1763), for example, one of the most notable figures in the field, began to teach in London in 1740. Over the next ten years he taught over 900 male students, providing tuition to some female students as well. Most of his lectures, dealing with all aspects of pregnancy and labour, were combined with practical demonstrations on what were referred to as "machines", simulating the female pelvis and the unborn child. (Peel 2004).

The new type of practitioner brought changes in the practices of assisting childbirth, promoted by general developments in the prevailing scientific thought-style. A new method of manual internal examination known as "touching" was introduced in order to examine the development of the fetus in the womb and to determine its position prior to delivery, and to turn the child in the womb, if needed. A new mechanical instrument, the obstetric forceps, was introduced in the early eighteenth century to help the delivery, particularly in the case of problems. With new discoveries concerning the anatomy of the pelvis, the birth position changed from the birthing chair to delivery in bed, preferably with the prospective mother lying on her side, facilitating examination and obstetric procedures for the obstetrician. There was a heated debate concerning these practices and the whole notion of men-midwives, giving rise to various kinds of treatises, pamphlets, broadsides and even plays drawing on the discourses of empirical science and ethics.

This "revolution in obstetrics" is also attested in the discourses of contemporary texts on childbirth. This is illustrated in examples under 6, containing extracts from two eighteenth-century English texts on midwifery, chosen deliberately to underline the differences in the discourses of knowledge in texts written by female and male midwives. Extract 6a is from *The Midwives Book*, written by a female midwife, Jane Sharp (*fl.* 1641–1671), who in the preface described herself as a "practitioner in the art of midwifery above thirty years". *The Midwives Book* is in fact the first printed midwifery manual written by a woman. First published in 1671, it became popular, and achieved four editions by 1725. The book addressed a wide range of issues considered relevant to human reproduction and found in several other contemporary and earlier midwifery manuals. These included conception and the promotion of fertility, signs of pregnancy and the care of the expecting mother, the management of normal deliveries and malpresentations, the treatment of venereal disease, and the post-natal care of both mother and baby. The book also contained information about the anatomy of the male and female reproductive organs, and a short discussion of children's diseases. According to Sharp, the book was based on her experience and her reading of foreign authors, but recent research has shown that she in fact borrowed from several works of Nicholas Culpeper (1616–1654), the best-known popular medical author and health educator of the seventeenth century. (King 1995, Moscucci 2004, Curry 2004.)

Extract 6b comes from a comprehensive work by William Smellie. In the three volumes of *A Treatise on the Theory and Practice of Midwifery*, Smellie, the "master of British midwifery", laid the foundations of what became the new science of obstetrics. The work became a classic, making its author the best-known name in the field in Britain, so that "no-one could discuss the subject without referring to his book" (Wilson 1995: 125). The first two volumes appeared in 1752, the third in 1764. In this monumental work, also translated into French, German and Dutch, Smellie described the physiology of pregnancy and the mechanisms of normal and abnormal delivery far more precisely than any writer before him. In 1754 the treatise was supplemented with a volume of illustrations, *A Set of Anatomical Tables, with Explanations*, containing the most detailed and accurate demonstration of childbirth ever printed, in terms of both technique and anatomical description. (Peel 2004, Wilson 1995.)

The samples address female reproductive anatomy. Example 6a contains Sharp's sketchy description of the womb, drawing on Galen's ancient theory, where, for example, the anatomy of the womb was used to explain the determination of sex.[3] In contrast, Smellie's account of uterine anatomy in 6b represents the discourse of the new science. The description is detailed, paying attention to minute physical and spatial characteristics of the organs examined, with frequent indications of the shape, size, proportion, distance and location of the various components in precise measurements. Smellie's discourse is characterized by the use of more technical and Latinate terminology than Sharp's, and by explicit reference to observation and experiments – cutting and injecting, and measuring a Fallopian tube by trying to insert a hog's bristle into it.

(6a) **The Matrix or Womb hath two parts**, the great **hollow part** within, and the **neck** that leads to it, and it is a **Member** made by Nature for propagation of Children. [...] The **figure** of the **Concavity** of the Womb is **four-square, with some roundness, and hollow below** like a Bladder. [...] It hath but

3. For the history of explanations of the relationship between uterine anatomy and gender determination, see e.g. Pahta 1998.

one hollow Cell, yet this at the bottom is in some manner divided into two, as if there were two Wombs fastened to one neck. **For the most part Boys are bred in the right side of it, and Girls in the left**. [...] Perhaps it is no Error to say the Wombs are two, because **there are two cavities** like two hollow hands touching one the other, both **covered with one Pannicle**, and both **end in one Channel**; **No Man that sees a Womb can well discern it** [...] BUT to handle these things more particularly, **Galen saith**, that Women have all the parts of Generation that Men have, but Mens are outwardly, Womens inwardly.
(Sharp, *The Compleat Midwife's Companion*, [1671] 1725, pp. 31–34)

(6b) THE **Uterus** is **about three inches long from** the **Os Internum to the upper part of** the **Fundus**, and **one inch in thickness from the fore to the back part**. It is divided into the **Neck** and **Fundus**; the **length** of the neck being **an inch and three quarters**, while that of the fundus is **one inch and one quarter**. The **width** of the Uterus at the neck is **about one inch**, but at the fundus **twice as much**. The Uterus is smaller in young women.

The **outside-shape** of the Uterus in some measure resembles a flattened cucurbit, or that kind of spear which hath a long neck. [...] The **cavity** of the Fundus, is **in point of figure something between an oval and triangle**: **one of the angles commencing at the upper end** of the foresaid canal, and **the other two expanding the sides** of the Fundus, from which arise the **Fallopian tubes**. These tubes are **about three inches long**; and **so narrow at their entrance from the Uterus, as scarcely to admit an hog's bristle**. [...]

The blood-vessels of the womb, in the virgin or unimpregnated state, are very small, except **just at their approach to its sides, at the roots of the Ligamenta lata**: But, **as soon almost as they enter** its substance, they are **dispersed into such numbers of smaller branches** through the whole, that, **when it is cut, we can observe** but few, and those very small, **orifices**, much less any cavities that deserve the name of **sinuses**. Indeed, **when this part is minutely injected, it seems to be almost nothing but** a mass of vessels; a circumstance common to it with other parts of the body: And **anatomists are agreed**, that the greater number of **vessels visible in such nice injections**, are those thro' which the **serum** or **lymph** of the blood circulates in the living body; whence the **error loci** in an **Ophthalmia** is imitated by **subtile injections of coloured matter into the arteries of the dead subject**. (Smellie, *A Treatise on the Theory and Practice of Midwifery*, [1752] 1784, pp. 66–68)

These passages show why it was easy for men-midwives to criticize their female colleagues and to advertise themselves as more reliable assistants in childbirth. Men-midwives were able to support their claims to the profession with accurate, up-to-date knowledge and in the discourse of the prevailing scientific ideology.

The final example (7) from William Smellie's acclaimed treatise on the theory and practice of midwifery, describing the growth of the fetus in the womb, captures the two prominent discourse traits that characterize eighteenth-century texts constructing the new knowledge in the field of reproduction. On the one hand, the passage abounds in details, measurements and calculations, and implies careful observation of fetal development inside the womb. On the other hand, the epistemology of eighteenth-century science is reflected in the hesitations, and the final words of the account openly acknowledge the modalities concerning the eighteenth-century knowledge base in the science of reproduction, where many aspects of fetal development were still uncertain.

(7) **WHEN** the ovum descends into the uterus, **it is supposed to be about the size of a poppy-seed**, and **in the third month augmented to the bigness of a goose egg**. **Ten days after conception**, the child (**according to some authors**) **weighs half a grain**; **at thirty days**, is increased to the **weight of twenty-two grains**; **at three months**, **weighs betwixt two and three ounces; and at nine months, from ten to twelve, and sometimes sixteen, pounds**: by which **calculation it would appear** that the progress of the Fœtus is quickest in the beginning of its growth; for from the tenth to the thirtieth day (**according to this supposition**) it **increases to three and forty times its weight**. **All these calculations are uncertain**. (Smellie, *A Treatise on the Theory and Practice of Midwifery*, [1752] 1784, pp. 84–85)

7. Conclusion

In this study, I have examined eighteenth-century English medical texts and discourses on reproduction in their socio-historical context. My main aim has been to describe variation in the cognitive layer of discourse in the texts in relation to certain language-external variables, including the nature of their audience and the disciplinary context in which they were created and used, at the same time providing some insight into the variety of texts in which issues of reproduction were addressed. I have also identified and highlighted some discourse features in the texts that are characteristic of the period and the prevailing style of scientific thinking in which they were created, by comparing them with samples from earlier texts. In the analysis, I have paid special attention to discourses of knowledge, the notions by which the writers of the texts operate, and to aspects of evidentiality, i.e. to the sources of knowledge and to features indicating modality. The analysis confirms a clear trend in texts from logocentric, authority-dependent and low-modality modes of discourse, dominant in all socially embedded layers of knowledge, to a more complex and layered view in seventeenth- and eighteenth-century texts. The old discourse patterns continue to dominate in some social contexts, specifically in texts aimed at general audiences and female professionals, who did not have an established channel to access the new knowledge that was being created with the help of new tools and methods in the field. New patterns of discourse emerge, however, in the upper layers of the discipline. Here the professionals are no longer just transmitters of knowledge constructed by earlier generations. They are actively engaged in the process of constructing new knowledge by their experiments and observations and drawing on their own cognitive processes. Knowledge itself is also represented in a new, more dynamic light: it is not what someone says or writes, but something that can be discovered, seen, hypothesized, tested, doubted, and proved wrong or verified.

Primary sources

Aristotle's Last Legacy, Unfolding the Mystery of Nature in the Generation of Man. London, 1766.

Aristotle's Master-piece, Completed in Two Parts. Glasgow, 1776.

De Spermate. Ed. by Pahta, Päivi. *Medieval Embryology in the Vernacular: The Case of* De Spermate. Helsinki, 1998.

Douglas, William. *Letter to Dr. Smelle Shewing the Impropriety of his New-invented Wooden Forceps…* London, 1748.

Haller, Albrecht von. *First Lines of Physiology, by the Celebrated Baron Albertus Haller*. Translated from the correct Latin edition printed under the inspection of William Cullen, M.D. In two volumes. Edinburgh, 1786.

Haller, Albrecht von. *Sur la formation du coeur dans le poulet*, sur l'oeil, sur la structure du jaune &c. 2 vols. Lausanne, 1758.

Harvey, William. *Essays on the Generation of Animals*. London, 1653.

Lanfrank of Milan, *Chirurgia Magna*. Ed. by Fleischhacker, R.V. *Lanfrank's 'Science of Cirurgie'*. Early English Text Society O.S. 102. London, 1894.

Röesslin, Eucharius. *The Byrth of Mankynde newly translated out of Laten into Englysshe*. London: T. R., 1540.

Medieval Woman's Guide to Health: The First English Gynecological Handbook. Ed. by Rowland, Beryl. Kent, OH: Kent State University Press, 1981.

Sharp, Jane. *The Compleat Midwife's Companion: or, the Art of Midwifry Improv'd.* Fourth edition. In six volumes. London, 1725.

Smellie, William. *A Treatise on the Theory and Practice of Midwifery.* A new edition. In three volumes. Edinburgh, 1784.

Smellie, William. *The Philosophy of Natural History*. Edinburgh, 1790.

Wolff, Caspar Friedrich. Letters to Albrecht von Haller. Ed. by Roe, Shirley A. *Matter, Life, and Generation: 18th-century Embryology and the Haller-Wolff Debate*. Cambridge, 1981.

References

Adelmann, Howard B.
1966 *Marcello Malpighi and the Evolution of Embryology*. 4 vols. Ithaca, NY: Cornell University Press.

Baker, Paul
2006 *Using Corpora in Discourse Analysis*. London: Continuum.

Bazerman, Charles
1988 *Shaping Written Knowledge: The Genre and Activity of the Experimental Article in Science*. Madison, WI: The University of Wisconsin Press.

Blommaert, Jan
2005 *Discourse*. Cambridge: CUP.

Brockliss, L. W. B.
1990 The Embryological Revolution in the France of Louis XIV: The Dominance of Ideology. In: Dunstan, G. R. (ed.), 158–186.

Clarke, Gary N.
2006 A.R.T. and History, 1678–1978. *Human Reproduction* 21:1645–1650; doi:10.1093/humrep/del067.

Cody, Lisa Forman
2005 *Birthing the Nation: Sex, Science and the Conception of Eighteenth-century Britons*. New York: OUP.

Crombie, A. C.
1994 *Styles of Scientific Thinking in the European Tradition: The History of Argument and Explanation, Especially in the Mathematical and Biomedical Sciences and Arts*. 3 vols. London: Duckworth.

Curry, Patrick
2004 Culpeper, Nicholas (1616–1654), Physician and Astrologer. *Oxford Dictionary of National Biography*. Oxford: OUP. Online at http://www.oxforddnb.com/view/article/6882. Accessed 11 Jan 2008.

Dunstan, G. R.
1990 *The Human Embryo: Aristotle and the Arabic and European Traditions*. Exeter: University of Exeter Press.

Evenden, Doreen
2000 *The Midwives of Seventeenth-century London*. Cambridge: CUP.

Fissell, Mary E.
2004 *Vernacular Bodies: The Politics of Reproduction in Early Modern England*. Oxford: OUP.

French, Roger
1994 *William Harvey's Natural Philosophy*. Cambridge: CUP.

Green, Monica H.
1992 Obstetrical and Gynecological Texts in Middle English. *Studies in the Age of Chaucer* 14: 53–88.

Green, Monica H.
2008 *Making Women's Medicine Masculine: The Rise of Male Authority in Pre-Modern Gynaecology*. Oxford: OUP.

Gunnarsson, Britt-Louise
1997 On the Sociohistorical Construction of Scientific Discourse. In: Gunnarsson, B.-L., Linell, P. and Nordberg, B. (eds.), *The Construction of Professional Discourse*, 99–126. Longman: London and New York.

Gunnarsson, Britt-Louise
2006 Medical Discourse: Sociohistorical Construction. *Encyclopedia of Language and Linguistics*, 709–717. Elsevier.

Gunnarsson, Britt-Louise
2011 The linguistic construction of scientificality in early Swedish medical texts. In: Britt-Louise Gunnarsson (ed.), *Languages of Science in the Eighteenth Century*, 303–332. Berlin, New York: De Gruyter Mouton.

Hyland, Ken
2000 *Disciplinary Discourses: Social Interactions in Academic Writing*. Harlow: Pearson Education.

Keller, Eve
2000 Embryonic Individuals: The Rhetoric of Seventeenth-century Embryology and the Construction of Early-Modern Identity. *Eighteenth-Century Studies* 33: 321–348.

King, Helen
1995 "As if None Understood the Art that Cannot Understand Greek": The Education of Midwives in Seventeenth-century England. In: Nutton, Vivian and Roy Porter (eds.), *The History of Medical Education in Britain*, 184–198. Amsterdam and Atlanta, GA: Rodopi.

Kronick, David A.
1976. *A History of Scientific & Technical Periodicals: The Origins and Development of the Scientific and Technical Press, 1665–1790*. 2nd ed. Metuchen, NJ: Scarecrow Press.

Lieske, Pam (ed.)
2007 *Eighteenth-century British Midwifery*. London: Pickering and Chatto.

Moscucci, Ornella
2004 Sharp, Jane (fl. 1641–1671), Midwife. *Oxford Dictionary of National Biography*. Oxford: OUP. Online at http://www.oxforddnb.com/view/article/45823. Accessed 11 Jan 2008.

Müller-Wille, Staffan
2007 Plant Taxonomy: The Love of Plants. *Nature* 446: 268. doi:10.1038/446268a.

Needham, Joseph
1959 *A History of Embryology*. New York: Abelard-Schuman.

Nutton, Vivian
1990 The Anatomy of the Soul in Early Renaissance Medicine. In: Dunstan, G. R. (ed.), 136–157.

OED = *Oxford English Dictionary*. Oxford: OUP. Online at http://www.oed.com/.

Pahta, Päivi
1998 *Medieval Embryology in the Vernacular: The Case of* De Spermate. Helsinki: Société Néophilologique.

Pahta, Päivi
2006 *Ful Holsum and Profetable for the Bodi:* A Corpus Study of Amplifiers in Medieval English Medical Texts. In: Dossena, Marina and Irma Taavitsainen (eds.), *Diachronic Perspectives on Domain-specific English*, 207–228. Bern: Peter Lang.

Pahta, Päivi and Irma Taavitsainen
2011 An Interdisciplinary Approach to Early Modern English Medical Writing. In: Taavitsainen, Irma and Päivi Pahta (eds.), *Medical Writing in Early Modern English*, 1–8. Cambridge: CUP.

Peel, John
2004 Smellie, William (1697–1763). *Oxford Dictionary of National Biography*. Oxford: OUP. Online at http://www.oxforddnb.com/view/article/25752. Accessed 10 Jan 2008.

Porter, Roy
1996 Medical Science. In: Porter, Roy (ed.), *The Cambridge Illustrated History of Medicine*, 154–201. Cambridge: CUP.

Porter, Roy and Lesley Hall
1995 *The Facts of Life: The Creation of Sexual Knowledge in Britain, 1650–1950*. New Haven: Yale University Press.

Roe, Shirley A.
1981 *Matter, Life, and Generation: 18th-century Embryology and the Haller-Wolff Debate*. Cambridge: CUP.

Schellenberg, Renata
2011 Scientific literacy in eighteenth-century Germany. In: Britt-Louise Gunnarsson (ed.), *Languages of Science in the Eighteenth Century*, 91–106. Berlin: De Gruyter Mouton.

Selosse, Philippe
2011 The Linnaean botanical nomenclature(s) as "a point of view on the world". From theory of ideas to theory of succedaneum. In: Britt-Louise Gunnarsson (ed.), *Languages of Science in the Eighteenth Century*, 157–168. Berlin, New York: De Gruyter Mouton.

Taavitsainen, Irma
2006 Medical Discourse: Early Genres. *Encyclopedia of Language and Linguistics*. Elsevier. 688–694.

Taavitsainen, Irma and Päivi Pahta
1997 Corpus of Early English Medical Writing 1375–1750. *ICAME Journal* 21: 71–78. Online at http://icame.uib.no/ij21/.

Taavitsainen, Irma and Päivi Pahta (eds.)
2004 *Medical and Scientific Writing in Late Medieval English*. Cambridge: CUP.

Taavitsainen, Irma and Päivi Pahta (eds.)
2010 *Early Modern English Medical Texts: Corpus Description and Studies*. Amsterdam and Philadelphia: John Benjamins.

Taavitsainen, Irma, Päivi Pahta, Noora Leskinen, Maura Ratia and Carla Suhr
2002 Analysing Scientific Thought-styles: What Can Linguistic Research Reveal about the History of Science? In: Raumolin-Brunberg, Helena, Minna Nevala, Arja Nurmi and Matti Rissanen (eds.) *Variation Past and Present.* Helsinki: Société Néophilologique. 251–270.

Taavitsainen, Irma, Päivi Pahta and Martti Mäkinen (compilers)
2005 *Middle English Medical Texts*. CD-ROM with MEMT Presenter software by Raymond Hickey. Amsterdam and Philadelphia: John Benjamins.

Valle, Ellen
1999 *A Collective Intelligence: The Life Sciences in the Royal Society as a Scientific Discourse Community 1665–1965*. Turku: University of Turku.

Wiesner, Merry E.
1986 Early Modern Midwifery: A Case Study. In: Hanawalt, Barbara A. (ed.), *Women and Work in Pre-Industrial Europe*, 94–113. Bloomington: Indiana University Press.

Wilson, Adrian
1995 *The Making of Midwifery: Childbirth in England, 1660–1770*. Cambridge, MA: Harvard University Press.

Subject index

Bei Fragen zur Produktsicherheit wenden Sie sich bitte an:
If you have any questions regarding product safety,
please contact:

Walter de Gruyter GmbH
Genthiner Straße 13
10785 Berlin
productsafety@degruyterbrill.com